D0490258

Sustainable Industrial Design and Waste Management

Sustainable Industrial Design and Waste Management

Cradle-to-cradle for Sustainable Development

Dr Salah M. El-Haggar, PE, PhD

Professor of Energy and Environment
Department of Mechanical Engineering
The American University in Cairo

AMSTERDAM • BOSTON • HEIDELBERG • LONDON • NEW YORK • OXFORD
PARIS • SAN DIEGO • SAN FRANCISCO • SINGAPORE • SYDNEY • TOKYO
Academic Press is an imprint of Elsevier

6447552

Elsevier Academic Press
30 Corporate Drive, Suite 400, Burlington, MA 01803, USA
525 B Street, Suite 1900, San Diego, California 92101-4495, USA
84 Theobald's Road, London WC1X 8RR, UK

This book is printed on acid-free paper. ∞

Copyright © 2007, Elsevier Inc. All rights reserved.

No part of this publication may be reproduced or transmitted in any form or by
any means, electronic or mechanical, including photocopy, recording, or any
information storage and retrieval system, without permission in writing
from the publisher.

Permissions may be sought directly from Elsevier's Science & Technology Rights
Department in Oxford, UK: phone: (+44) 1865 843830, fax: (+44) 1865 853333,
e-mail: permission@elsevier.co.uk. You may also complete your request on-line
via the Elsevier homepage (http://elsevier.com), by selecting "Customer Support"
and then "Obtaining Permissions".

Library of Congress Cataloging-in-Publication Data
Application submitted

British Library Cataloguing in Publication Data
A catalogue record for this book is available from the British Library

ISBN-13: 978-0-12-373623-9

For all information on all Elsevier Academic Press publications
visit our website at www.books.elsevier.com

Typeset by Charon Tec Ltd (a Macmillan company), Chennai, India
www.charontec.com

Printed in the United States of America

07   08   09   10   9   8   7   6   5   4   3   2   1

Working together to grow
libraries in developing countries

www.elsevier.com | www.bookaid.org | www.sabre.org

ELSEVIER   BOOK AID   Sabre Foundation
            International

Contents

Acknowledgments   ix

About the author   xi

Introduction   xiii

1 CURRENT PRACTICE AND FUTURE SUSTAINABILITY   **1**

   1.1   Introduction   1
   1.2   Waste management   1
   1.3   Treatment   3
   1.4   Incineration   6
   1.5   Landfill   11
   1.6   Zero pollution and 7Rs rule   12
   1.7   Life cycle analysis and extended producer responsibility   13
   1.8   Cradle-to-cradle concept   16
   Questions   18

2 CLEANER PRODUCTION   **21**

   2.1   Introduction   21
   2.2   Promoting cleaner production   22
   2.3   Benefits of cleaner production   23
   2.4   Obstacles to cleaner production and solutions   24
   2.5   Cleaner production techniques   25
   2.6   Cleaner production opportunity assessment   29
   2.7   Cleaner production case studies   30
   Questions   84

3 SUSTAINABLE DEVELOPMENT AND INDUSTRIAL ECOLOGY   **85**

    3.1   Introduction   85
    3.2   Industrial ecology   86
    3.3   Industrial ecology barriers   88
    3.4   Eco-industrial parks   91
    3.5   Recycling economy/circular economy initiatives   93
    3.6   Eco-industrial parks case studies   97
    Questions   124

4 SUSTAINABLE DEVELOPMENT AND ENVIRONMENTAL REFORM   **125**

    4.1   Introduction   125
    4.2   Sustainable development proposed framework   126
    4.3   Sustainable development tools, indicator, and formula   133
    4.4   Sustainable development facilitators   134
    4.5   Environmental reform   135
    4.6   Environmental reform proposed structure   137
    4.7   Mechanisms for environmental impact assessment   141
    4.8   Sustainable development road map   147
    Questions   148

5 SUSTAINABILITY OF MUNICIPAL SOLID WASTE MANAGEMENT   **149**

    5.1   Introduction   149
    5.2   Transfer stations   155
    5.3   Recycling of waste paper   158
    5.4   Recycling of plastic waste   163
    5.5   Recycling of bones   172
    5.6   Recycling of glass   173
    5.7   Foam glass   175
    5.8   Recycling of aluminum and tin cans   179
    5.9   Recycling of textiles   179
    5.10 Recycling of composite packaging materials   180
    5.11 Recycling of laminated plastics   187
    5.12 Recycling of food waste   189
    5.13 Rejects   194
    Questions   196

6 RECYCLING OF MUNICIPAL SOLID WASTE REJECTS   **197**

    6.1   Introduction   197
    6.2   Reject technologies   198

6.3	Product development from rejects	201
6.4	Construction materials and their properties	202
6.5	Manhole	215
6.6	Breakwater	217
6.7	Other products	219
	Questions	222

7   SUSTAINABILITY OF AGRICULTURAL AND RURAL WASTE MANAGEMENT — **223**

7.1	Introduction	223
7.2	Main technologies for rural communities	224
7.3	Animal fodder	226
7.4	Briquetting	227
7.5	Biogas	232
7.6	Composting	233
7.7	Other applications/technologies	234
7.8	Integrated complex	239
7.9	Agricultural and rural waste management case studies	243
	Questions	260

8   SUSTAINABILITY OF CONSTRUCTION AND DEMOLITION WASTE MANAGEMENT — **261**

8.1	Introduction	261
8.2	Construction waste	262
8.3	Construction waste management guidelines	263
8.4	Demolition waste	272
8.5	Demolition waste management guidelines	273
8.6	Final remarks	279
8.7	Construction waste case studies	279
	Questions	292

9   SUSTAINABILITY OF CLINICAL SOLID WASTE MANAGEMENT — **293**

9.1	Introduction	293
9.2	Methodology	294
9.3	Clinical waste management	295
9.4	Disinfection of clinical wastes	298
9.5	Current experience of clinical wastes	302
9.6	Electron beam technology	303
9.7	Electron beam for sterilization of clinical wastes	304
	Questions	306

10   SUSTAINABILITY OF INDUSTRIAL WASTE MANAGEMENT   307

10.1   Introduction   307
10.2   Cement industry case study   308
10.3   Iron and steel industry case study   317
10.4   Aluminum foundries case study   326
10.5   Drill cuttings, petroleum sector case study   339
10.6   Marble and granite industry case study   346
10.7   Sugarcane industry case study   350
10.8   Tourist industry case study   362
Questions   368

References   371
Index   387

Acknowledgments

I extend my heartfelt gratitude to everyone at Elsevier Ltd who was involved in the publication of this book. Without their help, devotion, and dedicated efforts, this book would not have come to fruition.

My sincere appreciation goes to all my graduate and undergraduate students at the American University in Cairo, who have always been an integral part of my research projects and who provided substantial assistance in the preparation of this book. Special thanks go to: Ahmed Elshall, Amal Mousa, Dalia Sakr, Dina Abdelalim, Eiman Hamdy, Hala Abu Hussein, Ishaq Adeleke, Islam El-Adaway, Lama El-Hatow, Marwa El-Ansari, Moataz Farahat, Mohamed Abu Khattowa, Mohamed El Gowini, Mona Hamdy, Passant Abou Yousef, Yasser Ibrahim, Yasser Kourany, Zainab Hermes as well as many other undergraduate and graduate students.

I would also like to thank the Ministry of State for Environmental Affairs in Egypt, the Egyptian Environmental Affairs Agency and the DFID-SEAM Program for their promotion of cleaner production in Egypt's various industrial sectors. Special appreciation to Mr Phil Jago, SEAM Program manager and cleaner production team leader for his effort in promoting cleaner production through demo projects where a number of case studies are demonstrated in Chapters 2 and 10.

Last but not least, I am eternally indebted to my wife Sadika, and my children and grandchildren. I am at a loss for words to adequately express my gratitude and appreciation to them all. They provided me not only with endless moral support, but also with the tranquil and appropriate environment, which made it possible for me to finish this work.

About the author

Dr El-Haggar has more than 30 years of experience in energy and environmental consulting and university teaching. Dr El-Haggar was a visiting professor at Washington State University and at the University of Idaho, and is presently the Professor of Energy and Environment at the Mechanical Engineering Department at the American University in Cairo.

Dr El-Haggar has received more than 18 academic honors, grants, and awards. He was awarded the outstanding teaching award from AUC in 1995 as well as a number of outstanding trustees awards. In addition Dr El-Haggar has 118 scientific publications in environmental and energy fields, 34 invited presentations, 50 technical reports and 12 books.

Dr El-Haggar's environmental consulting experience includes more than 40 environmental/industrial auditing for major industrial identities, 20 compliance action plans, nine environmental impact assessments in addition to extensive consulting experience in environmental engineering, environmental auditing, costal zone management, environmental impact assessment (EIA), environmental management systems (EMS), energy management, hazardous and non-hazardous waste management, recycling, pollution prevention and waste minimization, zero pollution, biogas/solar/wind technology, community/desert development, solid and industrial waste, and environmental assessment for the local government and private industries. Dr El-Haggar is a member/board member of 14 national and international societies in the areas of mechanical engineering, environmental engineering and community development.

Dr El-Haggar has been working in environmental technologies since 1987. His paramount objective is to transform waste into useful products. He developed a very simple theory called the 7Rs rule that applies cradle-to-cradle concepts to waste handling and management. Dr El-Haggar was able to develop different technologies for recycling unrecyclable waste such as Tetra-Paks, dippers, municipal solid waste rejects, etc. He has published two series of books on *Cleaner Production Technologies* and *Fundamentals and Mechanisms for Sustainable Development*. Dr El-Haggar has also written a number of chapters in several books on the environment.

Introduction

Some people see things as they are and ask why?
I dream of things that have never been and ask why not?

Senator Robert Kennedy

Ever since the Earth Summit of Rio de Janeiro back in 1992, much has been written and continues to be written about "sustainability". Throughout this time, however, we seem to have lost a direction for measuring sustainability. We need first to think about how we can develop sustainable projects and industries, and then think about how we can develop indicators to measure the sustainability or percentage of sustainability within these projects and industries. These two issues will help us draft plans for further sustainable developments. This book will discuss such indicators and the tools necessary for sustainable development. In this way, it suggests a formula for sustainability.

The real environmental and economical problem of the 20th century is that scientific and technological developments have increased the human capacity to extract resources from nature, process them, and use them, but have not offered parallel and similar insight into how these resources can be returned to their environmental origin or how they could be entered into a new cycle of extraction, processing, and use. Much of the resources extracted from nature are used in unsustainable activities and end up as waste. This can be described as a cradle-to-grave scenario in which the resources have a "lifetime" and are disposed of after they are used, ending up in a "grave" (a landfill, for example). If this were to continue unabated, we may end up completely depleting our natural resources. The only way to evade this dead end is to develop newer production and processing techniques that use up resources in an alternative cradle-to-cradle scenario.

This book is original in that it presents cradle-to-cradle production and processing alternatives to most of the traditional industries common today. These alternatives are not only environmentally friendly, but are also economically advantageous (either cutting costs or increasing profits). This book is replete with creative ideas and innovative technologies that, when implemented, would lead to cradle-to-cradle production and manufacturing in all industrial sectors.

Perhaps the major problem industries have with current environmental protection regulations is their cost and return. Pollution control and treatment, and environmental protection procedures, are all considered very expensive activities and, as such, they are seen as economic burdens and impediments to further industrial development.

Indeed proper waste handling and management is posing a complex problem for the entire world. On the one hand, it can be highly costly, and on the other hand, improper handling of waste can have harmful effects on life and habitat and at the same time lead to depletion of our natural resources.

Because of the universality of this problem, any comprehensive solution should be appropriate and applicable in both developed (industrial) and underdeveloped nations. And for any solution to be sustainable, it should promise economic benefits, require available or obtainable technology, and comply with the social and environmental norms within a given nation.

The main objective of this book is to conserve our natural resources by attempting to reach a 100% utilization of all types of waste. It offers alternative production and waste management techniques that employ cradle-to-cradle concepts and the methodologies of cleaner production and industrial ecology. It is filled with case studies that demonstrate the applicability of these techniques in most industrial sectors such as textile, food, oil and soap, etc. Case studies were also implemented in the heavy industries such as petroleum, iron and steel, cement, etc. Touristic activities are also included because they are considered an industry that uses up natural resources and generates waste.

The traditional waste management hierarchy implemented in most countries, which involves reduction, reuse, recycling, recovery, treatment, and disposal, should now be modified to exclude treatment (especially thermal and chemical) and disposal in landfills. Waste treatment converts harmful waste into less harmful waste, but produces in the process an effluent that itself becomes waste and must be disposed of in a landfill. A "disposal" of anything means depletion of our natural resources, and may also lead to environmental pollution (in the air, water, and soil). In contrast, recovery, as used in the hierarchy above, attempts to convert waste into energy. It is a very expensive procedure that cannot be afforded by most countries. And thus throughout this book, recovery will mean material recovery – for example, attempting to separate waste oil from water employing gravity using a gravity oil separator (GOS) technique, or employing air bubbles using a dissolved air flotation (DAF) technique. This book thus suggests a new hierarchy for waste management that would apply cradle-to-cradle concepts in order to conserve natural resources.

Natural resources are becoming a very crucial issue for sustainable development because finding new sources of raw material has proven to be very costly and difficult. Waste disposal has very significant impacts on the environment since it may cause contamination in the air, soil, and/or water. In order to make waste management more sustainable, it should shift from cradle-to-grave systems to ones that apply cradle-to-cradle concepts. These systems should also reduce or completely eliminate any disposal stages.

Chapter 1 of this book will cover the common waste management procedures currently practiced worldwide and will discuss their impacts on future sustainability and conservation of natural resources. The life cycle of waste in these procedures will be analyzed to demonstrate that it follows a cradle-to-grave approach. We will then examine the impact these procedures have on environmental protection and conservation of natural resources. Subsequently, the cradle-to-cradle concepts will be discussed in detail with a listing of their pros and cons. We will explain the role of the government and civil society in effecting these cradle-to-cradle concepts for the conservation of our natural resources using the principle of extended producer responsibilities. We will introduce a new term in environmental engineering – "sustainable treatment" – as well as a new hierarchy for waste management, which will apply cradle-to-cradle concepts. The following chapters will serve as applications and implementations of this definition.

Chapter 2 will introduce the concept of cleaner production (CP), its techniques, and its benefits. Obstacles or barriers to cleaner production will be discussed and solutions will be offered. This chapter attempts to successfully develop cleaner production opportunities and assess their implementations. A discussion will follow of various case studies in the different industrial sectors (food, textile, oil and soap, etc.), with elaborate cost/benefit analyses. The case studies will demonstrate and assess different cleaner production opportunities and implementation techniques.

Chapter 3 is about sustainable development and industrial ecology. It will discuss the principles of industrial ecology and will attempt to integrate our industrial activities within a natural ecosystem. Barriers to industrial ecology will also be discussed in all their dimensions: technical, marketing and awareness, financial, and barriers involving regional strategy and regulations. Eco-industrial parks (EIP) will be discussed in further detail with case studies implemented in different parts of the world demonstrating EIP applications using a top-down scheme, bottom-up scheme, or combinations of both. These case studies are intended to guide the readers in developing their own EIP implementation schemes in their own country or community. It also hopes to equip the readers with the methodologies for converting existing industrial estates into environmentally friendly ones – eco-industrial parks.

Chapter 4 on sustainable development and environmental reform will employ the first three chapters to develop a framework for sustainable development and environmental reform. Sustainable development tools and methodologies will be discussed such as the environmental management system (EMS), cleaner production (CP), environmental impact assessment (EIA), and environmental information technologies (EIT). This chapter will then move on to suggest an integration of cleaner production and environmental management systems to promote and manage cleaner production implementation throughout the different industrial sectors. This is a proposed modification to the ISO 14001 standard to be discussed by the ISO Technical Committee in its next round. This chapter will also propose an environmental reform structure and present a detailed discussion of all the relevant elements such as regulation, environmental impact assessment (EIA), environmental management system (EMS), cleaner production (CP), and industrial ecology (IE).

Chapter 5 will tackle the issue of municipal solid waste management sustainability (MSWMS). It is the most challenging chapter as it attempts to apply all the principles covered in the previous four chapters to reach practical cradle-to-cradle implementations. The fundamental issue in this chapter is limiting the use of landfills (i.e. reducing disposed waste) or completely eliminating disposed waste from MSWMS. Different techniques for recycling MSW will be presented such as recycling food waste, bones, tin cans, plastics, glass, and textiles. The recycling of composite material, used in packaging, will also be discussed.

The remaining kinds of waste (rejects), which cannot be recycled by any technique, will be discussed in Chapter 6. Chapter 6 is a completion for MSW, which allows the full and practical realization of a cradle-to-cradle model. This chapter will discuss technology developments to recycle unrecyclable wastes (rejects) as well as product developments to meet or match the needs of a given community. The properties of the resulting new materials and suggestions for its suitable applications will also be presented.

Chapter 7 on the sustainability of agricultural and rural waste management is very important for most developing countries as well as some developed countries. The unsustainable nature of agricultural and rural waste results in environmental pollution and may ultimately lead to complete depletion of our natural resources. Different technologies for handling this type of waste, such as composting, animal fodder, briquetting, biogas, construction materials, silicon carbide, etc., will be discussed in this chapter. These technologies are appropriate for and applicable in both developed and underdeveloped countries. Two different case studies are included in this chapter. The first involves converting soil conditioners into organic fertilizers for organic farming by composting agricultural and rural waste. The second will combine all agricultural and municipal solid waste, as well as municipal liquid waste, into one complex called an eco-rural park.

Chapter 8 will discuss the sustainability of construction and demolition waste and will explain the relevant guidelines to owners and contractors. This chapter includes three case studies. The first case study uses the 7Rs rule as a guideline for handling construction waste in a manner that applies cradle-to-cradle concepts. The second case study demonstrates how much money is typically spent on getting rid of construction waste. The third and final case study demonstrates how cradle-to-cradle implementations on construction waste can be advantageous and beneficial.

Chapter 9 on the sustainability of clinical solid waste management is the most critical chapter in this book because clinical wastes can be very

hazardous, are generated from very sensitive resources and should thus be handled and treated in a very sensitive manner. The chapter will discuss the most popular technologies used in clinical waste treatment and compare and contrast the advantages and disadvantages of each one. Technologies discussed in this chapter include incineration, autoclave (steam sterilization), chemical disinfection, microwave disinfection, pyrolysis, gasification, plasma systems, and irradiation. Current clinical waste management practices applied in both the developed and developing countries will be examined, and a final discussion will be given about the use of electron beam technology in sterilizing clinical waste and how that could be applied to achieve a cradle-to-cradle implementation.

The final chapter, Chapter 10, will discuss sustainability of industrial solid waste management. It attempts to define an outline for transforming the different traditional industries into more environmentally friendly ones that apply cradle-to-cradle concepts. Some industrial sectors discussed in this chapter have not been mentioned before in previous chapters such as the sugarcane industry, aluminum foundry, iron and steel, marble, petroleum, cement, and tourism.

In conclusion, this book advocates sustainable development and the conservation of our natural resources without inflicting harm on the environment. It discusses most types of waste generated in most industries and within our communities, and suggests techniques for utilizing it according to cradle-to-cradle concepts and avoiding the use of landfills, incineration, and treatment in general. I hope the reader finds it as stimulating and enjoyable to read as it was for me to write. Any questions, comments or suggestions – positive or negative – for further improving this book would be highly appreciated. Please feel free to contact the author at: elhaggar@aucegypt.edu

Chapter 1

Current Practice and Future Sustainability

1.1 Introduction

One of the major problems facing the world today is the environmental protection cost and return. The current practice of pollution control, treatment and environmental protection can be considered very expensive activities where people consider it a burden for development. There is a worldwide misconception that "environmental protection comes at the expense of economic development or vice versa". This is not true if sustainable development is achieved. Sustainable development promotes economic growth given that this growth does not compromise the management of the environmental resources. The traditional approach for clinical waste, agricultural waste, industrial and municipal solid waste, industrial and municipal liquid waste, etc. can be considered disastrous worldwide because it is depleting the natural resources and may pollute the environment if it is not treated/disposed of properly. Any solution should suit not only the developed countries but also the developing countries should include the economical benefits, technological availability, environmental and social perspectives otherwise they will never be sustainable. The objective of this book is to conserve the natural resources by approaching 100% full utilization of all types of wastes by a cradle-to-cradle concept through sustainable treatment.

1.2 Waste Management

Waste generations vary from one country to another, but many previous studies indicated that as gross domestic product (GDP) per capita increases, per capita municipal solid waste (MSW) generation and other types of wastes also increases. So, waste management is a must for conservation of natural

1

resources as well as for protecting the environment in order to approach sustainable development.

The selection of a combination of techniques, technologies and management programs to achieve waste management objectives is called integrated waste management (IWM). The hierarchy of actions to implement IWM is reduction, reuse, recycle, treatment and final disposal (Tchobanoglous *et al.*, 1993). Different sources use different terms and categories to describe the waste management hierarchy. The USEPA 1989 publication "The Solid Waste Dilemma: An Agenda for Action" states that their hierarchy for waste management is source reduction, recycling, waste combustion and landfilling. Others would list source prevention, source reductions and reuse as two categories, while most of the literature combines them under source reduction. The New Jersey Department of Environmental Protection includes recycling, on-site composting and reusing at the source under source reduction. However, reviewing diverse literatures reveals that the traditional waste management hierarchy is dominantly reducing, reusing, recycling, recovery, treatment, and disposing. Incineration might be included within treatment because it is thermal treatment, or within recovery as waste-to-energy recovery, or can be discussed as an independent item as will be discussed in this chapter.

Reducing: Reduced material volume at the source can be enforced through extended producers and consumers polices (e.g. less unnecessary packaging for products). Indeed, changing the consumer's practices is part of the source reduction concept. Reducing the raw material at the source will conserve the natural resources for other uses. Fortunately, statistics show that these trends are declining in developed countries. For example, the total source reduction in the USA, which includes prevention and reuse, increased from less than one million tons in 1992 to more than 50 million tons in 1999 (USEPA, 1999).

Reusing: Reuse means to continue using the product in its original or in a modified form. Reuse of materials involves extended use of a product (retrading auto tires) or use of a product for other purposes (tin cans for holding nails, glass bottles for holding water in refrigerators). Reusing the product does not return the material to the industry for remanufacturing or recycling. Reuse can be considered another aspect of source reduction which could be carried out not only by consumers but also by producers. Chemicals used in the tanning industry could be reused by installing an on-site chromium recovery unit. Source reduction and reusing can be encouraged through numerous regulations and programs such as the Pay-As-You-Throw program developed by USEPA as well as other programs.

It is clear that source reduction does not only include reduction in the use of material, but includes as well the activities that increase product durability and reusability. Source reduction, which includes source prevention and reuse, is the best option in waste management because it preserves natural resources and reduces pollution, and waste landfilling or incineration. The less preferred option in waste management is recycling.

Recycling: What cannot be reduced at the source is pumped in the waste stream. The above discussion shows that reuse has much to do with cultural habits and this is also the case with recycling but recycling involves additional technical know-how and could involve some capital investment. Recycling is the process of converting these wastes to raw material that can be reused to manufacture new products.

Through regulations governments have a great role to play in promoting recycling. Such regulations are even emerging in developing countries. For example, the Republic of Korea explicitly prescribes the Extended Producer Recycling system under the Resources Conservation and Recycling Promotion Law, amended in 2003 (IGES, 2005). In India and the Philippines, laws on the management of MSW have been enacted recently and the importance of material cycles is clearly mentioned in the laws (IGES, 2005).

Recovery: Recovery of materials or energy can take numerous forms. It is clear that material recovery is a limited activity worldwide and is mainly concerned with the recovery of energy from burning wastes. For example, the Oregon Department of Environmental Quality in the USA states that "construction and demolition wastes makes up the majority of the wastes being processed at MSW Recovery Facilities, followed by 'dry' commercial and industrial loads; virtually no recovery from residential garbage route trucks occurs" (ODEQ, 1997).

Recovery differs from recycling in that waste is collected as mixed refuse, and then various processing steps remove the materials. Separating oil from waste water effluent by a gravity oil separator (GOS) in the oil and soap industry is material recovery from waste. This material is then sold to another type of soap industry or returned to the industrial process within the same factory. The difference between recycling and recovery, the two primary methods of returning waste materials to industry for manufacturing and subsequent use, is that the latter requires a process to remove the material from the waste while the former does not require any processes for separation, sorting can be done manually.

1.3 Treatment

Treatment or end-of-pipe treatment or pollution control is one of the very important technologies for the traditional waste management hierarchy and environmental compliance for any industry. There is a variety of traditional treatment technologies for wastes to choose from depending on several factors such as physical form of the waste (solid, liquid, or gaseous), quantity of waste, characteristics, combined or segregated wastes, degree of treatment required, etc. The treatment technologies can be categorized into physical, chemical, thermal, or biological treatment. Combinations of treatment technologies are often used to develop the most cost-effective, environmentally acceptable solutions for waste management.

Physical treatment: Physical processes for waste treatment include screening, sedimentation and clarification, centrifugation, flotation, filtration, sorption, evaporation and distillation, air or steam stripping, membrane-based filtration processes, etc. These processes are mostly applied to liquid hazardous wastes, and involve the separation of suspended or colloidal solids from the liquid phase. The selection of the technology depends mainly on the concentration and characteristics of the suspended solids relative to the liquid phase. Physical processes segregate the waste from one form to another, reduce the volume, and concentrate the solids to facilitate further treatment or further actions. Whenever a waste containing liquids and solids is to be treated, physical separation of the solids from the liquid should be considered first because it is generally cost-effective to treat a low volume, high concentration waste. Usually physical treatment is used in combination with other treatment technologies for optimum waste treatment and disposal.

Chemical treatment: Chemical treatment involves the use of chemical reactions to transform harmful waste into less harmful, or non-harmful waste, or make it less mobile in the environment. Many different types of chemical treatment processes are used in waste management such as neutralization, precipitation, coagulation, flocculation, oxidation, reduction, etc. Chemical treatment can have some advantages such as volume reduction and promoting resource recovery from wastes. Because it can be employed for resource recovery, and to produce useful byproducts and environmentally acceptable residues, chemical treatment should be considered before sending an untreated hazardous waste to an off-site landfill for disposal. Also, since liquid wastes should not be disposed of in a landfill without prior treatment, chemical treatment is often used to make it either non-hazardous, or at least chemically convert it to a solid or semi-solid, which makes the contaminants chemically stable and not very mobile in the landfill environment.

Biological treatment: Biological treatment can be used for organic liquid wastes or organic solid wastes such as municipal wastewater, landfill leachate, contaminated soil, etc. Biological treatment may be categorized, according to the oxygen utilization, into aerobic and anaerobic processes. In the aerobic process, oxygen is required to decompose organic matter as the aerobic bacteria needs to grow and multiply. The anaerobic process uses anaerobic bacteria, in an oxygen deficient atmosphere, to decompose organic matter. Aerobic organisms are most commonly used to treat industrial and municipal wastewater. Anaerobic systems are usually used for the treatment of concentrated organic waste or organic sludges. Technologies have been developed in which anaerobic bacteria can be used to treat complex toxic organics such as solvent contaminated groundwater. Aerobic bacteria is used commonly for the treatment of petroleum contaminated soils and sludge.

Sustainable treatment: Sustainable treatment is a new term and is defined as "the type of treatment or combination of different types of treatments able to recover the raw material in order to conserve the natural resources on the condition that there is a full utilization or recycling of all effluents from the treatment facility". The first example of sustainable treatment is through mechanical treatment or recycling to convert waste into raw material and produce other products as explained above and will be explained in detail later on. The second example of sustainable treatment is through biological treatment to convert organic waste into a safe byproduct such as composting. Composting is an aerobic biological treatment process to convert organic waste into soil conditioner or organic fertilizer as will be discussed in detail in Chapters 5 and 7. The third example of sustainable treatment is through physical treatment to separate the waste streams from each other, for example the gravity oil separator (GOS) separates the oil and grease from oily water by gravity. The oil and grease can be recycled into the industrial process and the water with some remaining oil can be further treated physically through a dissolved air flotation (DAF) unit to separate the remaining oil and recycle both the remaining oil and water into the industrial process. The fourth example of sustainable treatment is through chemical treatment to separate raw material from waste by precipitation such as chromium recovery from liquid waste effluent produced by the tanning industry through pH control.

It is necessary for any establishment to treat its waste so that it complies with environmental protection regulations. Some industries resisted compliance in order to avoid costs. Now, new industries are accepting waste treatment as an integrated part of production cost. The added costs must then be passed on to consumers or deducted from the profits of the firm depending on market competition.

Through the traditional waste management hierarchy, hazardous waste should be treated before final disposal according to international regulations. Therefore, treatment means converting harmful waste into less harmful waste. In other words, treatment means converting waste from one form to waste in another form. The direct cost of waste treatment is more than just the expense of capital equipment and running cost (maintenance, operation and labor). This direct cost represents only a portion of the total cost. The other indirect cost may not be as easily identified and quantified. This includes the disposal cost and the cost related to adverse impact of the waste on the environment – contaminating air, water and land – as well as the equivalent cost of depleting the natural resources.

Some industries claim that it is not possible to have both jobs and capital spending for growth and at the same time, clean air and water. This statement is not true for industry, because wastes and emissions were originally raw material and should be treated as a byproduct not as a waste through reusing, recycling or recovery techniques – or sustainable treatment, a more generic term.

Treatment should be modified in the hierarchy of waste management for conservation of natural resources to sustainable treatment such as material recovery through physical treatment or biological treatment, etc., as explained above. In other words, what degree of treatment is required to arrive at the optimum outcome for material recovery without damaging the environment and depleting the natural resources? Thus, traditional treatment can be partially or completely eliminated for a new waste management hierarchy to sustainable treatment.

The optimum approach that industries can use to eliminate environmental damage completely is to weigh the pros and cons of each technique of the hierarchy. Economic indicators should be used through cost/benefit analysis as a primary criterion in making the decision but the health, safety and environment (HSE) intangible benefits, including the environmental monetary benefits of abating pollution, should be considered. The challenge of industry is to determine which techniques of the hierarchy, including treatment to some degree (if applicable), should be followed. Although technical parameters such as quantity and quality of waste are the primary factors, economical, political, social and psychological factors are also extremely important.

Innovative sustainable treatment technologies are required to solve the problem of industrial pollution through each of the cleaner production hierarchy techniques as will be explained later in detail in Chapter 2, such as:

- Reduction at the source by:
  - changing the raw material to one of better quality;
  - product modification.
- Reuse directly within a plant or indirectly by other industrial plants and/or recycle (on site) the waste stream resource.
- Marketing of stream resources (off-site reuse or recycling) and mixed with another industrial waste to produce a valuable product.
- Recovery of materials by sustainable treatment, for example the gravity oil separator (GOS) and dissolved air flotation (DAF) in the oil and soap industry to recover fat and grease and recycle the water in order for the effluent to comply with environment protection regulations.

On the contrary, the less the waste treatment provided by industry, the greater the cost of environmental damage. If industry does not provide waste treatment, environmental damage cost will be maximum. This will bring us to a very complicated formula, which is if no waste treatment the damage will be high, and if no proper disposal facilities the damage will be high too. Therefore, what is the solution for sustainable development without damaging the environment and depleting the natural resources? The solution is to approach a cradle-to-cradle concept through sustainable treatment as will be defined in sections 1.7 and 1.8 and implemented throughout the book.

1.4 Incineration

Hazardous and non-hazardous, solid and liquid wastes can be incinerated to convert them into ash. Incineration is the process of thermally combusting solid/liquid waste or, briefly, incineration is a thermal treatment process. There are various types of incinerators and the type used depends on the type of waste to be burnt. Conceptually, incinerators can be classified into the following common types of systems (Dasgupta and El-Haggar, 2003):

- Liquid feed incinerator
- Rotary kiln incinerator
- Grate-type incinerator
- Fluidized-bed incinerator

A liquid feed incinerator can handle liquid waste while a rotary kiln incinerator can handle both liquid wastes and solid wastes. A grate-type incinerator is used for large irregular-shaped solid waste and allows air to pass through the grate from below into the wastes. The fluidized-bed type incinerator is used for liquid, sludge and/or uniformly sized solid waste.

Liquid feed incinerator

A large number of hazardous liquid waste incinerators used today are of this type. The waste is burned directly in a burner or injected into a flame zone or combustion zone of the incinerator chamber through atomizing nozzles. The heating value of the waste is the primary determining factor for the location of the injection point of the liquid waste. Liquid injection-type incinerators are usually refractory-lined chambers, generally cylindrical in cross-section, and equipped with a primary burner (waste and/or auxiliary fuel fired). Often secondary combustors or injection nozzles are required where low heating value waste liquids are to be incinerated. Liquid incinerators operate generally at temperatures from 900°C to 1,500°C. Residence time in the incinerator may vary from milliseconds to as much as 3 seconds. The atomizing nozzle in the burner is a critical part of the system because it converts the liquid waste into fine droplets. Two fluid atomizers, using compressed air or steam according to fuel/waste type as an atomizing fluid, are capable of atomizing liquids. The atomizer design is therefore an important parameter of this system. The reasons for injecting the liquid waste as a fine spray are: (1) to break up the liquid into fine droplets, (2) to develop the desired pattern for the liquid droplets in the combustion zone with sufficient penetration and kinetic energy, and (3) to control the rate of flow into the combustion zone. In a good atomizer, the droplet size will be small providing greater surface area and resulting in rapid vaporization. The type of atomizer depends on the type of liquid waste and the combustion conditions required. Proper mixing of air with the atomized droplet is very important for complete oxidation. The method of injection of the liquid waste is one of the critical factors in the design and performance of these incinerators. If the liquids contain fine solids, the design must allow the particles to be carried to the gas stream without agglomeration affecting proper combustion.

Sufficient time must be provided to permit complete burnout of the solid particles in the liquid suspension. Inorganic particles carried in the liquid waste stream may become molten and agglomerate into molten ash. The combustor must be designed to collect the molten ash without plugging the flow passages of the incinerator. Primary and secondary combustion chambers are used in liquid feed incinerators. Primary combustion chambers are used to burn wastes, which have sufficient heating value to burn without auxiliary fuel. Secondary combustion chambers require auxiliary fuel to control the temperature and destroy the toxic emission. Sufficient air must be provided at all times to oxidize the organics in the combustion chamber. Incinerators can produce soot when burning under insufficient oxygen and poor air mixing conditions. Soot can clog up nozzles, and accumulate in the chamber, impairing burning conditions. The physical, chemical, and thermodynamic properties of the waste must be considered in the design of the incinerator.

Rotary kiln incinerator

The rotary kiln is often used in solid/liquid waste incineration because of its versatility in processing solid, liquid, and containerized wastes. The kiln is refractory lined. The shell is mounted at a 5 degree incline from the horizontal plane to facilitate mixing the waste materials. A conveyor system or a ram usually feeds solid wastes and drummed wastes. Liquid hazardous wastes are injected through a nozzle(s). Non-combustible metal and other residues are discharged as ash at the end of the kiln. Rotary kilns are also frequently used to burn hazardous wastes.

Rotary kiln incinerators are cylindrical, refractory-lined steel shells supported by two or more steel trundles that ride on rollers, allowing the kiln to rotate on its horizontal axis. The refractory lining is resistant to corrosion from the acid gases generated during the incineration process. Rotary kiln incinerators usually have a length-to-diameter (L/D) ratio between 2 and 8. Rotational speeds range between 0.5 and 2.5 cm/s, depending on the kiln periphery. High L/D ratios and slower rotational speeds are used for wastes requiring longer residence times. The kilns range from 2 to 5 meters in diameter and 8 to 40 meters in length. Rotation rate of the kiln and residence time for solids are inversely related; as the rotation rate increases, residence time for solids decreases. Residence time for the waste feeds varied from 30 to 80 minutes, and the kiln rotation rate ranges from 30 to 120 revolutions per hour. Another factor that has an effect on residence time is the orientation of the kiln. Kilns are oriented on a slight incline, a position referred to as the rake. The rake typically is inclined 5° from the horizontal.

Hazardous or non-hazardous wastes are fed directly into the rotary kiln, either continuously or semi-continuously through arm feeders, auger screw feeders, or belt feeders to feed solid wastes. Hazardous liquid wastes can also be injected by a waste lance or mixed with solid wastes. Rotary kiln systems typically include secondary combustion chambers of afterburners to ensure complete destruction of the hazardous waste. Operating kiln temperatures range from 800°C to 1,300°C in the secondary combustion chamber or afterburner depending on the type of wastes. Liquid wastes are often injected into the kiln combustion chamber.

The advantages of the rotary kiln include the ability to handle a variety of wastes, high operating temperature, and continuous mixing of incoming wastes. The disadvantages are high capital and operating costs and the need for trained personnel. Maintenance costs can also be high because of the abrasive characteristics of the waste and exposure of moving parts to high incineration temperatures.

A cement kiln incinerator is an option that can be used to incinerate most hazardous and non-hazardous wastes. The rotary kiln type is the typical furnace used in all cement factories. Rotary kilns used in the cement industry are much larger in diameter and longer in length than the previously discussed incinerator.

The manufacture of cement from limestone requires high kiln temperatures (1,400°C) and long residence times, creating an excellent opportunity for hazardous waste destruction. Further, the lime can neutralize the hydrogen chloride generated from chlorinated wastes without adversely affecting the properties of the cement. Liquid hazardous wastes with high heat contents are an ideal supplemental fuel for cement kilns and promote the concept of recycling and recovery. As much as 40% of the fuel requirement of a well-operated cement kiln can be supplied by hazardous wastes such as solvents, paint thinners, and dry cleaning fluids. The selection of hazardous wastes to be used in cement kiln incinerators is very important not only to treat the hazardous wastes but also to reap some benefits as alternative fuel and alternative raw material without affecting both the product properties and gas emissions. However, if hazardous waste is burned in a cement kiln, attention has to be given to the compounds that may be released as air emissions because of the combustion of the hazardous waste. The savings in fuel cost due to use of hazardous waste as a fuel may offset the cost of additional air emission control systems in a cement kiln. Therefore with proper emission control systems, cement kilns may be an economical option for incineration of hazardous waste.

Grate-type incinerator

Grate-type incinerators are used for incineration of solid wastes. These types of incinerators are suitable for large irregular-shaped wastes, which can be supported on a stationary or moving alloy grate, which allows air to pass through the grate from below into the waste. Generally, grate-type incinerators have limited application for hazardous waste incineration because the high temperature required in the chamber may affect the material of the grate. The primary furnace is followed by a secondary combustion chamber, where additional air and fuel are added, to complete destruction of all toxic emissions.

Fluidized-bed incinerator

This type of incinerator is used for liquid, sludge, or uniformly sized solid waste. The fluidized-bed incinerator utilizes a fluidized bed consisting of sand or alumina on which combustion occurs. Air flow is applied from below which has sufficient pressure to allow it to fluidize the bed of sand, or hold it in suspension, as long as the velocity of the air is not so great that it transports the sand out of the system. This is a fluidized bed in which the particles of the bed are in suspension, but not in flow. Waste is added/injected into this fluidized bed. The air with which the bed is fluidized is heated to at least ignition temperature of the waste, and the waste begins to burn (oxidize) within the bed. Most of the ash remains in the bed, and some exits the incinerator into the air pollution control system. Heat also exits with the flue gases and can be captured in a boiler or used to preheat combustion air.

In almost all types of waste incinerators, primary and secondary combustion chambers are necessary to complete the combustion and oxidation to achieve the required destruction and removal efficiency (DRE). The primary chamber's function is to volatilize the organic fraction of the waste. The secondary chamber's function is to heat the vaporized organics to a temperature where they will be completely oxidized.

Advantages of incineration:

- It is applicable to all kinds of waste.
- Incinerators are made to avoid air pollution through air pollution control units.
- The ash resulting from the combustion occupies only around 10% of the solid waste volume.
- Energy may be recovered from incineration through many ways, for example gas to water heat exchange. The water is converted to steam, which may then be used to generate electricity through steam turbines.

Disadvantages of incineration:

- Incinerator construction requires high capital cost.
- Incineration operation and management require high cost and skilled workers.
- Wastes require energy to be burnt.
- The air pollution control systems are very expensive. On the other hand, the emissions and the ash resulting from incineration are extremely dangerous. If not properly controlled, they cause air pollution which can have dangerous effects on human health.

1.5 Landfill

Solid waste and/or ash produced from the incineration process should be landfilled. Most people think of landfills as dumpsites, just an open hole in the ground where solid waste is put with all kinds of animals and insects wandering around. This is not true. A landfill is a very complicated structure, very carefully designed and positioned either into or on top of the ground in which solid waste is isolated from the surrounding environment. There are many steps before starting to construct a landfill. The first step is to choose a suitable site. Site location is one of the very important steps to avoid any impacts on the surrounding environment. After choosing the appropriate location for a landfill, the designing process starts. Beside the design of the lining and coverage of the landfill, a leachate collection system, biogas collection system and a storm water drainage system should also be designed and planned for implementation during operation.

The bottom liner isolates the solid waste from the soil preventing groundwater contamination. The liner is usually some type of durable, puncture-resistant synthetic plastic (polyethylene, high-density polyethylene, polyvinyl chloride). In landfills constructed below surface level, a side liner system is used which maintains mechanical resistance to water pressure, drainage of leachate and prevention of lateral migration of biogas.

The landfill size is directly related to the capacity and the lifetime of the landfill. To maximize the landfill's lifetime, solid waste is compacted into areas, called cells. Each cell contains only one day's solid waste. Each cell is covered daily with 15 cm of compacted soil. This covering seals the compacted solid waste from the air and prevents pests (birds, rats, mice, flying insects, etc.) from getting into the solid waste.

It is impossible to totally exclude water from the landfill. The water percolates through the cells in the landfill. As the water percolates through the solid waste, it picks up contaminants (organic and inorganic chemicals, metals, biological waste products of decomposition). This water with the dissolved contaminants is called leachate and is typically acidic. It is then collected into a pond by means of perforated pipes and sent for treatment.

Bacteria in the landfill break down the solid waste in the absence of oxygen (anerobic) because the landfill is airtight. A byproduct of this anerobic breakdown is landfill gas, which contains approximately 60% methane and 40% carbon dioxide with small traces of other gases. This presents a hazard because the methane can explode and/or burn if it is not collected and utilized. So, the landfill gas must be removed or collected for utilization through a pipe network embedded within the landfill. After the landfill has been closed a layer of soil is put above it to prepare it for landscaping.

Disadvantages of landfill:

- Landfill construction requires high capital cost.
- Mismanagement of landfill may cause soil contamination, water contamination as well as air pollution.

- Landfill operation requires high capital cost.
- The leachate collection and treatment facility and gas collection system might require huge capital.
- Natural resources are depleted.

1.6 Zero Pollution and 7Rs Rule

The rate of waste generated is increasing with population rate and social standards, i.e. the more advanced and wealthy societies (individuals) produce more waste. The amount of municipal waste for 2004 according to the Global Waste Management Market Report (Research and Markets, 2004) is about 1.82 billion tons with a 7% increase compared to 2003. The amount of waste is estimated to rise by 31% in 2008 to reach nearly 2.5 billion tons. This huge amount would cover the total continent of Australia at a thickness of 1 mm. Meanwhile, finding new sources of raw material is becoming costly and difficult. Concurrently, the cost of safe disposal of waste is escalating exponentially and even locating waste disposal sites is becoming more difficult. As a result, a new hierarchy for waste management to approach full utilization of waste is a must, which starts from reduction at source, reuse, recycle and sustainable treatment for possible material recovery for conservation of natural resources.

In the past, people's dream was to turn sand into gold. Today, the dream is to turn waste and pollution into gold. This was a dream until the new hierarchy to approach zero pollution was developed at the American University in Cairo (El-Haggar, 2001c) and the 6Rs Golden Rule was initiated (El-Haggar, 2003c). That is, the rule aims at Reducing, Reusing, and Recycling waste. The fourth R of the 6Rs Golden Rule emphasizes the Recovering of raw materials from waste through sustainable treatment. The last 2Rs are Rethinking and Renovation where people should rethink about their waste before taking action for treatment and develop renovation-innovative techniques to solve the problem. Another R should be added at the top to the previous 6Rs, which is Regulation; without Regulation nothing will be implemented. It is very important to add Regulation to the system and enforce the 6Rs Golden Rule into the management system. So the 7Rs Golden Rule encompasses Regulation, Reducing, Reusing, Recycling, Recovering, Rethinking and Renovation as the basic tools for zero pollution (El-Haggar, 2004a). The rule provides a methodology to manipulate current activities to approach zero pollution and avoid landfill, incineration and/or traditional treatment. This approach is based on the concept of adapting the best practicable environmental option for individual waste streams and dealing with waste as a byproduct. This 7Rs Golden Rule for zero pollution can be considered the Sustainable Waste Management Hierarchy for Zero Pollution and the Industrial Ecology Hierarchy for Zero Pollution as will be explained in Chapter 3. Fortunately, with full success, the theory was practically applied in a pilot scale to most of the industrial sectors (El-Haggar, 2000a, 2002, 2003a), numerous projects (El-Haggar, 2003b) and rural communities (El-Haggar, 2001b) as will be explained throughout this book.

Capital investment, running cost as well as adverse environmental impacts of landfills, incineration and treatment heartens the implementation of the 7Rs Golden Rule. It is very simple, natural and not a newborn theory. Fundamentally, the theory depends on all kinds of recycling (on-site recycling, off-site recycling, sustainable treatment for possible recycling, etc). This is mainly because recycling is considered a pivotal income generated activity that conserves natural resources, protects the environment and provides job opportunities.

Zero pollution can be defined as the pollution generated from any manmade activity and should be within the allowable limits stated by the national or international environmental regulations. The concept of zero pollution is not new; Professor Nemerow (1995) developed a methodology for the future of industrial complexes or parks to approach zero pollution. Most of the countries (developed and developing countries) are working toward zero pollution not only in industrial sectors but also in all other sectors.

From the above analysis, it is clear that traditional treatment, incineration and final disposal through landfilling processes for solid and liquid waste require a huge capital and might cause environmental problems if it is not managed and operated properly. If the treatment/disposal is not done properly, or the treatment/disposal facility is not well designed and constructed, the adverse impacts will be significant. Who will pay for the treatment/disposal facility? Who will manage the treatment/disposal facility? Who will run the treatment/disposal facility? Government, society or industry? In conclusion, sustainable treatment has proved to be the most suitable solution and can be implemented to solve the problem of waste not only in developed countries but also in developing countries compared with incineration and/or landfill according to the 7Rs Golden Rule using the concept of cradle-to-cradle that will be discussed in the next section.

1.7 Life Cycle Analysis and Extended Producer Responsibility

The USEPA has defined life cycle assessment/analysis (LCA) as a method to evaluate the environmental effects associated with any given industrial activity from the initial gathering of raw materials from the earth to the point at which all residuals are returned to the earth, a process also known as cradle-to-grave.

LCA results will not be promising as long as the evaluation is done for industrial activities that adopt a cradle-to-grave flow of materials. Unfortunately most manufacturing processes since the industrial revolution are based on a one-way, cradle-to-grave flow of materials; initially by extracting raw materials, followed by processing, producing goods, selling, utilization

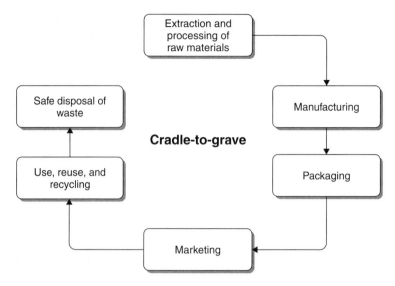

FIGURE 1.1 Traditional life cycle analysis

by consumers and finally disposal and waste generation as shown in Figure 1.1. The technological advancements in manufacturing processes and the constantly increasing variety of materials and products have led to a continuous rise in the amounts of waste generated. The cradle-to-grave flow of materials has proven to be just enough to protect the environment but inefficient due to the depletion of natural resources.

Several organizations have developed methods for LCA each using a different analytic approach to this complex activity. Regardless of the approach, several generic difficulties challenge LCA, including poor quality data, weak reasons or procedures for establishing analytic boundaries, and diverse values inherent in comparing environmental factors with no common objective, quantitative basis. The current selection of products undergoing LCA has been chaotic; some products have been strongly scrutinized while others have been totally neglected.

LCA is a tool for identifying the impacts of a product on the environment throughout its entire life cycle or, in other words, from cradle-to-grave, including the extraction of raw materials; the processing of raw materials in order to fabricate a product; the transportation and distribution of the product to the consumer; the use of the product by the consumer; and finally the disposal of the product's materials after its use (TecEco Pty. Ltd., 2006) as shown in Figure 1.1. The main components of a life cycle analysis or assessment should include the identification and quantification of not only the waste generated through the entire life cycle but also the raw materials and energy requirements throughout the entire life cycle and their environmental impacts. Comparing existing methods for LCA gives insight into the conceptual framework used by researchers. The code of practice for LCA stands out currently as the most widely recognized procedural model. The code divides LCA into four distinct components: (1) scoping; (2) compiling quantitative data on direct and indirect materials/energy inputs and waste emissions; (3) impact assessment; and (4) improvement assessment. While variations exist, the theme of taking an inventory and performing an assessment based on collected data is common to all LCA approaches dating back to the early 1970s (Wernick and Ausubel, 1997).

There have been a lot of different methods developed by researchers to obtain LCA. Though some methods for LCA receive approval for thoroughness and analytic consistency, these same methods have been criticized as requiring too much data, time, and money when each is in short supply. As an alternative method for assessing the environmental impact of products, researchers at AT&T have devised the Abridged Life Cycle Assessment Matrix, a method that couples quantitative environmental data with qualitative expert opinion into an analysis that conveys the uncertainty and multidimensionality of LCA and also yields a quantitative result (Wernick and Ausubel, 1997).

A lot of work has been done to develop methodologies, guidelines, benefits, etc. for LCA according to the "cradle-to-grave" concept to protect the environment throughout the life cycle of the product. Nowadays, individuals and organizations adhere to the International Organization for Standardization (ISO) which has developed a series of international standards to cover LCA in a more global sense such as ISO14040 (LCA-Principals and guidelines), ISO14041 (LCA-Life Inventory Analysis), ISO14042 (LCA-Impact Assessment), and ISO14043 (LCA-Interpretation). All ISO 1404X related to LCA based on "cradle-to-grave" approach for environmental protection.

It is now time to change the LCA-ISO standard from a "cradle-to-grave" concept to protect the environment to a "cradle-to-cradle" concept to protect not only the environment but also the natural resources as will be discussed in the next section. This might require an added responsibility to the producer according the principle of extended producer responsibility.

The concept of producer responsibility was introduced to solve the problem of waste recycling from the beginning "design phase" to give full responsibility to the producer to select recyclable material in the products. The concept of producer responsibility was first introduced by Riddick (2003) and requires all producers to be responsible for any environmental impacts their products may have throughout the product's wide life cycle and not just at the end. In reference to the Organization for Economic Co-operation and Development (OECD), Riddick defined producer responsibility as "an environmental policy approach in which a producer's responsibility, physical, and/or financial, for a product is extended to the post consumer stage of a product's life cycle". Applying this concept shifts the physical and financial responsibilities of managing these wastes from the government to the producer. Riddick explained that since producers are fully responsible for designing products, they must also become responsible for dealing with the damage their products have caused to the environment. This will give producers the incentive to change the design of their products and take into consideration the "recyclability" of these products when they are disposed of as waste. The number of components as well as the amount and types of material used can then be reduced. According to Riddick, "Producer Responsibility can also address cradle to grave environmental problems" by encouraging reductions in the use of natural resources, extension of the product's lifetime, and proper management of the product at the end of its useful life through recovery or recycling. A key objective of this principle is to include the cost of use and disposal of products together with the cost of manufacturing in the total cost of the consumer product. He further explained that there are a range of policy instruments that governments can use to apply the principle of producer responsibility mechanism; it depends on the priority of the government whether it is sustainable development, product policy, or waste management.

Extended producer responsibility (EPR) was also defined by McKerlie *et al.* (2005) as "a policy measure that recognizes the producer's role in reducing the impacts of their product throughout its entire life cycle, including waste management or recovery at end-of-life". He illustrated the principle of EPR in a form of a cycle that encompasses both the upstream and downstream phases of a product's life cycle. McKerlie *et al.* claimed that the producers have the greatest ability to develop the product and induce changes in its different stages in order to achieve environmental improvements. He added that enforcing the "polluter pays principle" encourages product innovation and pollution prevention activities. In general, effective EPR programs can influence the development of more sustainable materials management systems and encourage Design for Environment practices such as dematerialization, the elimination of toxic chemicals in products, and the reuse of products and packaging. They stated that EPR has encouraged more sustainable packaging design and resulted in significant waste reduction. Design changes such as an increase in the use of reusable packaging, a reduction in the use of composite and plastic packaging, and changes to container shapes and sizes have led to a reduction of about 66% in weight of packaging waste to be landfilled and incinerated in Germany. Although applying producer responsibility in the UK increased the percentages of packaging waste recovery and recycling, it led to some undesirable consequences: the amount of packaging in the market was not minimized, recycling was favored over reuse, disposal of non-packaging wastes was favored over their recycling in addition to problems regarding waste collection (Riddick, 2003). Riddick finally argued that the application of producer responsibility needs further modifications and refining.

1.8 Cradle-to-cradle Concept

LCA is a very important tool to guarantee there is no harmful impacts on the environment starting from extracting the raw material (cradle) all the

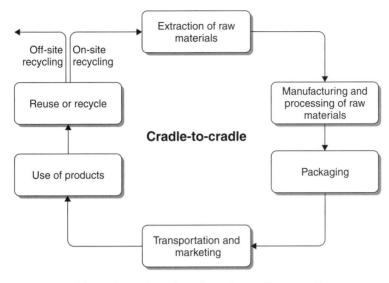

FIGURE 1.2 New life cycle analysis based on the cradle-to-cradle concept

way to the final disposal in a landfill (grave). In other words, the product's design should be selected according to a safe disposal process. This process will protect the environment but will deplete natural resources. If the product is designed according to the cradle-to-cradle concept, materials can be reused, no wastes get produced or can be recycled and, accordingly, there are no negative impacts on the environment generated within the closed loop of the life cycle of the product as shown in Figure 1.2. This can be achieved by having industries change their products from having a cradle-to-grave design where the product will eventually get disposed of in a landfill at the end of its life to a cradle-to-cradle designed product where the materials are circulated in a closed loop without losing any natural resources.

Due to development of strict government regulations more attention has been directed towards eco-efficiency, which implies minimizing waste disposal, pollution and natural resource depletion. Although some environmental, economic and health benefits have been achieved due to eco-efficiency, sustainable development cannot be based on improving the efficiency of a fundamentally destructive cradle-to-grave system. It can only be achieved in the presence of an effective system that fundamentally promotes sustainability.

The environmental and health impacts as well as depleting the natural resources as a result of using traditional treatment, incineration and/or final disposal through landfill are becoming more dangerous and disastrous for sustainable development. Establishing or approaching a new LCA based on the concept of cradle-to-cradle instead of cradle-to-grave by a full utilization of raw material, water and energy using the 7Rs Golden Rule (El-Haggar, 2004b) is a must for sustainable development.

Braungart and McDonough (2002) proposed a shift from a cradle-to-grave approach where waste products are disposed of in a landfill to a cradle-to-cradle approach where waste can be used for the production of other products. They recommended the "eco-effective" recycling approach to enable material reuse with high quality. They added that combining different materials in one product prevents the products from being fully recycled. Accordingly, product designers need to plan for the reuse of their products in order to prevent waste generation. This shift in products' design approach will require an added responsibility of the producer – EPR as discussed above.

The cradle-to-cradle concept promotes sustainable development. It is a system of thinking based on the belief that human endeavors can emulate nature's elegant system of safe and regenerative productivity, by transforming industries to sustainable enterprises and eliminating the concept of waste.

Natural ecosystems are based on principles, which can be adopted by humans in industry. For example, no waste generation; in natural ecosystems an organism's waste is consumed by others. This can be applied in industry such that one industry's wastes are another's raw material – "industrial ecology" will be discussed in Chapter 3. This is the fundamental concept of eco-industrial parks, where industries are grouped together to have a continuous flow of material and no waste generation as in the case of an eco-industrial park in Kalundborg, Denmark, which will be discussed in detail in Chapter 3.

Adopting cradle-to-cradle principles creates a cyclical flow of materials, as opposed to the one-way cradle-to-grave concept. The materials consumed in industry resemble the nutrients that flow cyclically in natural ecosystems and can circulate in one of two metabolisms, biological or technical.

According to the cradle-to-cradle concept, products are designed to be made of materials that can be safely manufactured, used, recovered, and reused while still maintaining their high value throughout their life cycle. This way valuable used material can be continuously cycled in closed loops and transformed for reuse as other products. By applying the principle of cradle-to-cradle design and transforming industrial systems to a closed loop system of material flow, not only will this design save the environment from waste generation and negative impacts, but industries can even benefit from the continuous availability of products made of high value material even after the useful stage of the product's life.

Questions

1. Discuss the differences between treatment and sustainable treatment.
2. Discuss the differences between the traditional waste management hierarchy and the sustainable waste management hierarchy.

3. Is the traditional landfilling process sustainable? Why?
4. What are the differences between the cradle-to-cradle concept and the cradle-to-grave concept?
5. What are the advantages and disadvantages of waste disposal through landfill?
6. Why is incineration as a thermal treatment process not considered a sustainable treatment?
7. What is the difference between recycling and material recovery?

Chapter 2

Cleaner Production

2.1 Introduction

The term cleaner production (CP) was launched 1989 by the United Nations Environment Program (UNEP) as a response to the question of how to produce in a sustainable manner. Its core element is prevention vs clean-up or end-of-pipe treatment to environmental problems. Resources should be used efficiently thus reducing environmental pollution and improving health and safety. Economic profitability together with environmental improvement is the aim. Cleaner production typically includes measures such as good housekeeping, process modifications, eco-design of products, and cleaner technologies, etc. The United Nations Environment Program defines cleaner production as "the continuous application of an integrated, preventative environmental strategy to processes, products and services to increase eco-efficiency and reduce risks to humans and the environment" (UNEP, 1997).

Cleaner production is a preventive approach to environmental management; it is a term that encompasses eco-efficiency and pollution prevention concepts with risk reduction to humans and the environment. Cleaner production is not against economic growth, it just insists that the growth is sustainable. Therefore cleaner production is a "win–win" strategy. It protects the environment, conserves our natural resources while improving the operating industrial efficiency, profitability, company image and competitiveness. Cleaner production focuses on conservation of natural resources such as water, energy and raw materials and avoids the end-of-pipe treatment. It involves rethinking for products, processes and services to move towards sustainable development.

By considering production processes, cleaner production includes conserving raw materials and energy, eliminating toxic raw materials, and reducing the quantity and toxicity of all emissions and wastes before they leave a process. For products, the strategy focuses on reducing impacts along the entire life cycle of the product. Cleaner production is achieved by applying know-how, by improving technology, and by changing attitudes. Changing

Leabharlann
6447552

attitudes is the most challenging and the most important step in applying the cleaner production concept.

The conceptual and procedural approach to production which demands all phases of the life cycle of products, must be addressed with the objective of the prevention or minimization of short- and long-term risks to humans and to the environment.

One factor in defining cleaner production is therefore the reduction in production costs that results from improved process efficiencies. In terms of investment the key difference is that investment in end-of-pipe technologies "treatment" is nearly always additional investment, whereas investment in cleaner production always pays. This has obvious implications for employment and production cost.

A useful definition of cleaner production needs to take account of the distinction between technologies and processes. For example, a process may be made "cleaner" without necessarily replacing process equipment with cleaner components – by changing the way a process is operated, by implementing improved housekeeping or by replacing a feedstock with a "cleaner" one. Cleaner production may or may not, therefore, entail the use of cleaner technologies. Investment in cleaner production via the implementation of clean technologies is clearly easier to identify than investment in cleaner production by any other means. Whatever the method employed to make production cleaner, the result is to reduce the amount of pollutants and waste generated and reduce the amounts of non-renewable or harmful inputs used.

Most of the developed and developing countries are working toward zero pollution not only in industrial sectors but also in vehicle emissions to reduce gaseous emissions to allowable limits and in other sectors such as construction and agricultural. To approach zero pollution, industry should prevent all pollutants from its effluent. The cleaner production hierarchy to eliminate all pollutants and approach zero waste/pollution should start from raw material selection through to recycling and all the way to product modifications in order to avoid end-of-pipe treatment as will be explained throughout this chapter.

2.2 Promoting Cleaner Production

The UNEP launched the Cleaner Production Program in 1989 in response to the need to reduce worldwide industrial pollution and waste. Positive future expectations exist for the spread of the cleaner production concept, as it combines maximum effect for the environment with significant economic savings for any business.

Although it is up to industry to adopt a cleaner production concept, the role of government is required to encourage industries to begin their own CP programs. The tools that governments use regarding their country's needs and circumstances fall into:

Adopting regulations: The regulations specify the environmental goals, methodology of achieving them, and technology used.

Economic instruments: Economic instruments are used to make the costs of pollution more expensive than the costs of cleaner production. Two forms of these instruments exist; those that provide rewards and those that penalize.

Provide support measures: Government can provide support in five main areas:

- Providing information about cleaner production.
- Assisting in the development of management tools in the industry.
- Organizing training workshops.
- Promoting the concept in engineering schools, universities and research institutes.
- Providing a case study with cost/benefit analysis in different industrial sectors.

Provide external assistance: Assistance can take several forms; financial aid, development of case studies in different sectors, technology transfer, and exchange of expertise.

Provide guidelines for implementing CP: Three main steps are required to initiate cleaner production successfully: (1) structured methodology, (2) management commitment, (3) operator's involvement.

2.3 Benefits of Cleaner Production

Cleaner production can reduce operating costs, improve profitability and worker safety, and reduce the environmental impact of the business. Companies are frequently surprised at the cost reductions achievable through the adoption of cleaner production techniques. Frequently, minimal or no capital expenditure is required to achieve worthwhile gains, with fast payback periods. Waste handling and charges, raw material usage and insurance premiums can often be cut, along with potential risks. It is obvious that cleaner production techniques are good business for industry because it will:

- Reduce waste disposal cost.
- Reduce raw material cost.
- Reduce Health Safety Environment (HSE) damage cost.
- Improve public relations/image.

- Improve companies performance.
- Improve the local and international market competitiveness.
- Help comply with environmental protection regulations.

On a broader scale, cleaner production can help alleviate the serious and increasing problems of air and water pollution, ozone depletion, global warming, landscape degradation, solid and liquid wastes, resource depletion, acidification of the natural and built environment, visual pollution, and reduced bio-diversity.

2.4 Obstacles to Cleaner Production and Solutions

An industrial program in education must precede a successful reuse/recycling program by acquainting plant personnel with the potential value contained in the waste. Detailed qualitative analysis of wastes should be made available over relatively long periods of time (one year).

Establishing another industry correlating as much industrial wastes as possible with some other additives to produce a good quality product is the challenging point because industrial wastes vary in quality and quantity from time to time according to changes of products' types and amounts.

To overcome these obstacles, a government/industrial development agency assistance would be helpful in obtaining agreement for locating plants of these types with all the necessary licenses and support. In fact, the newly established industries based on industrial waste offer one of the most promising long-term solutions to today's environmental pollution problems as well as to many future industrial economic problems as a result of any damage that might occur.

Waste treatment may cost more than an establishing industrial plant based on waste as a raw material. This will lead to economic stability. Resources are limited and there is competition between users and consumers of these resources. What route should an industry follow? Cease production? Move to another site?

Cleaner production barriers

Although cleaner production techniques have evolved greatly over the past years, and have consequently been adopted by and implemented in many industries, there still remain two major "attitude problems" that present obstacles to an even more widespread realization of cleaner production.

The first attitude problem pertains to those who fear being seen as fools, going against tradition, being alone, being criticized, or making mistakes.

The second problem involves the "idea killers" who would say things like "let's think about it later", "we have already tried it", "it's not the right time", "you don't understand the problem", "talk to John, it's not my field", etc.

2.5 Cleaner Production Techniques

The key difference between cleaner production and other methods like pollution control is the choice of the timing, cost and sustainability. Pollution control follows a "react and treat" rule, while cleaner production adopts "prevention is better than cure". Cleaner production focuses on before-the-event techniques that can be categorized as shown in Figure 2.1 as follows:

- Source reduction:
  - Good housekeeping
  - Process changes:
    - Better process control
    - Equipment modification
    - Technology change
    - Input material change.
- Recycling:
  - On-site recycling
  - Useful byproducts through off-site recycling.
- Product modification.

Implementing cleaner production techniques while performing the cleaner production audit (as will be explained later) will generate options for cleaner production opportunities within the industry. It can be seen that there is no solid boundary between any of the above-mentioned cleaner production techniques. Some of the cleaner production opportunities might require one

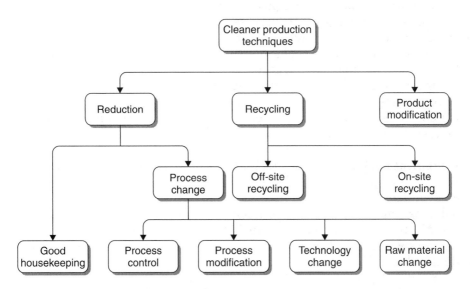

FIGURE 2.1 Cleaner production techniques

or more cleaner production techniques to be able to approach "sustainable development" as will be explained.

Good housekeeping: Also referred to as good operating practices and depends on reducing wastes from the source. It implies all the measures that a company can take to minimize waste and emissions. Good housekeeping involves either reducing potential wastes or conserving natural resources through optimizing the process and eliminating spill, leakage, overheating or any fault that would result in unnecessary losses. This can be achieved by following a regular preventive maintenance program for the production process or raising staff environmental awareness or through an incentive mechanism. Meanwhile, good housekeeping could be implemented at a very low cost or no cost through procedural instructions in production, maintenance, storage and material handling. The UNEP lists several good housekeeping measures that industries could adopt to enhance efficiency, which could be summarized in the following points (UNEP, 2000/2001):

- Minimize wastes and emissions by repairing all leakages from equipments and spillage. Keep taps closed when not in use.
- Separate the hazardous wastes from the non-hazardous to avoid mixing the two to decrease the volume of the hazardous wastes.
- Reduce the loss of input materials due to mishandling, expired shelf life and proper storage conditions.
- Perform employee training and incentives to encourage all employees to continuously strive to reduce wastes.

Better process control: Good practice is achieved by an effective and efficient planning and regulating process. It ensures that the conditions of the process are optimal with respect to resource consumption, production and waste generation. Process conditions such as temperature, pH, pressure, water level, time, etc. should be measured, monitored and maintained at possible optimum conditions. It requires continuous monitoring and management. Good control for the operating parameters will save energy, water and raw material as well as increasing the efficiency of the industrial process avoiding excessive wastes and emissions.

Equipment modification: This technique would include any modification in the existing equipment in order to improve the process and account for more efficient (optimum) utilization of raw materials, water or energy as well as reducing emissions to the environment. For example, switching energy from using heavy liquid oil "industrial fuel" to natural gas for boilers or industrial furnaces will have a lot of economical and environmental benefits. Heavy liquid oil has a high percentage of sulphur and other heavy metals and impurities. Sulphur generates sulphur oxides while heavy metal generates metal oxide. Both chemical compounds are very dangerous to human health. Heavy liquid oil produces high temperature gases causing the formation of NO_x, which is hazardous to the environment. In addition heavy oil has low combustion efficiency producing a high percentage of hydrocarbon with the exhaust. However, natural gas does not contain (or contains a very low percentage of) sulphur, heavy metals and impurities and emits lower levels of NO_x and CO_x. Using natural gas instead of heavy oil or other fuels will lessen the pollution because natural gas is more environmentally friendly. Moreover it will increase the efficiency of the combustion process. In order to achieve this switch from heavy oil or other types of liquid/solid fuel to natural gas, a change in combustion equipment is a must.

Another example for this technique is Wiggins Ltd, a company that operate a paper mill with a large output of waste water with consequent high charges for its disposal in the public sewage network. As a result a cleaner production industrial audit decided to replace the water lines, recycle cooling water, and install backwater tanks. This was a partial change in the process, its control and operation. It resulted in a 77% reduction in water consumption with overall reduction in cost (Williams, 1998).

In yet another example dealing with the food sector where very high water consumption was observed and measured during the washing process, it is always recommended to install high-pressure and low-volume spray nozzles to guarantee a good cleaning process with less water consumption. In the textile sector, with high energy and water consumption during wet processes, it is always recommended to use countercurrent flow rather than co-current flow to save water and energy as will be explained later in the case studies.

Technology change: Technology change is oriented towards processes modifications to reduce raw materials, water, energy as well as wastes and emissions, and ranges from minor changes that can take place in a process at low cost to the change of the whole production process that requires large capital investment. In other words, technology change involves replacing one process with another that could be less energy consuming or more efficient. For example, cement is produced by mixing limestone and clay together with some other additives in a dry or wet process, so using the dry process technology would consume less energy. Either production techniques can be used. To get the final product, the wet process requires evaporating water after mixing – this evaporation process consumes a large amount of energy – while the dry process requires less energy producing fewer air pollution emissions. Technology change using cleaner production techniques is always recommended as a last option.

Input material change: Input material change involves substituting one material for another which is less harmful to the environment, more feasible to use and has the same or better technical requirements. In the textile industry, for example, sulphur black dyeing, as will be explained later in detail, is widely used due to its low cost and good fabric quality. Sodium sulphate and sodium dichromate are both used in this process. They are toxic and hazardous chemicals. In addition they are dangerous to worker's health during handling and storage. The cleaner production opportunity assessment recommends substituting sodium sulphate with glucose and sodium dichromate with sodium perborate based on economical, technical and environmental feasibility. The new materials are safer to the environment and public health. They also improve the quality of fabric and the process's productivity.

On-site recycling: This involves returning the waste material either to the original process or to another process as input material (inside the same factory). In other words, on-site recycling means re-entering the waste into the process as a substitute for an input material or sent as useful byproducts or raw material for other processes within the same factory. For example, in the pulp and paper industry, fiber recovery from white paper, surplus pulp fiber, paper mill offcuts and damaged paper rolls are recycled into the pulping process. Another example, in the oil and soap industry the gravity oil separator (GOS) will separate the oil from the industrial wastewater stream and return it to the industrial process as explained above.

On-site recycling could be planned for and implemented through simple processes that might include equipment modifications and/or better process control in order to ensure that waste could be safely used as byproducts for the same industry with the required specifications.

Off-site recycling: This implies that the recycling process is done by another party that recycles the industrial wastes or at the post-consumer stage such as municipal solid waste. There are companies specialized in recycling the specific wastes. They buy certain types of waste, recycle them then sell them to other industries. All municipal solid waste recycling, which is explained in detail in Chapters 5 and 6, is considered off-site recycling.

Product modification: This is oriented towards change in the product design to use less raw material or energy or to minimize emissions. For example, reducing the metal content of soft drink cans by changing their design and the material from steel to aluminum. This change in can design made it lighter to handle and used less material to manufacture. Another example is a paper mills using unbleached toilet tissues instead of white tissue to save raw material and chemicals.

This product modification change is used to reduce not only waste generated but also materials consumed. Also, changing the design of the product such that it requires fewer packaging materials is considered within this category.

One element of sustainable development that is extremely important is the change in attitudes concerning the way development and the environment are seen. There are many prerequisites that must be fulfilled so that sustainable development can be practiced correctly. There must be willingness to try

new methods, sincere commitment to the cause, an open mind towards new ideas and tools, good team dynamics, as well as a structured methodology for the work. Current attitudes, which act as a main barrier preventing the implementation of practices of sustainable development through cleaner production, are fear and unawareness. Fear of the new, of criticism, and of breaking a dog-eared system is a common attitude. Also, basic unawareness coupled with this fear causes individuals to refuse ideas even before fully understanding them. The many stereotypes and incorrect information that continue to mislead many decision makers are a major barrier to sustainable development. However, this problem can be and is in the process of being remedied through the use of deliberate awareness campaigns that target various members of the community, primarily decision makers.

2.6 Cleaner Production Opportunity Assessment

The main goal of undertaking a CP assessment or CP audit in the industry is to be able to identify opportunities for cleaner production. The methodology for the cleaner production opportunity assessment is as follows:

- Team: Forming a cleaner production team from a CP expert and industrial expert(s) is the most important step toward the success of cleaner production opportunity assessment. Allocation of responsibilities within the audit team is very important to facilitate the reporting process.
- Pre-audit: The audit team request all available information regarding the water consumption, energy consumption, amount of liquid and solid wastes generated, raw materials used in the process "quantity, quality and pricing", products and byproducts, etc. Confirmation from the audit team to top management regarding confidentiality of the information. The most important factor is selecting a facilitator from the CP team side and the company side to act as a liaison point for information exchange.
- Surrounding environment: Determine the relationship of the factory with the surrounding environment and consider the potential impacts. These will include neighboring industries, surface water sources, groundwater sources, agricultural activities, roads, schools, residential areas, hospitals, etc.
- Operations and processes: Understand the operations and processes including the receiving area of raw material(s), handling and storage facilities, packaging area, etc. Site layout and plot plan including process flow diagram are very import for operations and processes.
- Inputs and outputs: List process steps with process flow diagram(s) indicating inputs, outputs as well as wastes and emissions. Use "best estimate" if specific data is not available.

- Wasteful processes: Identify wasteful process with cause analysis.
- Material and energy balance: Analysis of process steps with material and energy balance including cost of waste streams is very important to identify the missing wastes and emissions. Usually the inputs will exceed the outputs at this stage so it becomes necessary to look for missing data such as evaporation losses, leaks, inaccuracy in the original data, etc.
- Opportunities: Develop CP opportunities according to cleaner production techniques shown in Figure 2.1 and select workable opportunities.
- Priorities: Select CP solutions according to:
  - Technical feasibilities
  - Economic feasibility
  - Environmental aspects
  - Social aspects.

  Priorities can be assigned according to significant/severe environmental impacts to the working environment or surrounding environment or significant benefit to the company which will result from reduced cost or improved efficiency. The following is the recommended ranked improvement measure:
  - Low cost/high cost
  - High benefit/low benefit
  - Short payback/long payback.
- Implementation: Prepare for implementation through operator's involvement and management commitment. Management must understand and respect constructive criticism of current operation practices and must concur with, or correct, the assumption used in the audit. It is very important to involve management with the formulation of the action plans and thus the plan can be disseminated to all levels within the organization.
- Assessment: Monitor and evaluate results.
- Process sustainability: Maintain cleaner production.
- Sustainable development: Go back to re-identify another wasteful process and continue.

2.7 Cleaner Production Case Studies

The ultimate increase in demand and consumption rates resulted in a typical increase in production rates causing more depletion of natural resources, water and energy. Accordingly, wastes and pollution resulting from production activities became a real threat on both human health and the environment. This led the environmental aware communities to develop new approaches and methodologies in order to achieve sustainable development which maintains the balance between the economic and social development on one side and the conservation of natural recourses, water and energy for future generations on the other side. Among the tools that can achieve sustainable development is cleaner production (CP) concept and techniques covered in this chapter. CP calls for the continuous application of an integrated preventive environmental strategy to processes and products so as to reduce risks to humans and to the environment. Cleaner production provides procedures and steps that, if followed and applied with high commitment from top management, would result in significant economic benefits in addition to improving the quality of the final product, as well as conserving the environment and natural resources. These economic advantages of cleaner production are especially evident if compared with other environmental protection strategies such as end-of-pipe treatment.

The following case studies will demonstrate the use of CP techniques in different industrial sectors such as the food sector, the oil and soap sector, the textile sector, the wood sector, and the iron and steel sector. The CP opportunities were selected carefully to cover all CP techniques with cost/benefit analysis. Some details were given to some case studies to demonstrate not only the methodology but also the procedure to calculate the benefits. Different formats were given to the case studies to avoid standardizing the CP opportunities format. The main items that should be covered in all case studies were included such as: background information regarding the company, process description, CP opportunities, CP techniques implemented, and cost/benefit analysis.

Some CP opportunities may require prior investigation through laboratory research to check the achievements on a smaller scale in a controlled environment, and then apply the techniques in a pilot scale to guarantee the possibility of upsizing the research findings. If the pilot project succeeds, this will encourage implementing the prototype on a commercial scale within the factory or any other similar factory. One case study demonstrates moving from research scale to a pilot scale with brief feasibility for actual implementation. Another case study demonstrates moving from research scale all the way to commercial scale.

All case studies deal with CP opportunities and CP techniques. However, different approaches of CP are presented in each case study. Hopefully the reader will learn something new from each case study.

Food sector: Conservation of water and energy in preserved food companies

The following case study elaborates how cleaner production techniques can be applied in a food sector industry. The project was implemented with the contribution of the Egyptian Environmental Affairs Agency (EEAA) through the SEAM "Support for Environmental Assessment and Management" program in Edfina and Kaha Companies for preserved food to reduce the amount of energy and water consumption (SEAM 1999b). Edfina Company is one of the largest producers of preserved food in Alexandria, Egypt. It was built in

TABLE 2.1
Annual Production of Preserved Foods

Production line	Edfina (tons)	Kaha (tons)
Fruit juice	4,484	1,997
Jam	3,839	1,358
Canned vegetables	231	69
Canned beans	1,428	3,150
Frozen vegetables	812	253
Tomato paste	519	840
Other	258	109
TOTAL	11,571	7,776

1958 on 56,000 m^2. After its privatization, the workforce was reduced to 600 employees. Kaha Company was built in 1976 on 83,800 m^2 in Qalubiya, Egypt. The factory has a workforce of 650 employees and is privatized.

Process description
The production lines in both factories produce the following products:

- Fruit juices: Fresh fruits are received, sorted, washed and squeezed. Then the pulp is heated, screened and mixed with other ingredients. The mixture is heated, screened, bottled or canned then pasteurized.
- Jam: Fresh fruits are sorted, washed, peeled, cut, mixed with sugar then steam cooked and concentrated under vacuum. Concentrate is packed in tin cans or pots then sterilized.
- Frozen vegetables: Fresh vegetables are sorted, peeled and cut. Peeled vegetables are then sorted, frozen and packed.
- Canned beans: Green beans are sieved, sorted, dip and spray washed, then steam cooked, cooled, canned and sterilized.
- Tomato paste: Raw tomatoes are sorted, washed, pressed for juice and screened. Then the juice is concentrated under vacuum and heat treated. Paste is canned and sterilized.

The annual production of preserved food at both factories is presented in Table 2.1. The production levels, shown in the table, are based on full capacity.

Both factories have their own canning facilities, freezing units, water treatment facility, boiler station, quality control laboratories, refrigerators, cooling towers, garages and maintenance workshops.

Energy and water consumption
The types of energy used in both factories are electricity, heavy oil, diesel fuel and steam as shown in Tables 2.2 and 2.3. Electricity is used mainly for the

TABLE 2.2
Annual Electricity and Boiler Fuel Consumption

	Edfina	*Kaha*
Electricity (kWh)	5.95 million	2.78 million
Boiler fuel oil (tons)	2,419	1,890

TABLE 2.3
Steam Usage in Kaha Factory

	% of steam
Concentration	75
Sterilization	14
Cooking	11

TABLE 2.4
Water Consumption in Both Factories (per Annum)

	Edfina	*Kaha*
Water consumption	700,000 m^3	936,000 m^3
Water use		
Process and washing	41%	67%
Cooling	42%	32%
Domestic use	17%	

TABLE 2.5
Water Sources and Discharges

	Source of water	*Discharge (m^3/y)*	*Discharge destination*
Kaha	Groundwater	784,680	Drain canal
Edfina	Water from municipality	520,000	Public sewer network

machines in the industrial process, process control and lighting. Heavy oil is used mainly as a boiler fuel (Kaha was in the process of changing its boiler fuel from heavy oil to diesel because it is close to a residential area). Diesel is also used for vehicles, varnishing and the printing plant. Concerning the steam generation, at Edfina Company there are two boilers (12 ton/h) each and at Kaha Company four boilers (3 × 12 ton/h and 1 × 6 ton/h). The annual water consumption and use in both factories are shown in Table 2.4. Two sources of water were used as shown in Table 2.5: one from groundwater and one from municipality.

CP opportunities assessment

The cleaner production industrial audit for both factories concluded that water and energy consumption are very high in both factories. The cleaner production opportunities assessment for water and energy audits indicate the following opportunities:

Energy issues:

- Heat losses due to poor insulation of steam lines.
- Leakage of steam from the processes.
- Steam losses from steam traps due to poor maintenance and defective steam traps (the purpose of a steam trap is to let water that condensates in steam lines escape while stopping the flow of steam from escaping).
- Condensate is usually discharged to the drain.
- Lack of steam traps in some jacketed equipment.
- Steam losses in the process of can sterilization (done by direct steam injection in water).
- Power factors were less than 90% (the value of the utility contract without penalty).

Water issues:

- Water leakage in some of the processes.
- High water consumption due to open cooling cycles.
- High water consumption in vegetable washing.
- High water consumption in floor and equipment washing.
- Taps and hoses left running.
- Lack of cooling water recovery systems.

The CP audit selected a number of cleaner production solutions taking into consideration the technical, economical and environmental aspects. The energy audit was responsible for the fuel and steam consumption in both factories. Their major objective was to decrease energy loss and fuel consumption.

Energy saving measures

Insulation of bare steam pipes: The lack of insulation of the steam pipes resulted in significant heat loss. A Rockwool insulation of $80\,kg/m^3$ density was used to insulate 1,475 m of bare steam pipes at Edfina and 485 m at Kaha resulting in the savings shown in Table 2.6.

TABLE 2.6
Annual Savings from Insulating Steam Pipeline

	Edfina	Kaha
Steam saving (m^3/y)	5,394	2,123
Fuel saving (ton/y)	440	162

Replacement of leaking steam valves: Leaking steam valves cause steam loss. At Kaha factory 74 high-pressure leaking steam valves were identified (ranging in size from 1.9 cm to 15.2 cm). At Edfina factory 98 leaking steam valves were identified (ranging in size from 1.27 cm to 10.1 cm). It was assumed that every defective valve leaks 7.5 kg steam per hour. This resulted in the savings shown in Table 2.7.

Replacement of defective steam traps: Only some steam jacketed equipments were fitted with steam traps, while others had defective steam traps resulting in steam waste. Twenty-four steam traps were installed at Kaha and 35 steam traps were installed at Edfina resulting in the savings shown in Table 2.8.

Installation of temperature controller on sterilizers: The sterilization of juice and tomato paste was done by heated water with direct steam. Four temperature controllers with automatic pressure regulators were installed in each factory on steam lines used for sterilization to avoid steam loss. The temperature controller regulates the flow of steam according to water temperature. This resulted in the savings shown in Table 2.9.

TABLE 2.7
Annual Savings from Steam Valves

	Edfina	Kaha
Steam saving (m³/y)	1,055	799
Fuel saving (ton/y)	86	61

TABLE 2.8
Annual Savings from Steam Traps

	Edfina	Kaha
Steam saving (m³/y)	1,362	933
Fuel saving (ton/y)	111	71

TABLE 2.9
Annual Savings from Using Temperature Controller

	Edfina	Kaha
Steam saving (m³/y)	3,600	3,600
Fuel saving (ton/y)	294	343

TABLE 2.10
Annual Savings from Steam Condensate

	Edfina	Kaha
Steam saving (m³/y)	3,867	10,670
Fuel saving (ton/y)	29	74

Recovery of steam condensate: 2.75 tons of steam are required to produce one ton of tomato paste and 3.75 tons of tomato juice. After recovering the required heat (latent heat) from the steam, the steam will condense and discharge to the sewer or return to the boiler for water and energy conservation. By recycling this condensate the savings were achieved as shown in Table 2.10.

Improving boiler efficiency: By reducing the air to fuel ratio to 20–30% excess air, boiler efficiencies were improved at the two factories by an average of 3% resulting in fuel savings. Fuel savings were 85 ton/year for Edfina and 77 ton/year for Kaha.

Water savings measures
Concerning the water saving achieved in Edfina factory, the following measures were implemented.

Installation of water meters: Water meters were installed in 13 different locations to monitor water consumption. Monitoring is one of the very important tools for cleaner production to measure the consumption before and after modification to be able to calculate the amount of savings.

Installation of hose nozzles: A huge amount of water was needed to clean the tanks and wash the floors in the tomato paste section. On/off spray nozzles were fitted to control the water consumption. A water saving of 9,000 m³/y was achieved.

Improving the water collection system on the juice line: The cooling water from the juice line was collected in a tank to be recycled. This tank was smaller than the flow of water resulting in overflow. A larger tank with a new water pump was installed resulting in annual savings of 24,000 m³/y.

Installation of cooling tower for the bottled juice line: Juice bottles were sterilized at 90°C then cooled by water in an open cycle. A cooling tower was installed to recover and recycle the cooling water in a closed cycle. A water saving of 86,400 m³/y was achieved.

TABLE 2.11
Energy Savings – Cost/Benefits, Edfina Company

Action	Heavy oil savings (ton/y)	Costs of installation ($)	Annual savings ($)	Projected payback (months)
Insulation of steam pipes	440	21,792	14,049	19
Replacement/Installation of leaking steam traps	111	2,452	3,544	9
Replacement of leaking steam valves	86	8,244	2,746	36
Installation of pressure regulators	294	7,642	9,387	10
Recovery of steam condensate	29	5,821	1,589	44
Improved boiler efficiency	85	0	2,714	0
TOTAL	1,045	45,952	34,029	17

CP techniques implemented

From the above analysis we can classify the modifications achieved in both factories as:

- Good housekeeping: Leakages were repaired by fixing or replacing defective valves and steam traps. Water overflow and spillages were avoided.
- Technology change: By installing a cooling tower for the bottled juice line associated with collecting sump, pumps and piping instead of using water for cooling in an open cycle. Part of the process was changed and a new technology was used.
- Better process control: By controlling the steam flow by installing a temperature controller on the sterilizer and using a larger tank to avoid overflow and insulating the bare steam pipes to avoid heat losses and reducing the air fuel ratio to improve boiler efficiency.
- On-site recycling: By recycling the condensate resulting from the tomato paste process as well as recycling the cooling water in the bottled juice line.

Cost/benefit analysis

The cleaner production solutions proved to be very effective from the technical and environmental points of view. The cost/benefit analysis is summarized in Tables 2.11, 2.12 and 2.13. The payback period calculated in all case studies is called the "projected payback period" because it is based on investments and savings only without taking into consideration the running cost. This is due to the fact that the factory is operating with all necessary staff and infrastructure. The annual savings for energy are based on fuel savings.

TABLE 2.12
Energy Savings – Cost/Benefits, Kaha Company

Action	Diesel savings (ton/y)	Costs of installation ($)	Annual savings ($)	Projected payback (months)
Insulation of steam pipes	162	10,868	12,789	10
Replacement of leaking steam traps	71	2,540	5,605	6
Replacement of leaking steam valves	61	6,818	4,816	17
Installation of pressure regulators	343	7,925	27,079	4
Recovery of steam condensate	74	6,985	5,842	15
Improved boiler efficiency	77	0	5,974	0
TOTAL	788	35,136	62,105	7

TABLE 2.13
Water Savings – Cost/Benefits, Edfina Company

Actions	Water savings (m^3/y)	Costs of works ($)	Annual savings ($)	Payback (months)
Hose nozzles	9,000	860	1,580	7
Rehabilitation of the water collection system	24,000	1,480	4,210	5
Cooling tower for juice sterilizer	86,400	14,880	15,160	12
TOTAL	119,400	17,220	20,950	10

The benefits of the cleaner production solutions were:

- Steam savings of 15,278 m^3/y at Edfina and 18,125 m^3/y at Kaha were attained.
- Fuel oil consumption was reduced by 40% at Edfina and 34% at Kaha.
- Water consumption was reduced by 17% at Edfina.
- Wastewater volume and hence the load on the wastewater treatment facility were reduced.
- Energy savings implemented with an overall average payback period of 12 months.
- Water savings implemented with an average payback period of 10 months.

Referring to the above cost/benefit analysis, those cost savings were estimated on current production levels which were below full capacity and would increase 2–3 times when the two factories are running at full capacity. It is noticeable that when the annual saving is low with respect to the costs of installation, the payback period increases. This means it will take longer to recover the investment made. Some of the cleaner production solutions were implemented with no cost solutions like improving the boiler efficiency but did produce an annual saving of $2,714 in Edfina and $5,974 in Kaha. The other measures were meant to be low cost to be economical and technologically feasible. Most of the solutions have a payback period of around one year so that the investor reaps a quick return on his investments and continues with other CP opportunities.

Food sector: Process optimization at a sugar beet manufacturing facility

The process optimization at a sugar manufacturing facility in Tunisia is a clear example of the application of a source reduction technique. It employed good housekeeping and process modification to achieve cleaner production objectives (NCDENR 1997).

The Tunisian cleaner production program started in September 1993 with the aim of promoting the use of cleaner production practices to improve productivity and reduce pollution with an opportunity of financial savings. The United States Agency for International Development (USAID) funded the initiative, and currently funding is continued through the United Nations International Development Organization (UNIDO)/United Nations Environmental Program (UNEP). The program is implemented locally by the "Center de Production Plus Propre", CP3.

The assessed facility is a processing plant which produced white sugar from beets. It started production in June 1983. It employed 658 workers and operated a maximum of 70 working days during the beet harvesting season. The remaining days of the year were dedicated to maintenance works. The total capacity of the plant is 36,000 tons and produced 18,500 tons of white sugar in 1993.

Process description
The plant initially performs the following processes:

- Beet reception: Beet loads are sampled and analyzed for tare and sugar content; they are then unloaded and run through preliminary dry cleaning to a storage slab before being hydraulically transported through a cleaning station designed to separate the flow into clean beets, rocks, beet chips and vegetable matter.
- Beet processing: Beets are brought into the factory for slicing into strips called cossettes. These cossettes are heated to 85°C for scalding.
- Juice preparation: Juice produced by the diffuser is purified by two major treatment steps: first and second carbonation before it passes through ion exchange to remove possible remaining impurities.

- Juice concentration: Purified juice is concentrated in a four stage evaporator.
- Sugar crystallization: This is done in three stages. Water evaporation with steam from the evaporators drives the crystallization.
- Sugar processes: The mixture of sugar and mother liquor (masse-cuite or magma) are centrifuged to separate the sugar from the liquor. The white sugar is then dried and bagged.

This process, however, was not environmentally friendly due to gas emission, waste, energy, water and raw material losses, thus the company instituted the cleaner production program to resolve these issues. The pollution problems observed at the facility include:

- High BOD (biological oxygen demand) loadings in wastewater discharge.
- High COD (chemical oxygen demand) loadings in the wastewater discharge.
- High energy and water consumption.
- The use of harmful materials.
- Poor maintenance leading to excessive loss of raw materials as waste and low production efficiency.

CP opportunity assessment/techniques implemented
The cleaner production techniques used to improve the operations within the beet factory include:

- Chip and trash recovery system: Reactivating the existing system to recover pieces of whole beets and tips of beet roots containing sugar lost as wastes. This operation falls under the good housekeeping technique.
- Diffusion and juice purification: Controlling operation parameters to optimize production processes such as knife filling, diffuser temperature and stable operation conditions. This operation falls under the better process control technique.
- Heat exchange and evaporation: Monitoring and controlling operation parameters like the venting of non-condensable gases from the bottom of heaters and calandrias and controlling sugar end brix, to optimize evaporation and speed up production. This also falls under the better process control technique.
- Sugar end purity control: Modifying purity management steps of the sugar, lowering the purity of liquor (molasses) and reviewing crystallization management to optimize this operation; a process modification technique.
- Unknown losses: Improving maintenance on pump seals to stop leakages and improving operation methods to provide steady state condition; a good housekeeping technique.

Process modification/better process control and proper housekeeping as cleaner production techniques were employed to achieve the cleaner production objectives in this case study.

Cost/benefit analysis

The cleaner production recommendation resulted in the following environmental advantages:

- Sugar production yield increased by 6.5% by saving about 1,690 ton/y of sugar from being disposed off as waste in landfill.
- Energy consumption was reduced by 2.1% thus sparing about 150 ton/y of heavy fuel oil.
- About 2,000 ton/year of beet pieces were saved from being disposed of as waste to landfill.
- The most significant savings resulting from this case study were:
  - Process optimization through increased productivity and production yield accounting for more than 85% of savings
  - Resource conservation due to reduced material and energy consumption accounting for 15% of savings.

In all, the implementation of the cleaner production process led to an annual saving of US$1,400,000 for a total investment of no more than US$35,000, thus economically efficient.

Food sector: Recovery of cheese whey for use as an animal feed

This case study will show that when cleaner production and pollution control options are carefully evaluated and compared, the cleaner production options are often more cost effective. By implementing two simple techniques, which are on-site recycling and input material change, the cleaner production option will generate savings through reduced costs for raw materials for the animal farms, and reduce waste treatment cost for the factory and help it to comply with environmental regulations.

This case study was implemented in Misr Company for Dairy and Food in Damietta, Egypt (SEAM 1999a). The objective of the case study is to explain the methodology of implementing the cleaner production techniques, which starts by forming an environmental audit to identify the problem, propose solutions, test the proposed solution through laboratory and model experiments, prepare a feasibility study based on the data obtained from the model experiment, and finally assist the concerned parties in implementing the CP techniques on a prototype scale.

Process description

Misr Company for Dairy and Food, located in Damietta, Egypt, is a public sector factory with a workforce of 512. The factory annually processes

8,250 tons of raw milk to produce 3,100 tons of white and hard cheese and 200 tons of ghee. The farm that uses this byproduct (cheese whey) and in which the model experiment is to be conducted is the Animal Wealth Society Farm located 10 km away from the factory. The farm has 725 head of Holstein, Brown Swiss and Friesian breeds managed under a sophisticated dairy feedlot operation and quality control.

Waste analysis
The waste generated from this factory mainly consists of:

- Product losses
- Wash water
- Whey

Whey is a liquid byproduct originating from cheese manufacturing, which has tremendous biological impact. To produce hard cheese more than 83% of the milk is converted to sweet whey, and to produce white cheese more than 60% of the milk is converted to permeate whey. This first operation of sweet whey produced from Grade-A milk takes place from December to May, and the second process of ultra-filtration occurs through the year. Byproducts of these operations are sweet whey ($4,250 \, m^3/y$), deproteinized whey ($4,070 \, m^3/y$) and permeate whey ($1,900 \, m^3/y$).

Whey is a variable source of carbohydrates, a good supply of energy, and contains high quality protein and minerals. Only 4% of the whey is recovered as fat and protein by centrifugation and settling. The remaining whey – based on its nutritional true value, compared to other protein and energy sources for remnants such as roughage and mix feed – permeate whey was determined to be $12/ton. However, these volumes of different types of whey get mixed with other effluents to be discharged in the city sewer.

The annual factory effluent containing both sweet and permeate whey is $183,000 \, m^3$ with BOD (2,300 ppm), COD (4,050 ppm), TSS (540 ppm), TDS (2,290 ppm) and oil/grease (420 ppm). This highly biologically polluted water is discharged into the city sewer without prior treatment.

Therefore, to comply with environmental regulations the factory has to install an industrial wastewater treatment plant. This would involve a capital cost of at least $17,545 plus the running cost and would involve primary settling, tricking filters, coagulation, chlorination, and sludge digestion.

From the above analysis the problem is obvious. The factory has to pay a capital cost of $17,545 in the form of a wastewater treatment station plus the annual running cost in order to discharge whey that has a nutritional value of $74,000 per year. To illustrate the problem in the simple language of mathematics, the factory has to pay more than $17,545 in order to get rid of $74,000. Therefore, one of the eight clean production techniques explained above is needed to reverse this absurd equation. In other words, finding how whey could be recovered and used would reduce the cost of end-of-pipe treatment and save resources.

CP opportunity assessment

This structured implementation process is part of a systematic audit procedure. According to Johansson (1992), the environmental audit consists of four phases, which are informational group meeting, on-site meeting, plant walk-through, and alternative methods. In the first three phases specialized staff consisting of industrial consultants with the aid of a firm representative gather relevant information. Then the industrial experts develop a preliminary assessment for the possible solutions and alternatives. Some of these solutions may require prior investigation before being implemented as will be discussed during the implementation phase.

Investing in cleaner production to prevent pollution and reduce resource consumption is more cost effective than continuing to rely on expensive end-of-pipe solutions. Therefore, if the factory could conduct on-site recycling/recovery for the whey, this would decrease its end-of-pipe treatment cost and generate extra revenue. On the other hand, if the animal farm could use the whey as an animal feed for ruminants as an input material instead of water or molasses, this would reduce the use of natural resources in terms of water and dry feed at an extra cost to the farm.

CP techniques implemented

Experimental phase

A cleaner production industrial audit is performed but some corrective measurements need to be tested before implementation to insure that they satisfy the technical, financial and environmental requirements. This is done first on a laboratory scale by a research institute or qualified experts. In this case the characteristic of the whey produced by Misr Company for Dairy and Food was determined by conducting chemical, physical and nutritional analyses.

The aim of this laboratory work is to assess the nutritional value of the whey against that of the molasses. The total nitrogen (%) for sweet whey, whey permeate and cane molasses was found to be 1.3, 0.26 and 0.66 respectively, and the digestible energy (%) was 1.86, 1.7 and 1.4 respectively. This shows that whey has a higher nitration value than molasses.

Whey may cause health problems for animals if pH is below 4.5 or above 8.5, the microbial content is above $(1 \times 10^6/100\,ml)$ and the total coliform exceeds $(1/100\,ml)$ for calves and $(30/100\,ml)$ for cows. Therefore, the aim of this analysis is to check the whey against these values. The storage life of whey was determined and the whey was found to be safe if used under certain preservative conditions since pH may drop to 4.8 after storing fresh whey permeate for 24 hours.

Therefore, the third objective of this research work was to determine the optimum preserving conditions needed for permeate whey. The research work revealed that:

- Permeate whey from ultra-filtration does not need any further pasteurization.
- For permeate whey from hard cheese, formaldehyde (0.01%) and hydrogen peroxide (0.02%) were the most effective and economical preservatives that could be applied.

From the above theoretical findings we could conclude that whey has higher nutritional value than molasses and it is safe to use under certain conditions. However, this should be verified on the laboratory scale phase before implementing it on the pilot scale.

Pilot phase

The pilot feasibility experiment should be conducted based on detailed and structured plans and terms of reference by experts in order to ensure accurate and unbiased results. This phase involves numerous expenses and therefore funding agency and funding agreement should be secured.

The plan should state clearly the objective and the expected outcome of the pilot project. The selection of the place to conduct the pilot project should be based on criteria that match the objectives and outcomes of the pilot project. When working in Egypt, it is always recommended to choose places that operate under high professional standards because this would help eliminate any errors due to carelessness, or other problems that could affect the results of the experiment.

For this case study, the Animal Wealth Farm located 10 km from the factory was chosen from seven farms, based on its location, size, facilities and management proficiency. This farm follows a controlled breeding and diet system. The experiment was conducted on sheep with a total duration of eight weeks. The objectives of the experiment were to:

- Verify the safety of using the whey as food for animals.
- Assess the value of the whey in terms of growth rate and weight gain.
- Demonstrate to the farmers the potential benefits of using the whey as a substitute for water and molasses.
- Assess the feasibility of the process.

The experiment was conducted on six pen trials of five sheep aged between 8 and 10 months and weighing 37 to 39 kg. One pen was used as a control feeding only on water and the other five on diets as shown in Table 2.14.

TABLE 2.14
The Experimental Groups and Water as Control

Liquid feed	Feed efficiency kilogram feed/ kilogram weight gain	64-day gain (%)
Water	9.9	28.2
Permeate (100%)	7.5	33.9
Permeate (95%) + urea (5%)	8.9	28.9
Molasses (80%)	9.2	34.6
Molasses (7.5%) + urea (0.5%)	8.0	28.9
Permeate (50%) + molasses (4%) + urea (0.5%)	9.7	27.7

TABLE 2.15
The Traditional Feeding Ingredients in the Animal Farm

Feed type	Annual quantity (ton)	Total price ($)	Daily rate	Remarks
Municipal water	21,800	2,315	60 ton	
Concentrate Mix feedlot	1,825	175,440	7 kg/head	Cotton seed meal, yellow corn, wheat bran, salt, limestone, DM 89.5%, CP 17.3%
Roughage	900	63,160	3 kg/head	Clover hay, DM 87.4%, CP 13.6%

The results of the pilot experiments compared with the traditional feeding diets detailed in Table 2.15 show that:

- Permeate alone shows better feeding than when being supplemented with nitrogen.
- Permeate fed results in terms of weight gain in 64 days is 33.9%, which is almost the same as molasses (34.6%) and better than water (28.2%).
- Permeate has the highest yield efficiency of 7.5, which is kilogram whey fed per kilogram live weight gain, compared to molasses at 9.2.

Thus we could conclude that whey is technically acceptable, and safe as a food source for animals.

The pilot experiment is also important to determine problems that could be avoided in the prototype, for example developing a provisional business plan as well as making observations and documenting deductions to be accounted for in the prototype. Among the recommendations of the pilot experiment were:

- Quality assessment to minimize contamination and extend the storage life.
- Fresh delivery is highly recommended.
- Preservatives must be used in case of storage.
- Formaldehyde is not recommended in feeding latching cows.
- Feed to animals should be introduced gradually.
- Control amount of whey offered to prevent exhaustive consumption.
- Clean pipe at least once a week to avoid microbial contamination.
- Use corrosion resistant equipment, pipes and tanks in the transfer, feeding and storage.
- Fly control measures to be continuously implemented.

Cost/benefit analysis

Using whey is also financially acceptable. Based on the above findings and considerations, a primary cost/benefit analysis could be conducted.

Based on its nutritious value, permeate was determined to be $12/ton. However, the dairy factory will provide it for a cost as low as water at first until farmers realize its true value. In three years the factory may raise the price of the whey to $2.6/ton, which is still 6.8% of the price of molasses.

However, implementing this recycling technique, which consists of segregating whey and establishing a factory-to-farm transfer system, would require initial capital cost, and operation and maintenance costs.

The initial investment includes the following items:

- Installing whey corrosion resistant equipment, pipes and tanks in the transfer, feeding and storage of whey as the acid of the whey is corrosive.
- Purchasing monitoring equipment and meters for factory and farm.

The operation and maintenance costs include the following items:

- Provide training for factory and farm personnel.
- Whey preservation and quality assurance.
- Delivery of whey from the factory to the farm.
- Thorough cleaning of the pipe network at least once a week to avoid microbial contamination and off-flavors of the whey.
- Continuous implementation of fly control measures.

Based on the results obtained from the pilot experiments, a summary of the cost/benefit analysis is shown in Table 2.16.

From Table 2.16 it is clear that the factory would benefit from saving the cost of treatment and gaining revenue from selling molasses. The farm would benefit by saving the cost of feeding and increasing productivity.

TABLE 2.16
Cost/Benefit Analysis of Implementing CP Technique (Feeding Whey to 412 Beef Cattle)

Description	Year 1		Year 2		Year 3	
	Cash flow Factory ($)	Cash flow Farm ($)	Cash flow Factory ($)	Cash flow Farm ($)	Cash flow Factory ($)	Cash flow Farm ($)
Whey transfer and storage equipment	8,750	–				
Monitoring equipment and meters	(75)	(745)				
Operation cost of whey transfer to farm	(880)	–	(880)	–	(880)	–
Capital investment in whey WTP (Water Treatment Plant)	17,545	–	17,545	–	17,545	–
Sales of 6,000 m³ whey (E 1/ton)	1,050	(1,050)	7,895	(7,895)	15,790	(15,790)
Saving 40 kg water/ head/day	–	1,050	–	1,050	–	1,050
Saving 2 kg dry feed/ head/day	–	24,260	–	24,260	–	24,260
Net savings	26,390	23,515	24,560	17,415	32,455	9,520
Payback period (month)	<10	<1				

Note: Figures in brackets indicate an increase in cost.

Prototype phase

It is often claimed that cleaner production techniques do not exist, are impractical, false, infeasible, require sophisticated know-how to implement and maintain, or require huge capital investment. In other words, both the factory and the farm could claim that the aforesaid opportunities are false, and even if they are true, these techniques cannot be implemented. The pilot project confirmed that using whey to feed animals is technically, financially and environmentally acceptable. This means that implementing these techniques on the prototype scale is not only safe but also more feasible than traditional practices.

The cost/benefit analysis in Table 2.16 shows that the techniques are very simple and require small capital cost with a payback period of less than 10 months and one month for the factory and the farm respectively. Actually, implementing these techniques has to do more with attitudes and management approaches rather than capital investment and technology. However, for the second premise, which is that the techniques are false or infeasible, the structured implementation approach showed the feasibility and practicality of the proposed techniques.

At this point management commitment is needed to convert the provisional feasibility study prepared by the auditing team along with these recommendations into a detailed study that accounts for total budget and time required to implement these techniques on a prototype scale.

Outcomes
Factory outcome
Misr Dairy while implementing the proposed clean production technique of recovering the whey was able to make the following benefits and achievements:

- Recovery of more than 6,000 tons of byproducts that have a nutritional value of $12/ton at a low capital investment.
- Elimination of environmentally polluted water of BOD (415 tons), COD (522 tons), TSS (58 tons), TDS (218 tons) and oil/grease (60 tons) per year.
- Elimination of the capital cost needed in the end-of-pipe treatment.
- Up to 25% reduction in wastewater disposal.

Farms' outcome
The animal farms that carried out the proposed clean production technique of changing the input material from water to whey obtained the following results:

- Input material change through substituting water with high protein and energy value whey.
- 100% saving in water consumption.
- 75% saving in dry feed consumption.
- Reduction of manpower and equipment cost associated with dry feed.
- Improvement in feed palatability, texture, and dust control of feedlot rations.
- Increase in the safety factor of providing poor or variable quality diets.
- Provision of an economic and convenient method to feed urea supplements, vitamins, minerals, and feed additives.

Sustainability
From the above discussion, it is clear that implementing these two simple techniques will reduce waste disposal, reduce raw material cost, reduce pollution damage cost, improve the farm's performance, increase competitiveness

and aid the factory to comply with environmental regulations. This is enough to insure the continuous application of these techniques. However, these two approaches are so simple and are mainly based on attitudes and management approaches, so commitment from management to continue implementing a win–win process is a must. Also, continuous researches to develop these two techniques and investigating new alternatives are also needed as a part of any living process.

Final remark

It is often claimed that cleaner production techniques do not yet exist or if they do they are only obtained at an expensive cost, by big enterprises with qualified personnel. The above discussion shows that this statement is not true. Clean production is an affordable and conscious process that translates to savings, health and environmental safety and market advantages. Future application of eco-labeling would certainly give an edge to clean producers.

However, clean production is obtainable not with large funds but with improved management techniques and the correct attitude. The traits of this attitude are a willingness to change and continuously finding innovative solutions. Clean production is the will to engineer.

Food sector: Reduction of milk losses at dairy and food company

The Misr Company for Dairy and Food factory, located in Mansoura, Egypt, is one of nine owned by the public sector company Misr for Dairy and Food and is one of the largest producers of dairy products in Egypt. The Mansoura factory was built in 1965 and has a workforce of around 420. The factory annually processes an average of 7,200 tons of milk, producing mainly pasteurized milk, white cheese, blue cheese, yoghurt, sour cream, and processed cheese. The main objective of this case study is to reduce the amount of milk losses within different departments of the company (SEAM 1998).

Process description

The main process units present in the factory are:

- Milk receiving, preparation and storage.
- Milk pasteurization.
- White cheese manufacturing.
- Ghee manufacturing.
- Blue cheese manufacturing.
- Processed cheese manufacturing.
- Yoghurt and sour cream manufacturing.
- Mish manufacturing.

CP opportunity assessment

Cleaner production opportunities were identified by means of an industrial or cleaner production audit. The audit findings were as follows:

- Different solid wastes stored haphazardly in open areas and on roads, constituting a fire risk and negatively impacting the aesthetic value of the premises.
- Milk was wasted due to overflow during the filling for storage and service tanks.
- Milk leakages in the milk packaging and refrigeration units.
- Oils used in vehicle maintenance facilities were drained to factory sewers leading to drain blockage and foul odors.
- Excessive consumption of heavy oil in the boiler house due to poorly tuned boilers.

CP techniques implemented

- Good housekeeping was applied to the plant as a whole; improvements include the following:
  - Factory drainage, sewers and manholes were maintained and upgraded to eliminate blockage and overflow problems.
  - Plant roadways were paved to allow better traffic flow of factory vehicles.
  - Waste management: Application of an effective solid waste management system was shown in the following:
    - Garbage and packing wastes are trucked out and disposed of.
    - Solid wastes such as scrap iron and metal objects are sold in auctions.
- Onsite recycling: Oil, grease and lubricants are now collected instead of being disposed of to the sewer with the following benefits:
  - Approximately 0.75 tons of oil are accumulated monthly and sold at $50/ton.
  - Reduction of the pollutants in wastewater.
  - Prevention of serious blockage of sewers.
- Equipment/Process modification:
  - Since the boilers were consuming much heavy oil, the ratio of air to heavy oil intake was optimized to increase the efficiency of the boilers.
  - The refrigeration room was upgraded to prevent spoilage and loss.
  - The packaging unit was relocated to prevent handling losses.
  - 50% of whey reused at the cheese packing stage in place of fresh water.
  - Milk storage tanks were equipped with level sensors and stopcocks to prevent overflow particularly during the receiving stage.

TABLE 2.17
Cost/Benefit Analysis

Factory unit	Action	Capital and operation costs ($)	Yearly savings ($)	Payback period (month)
All	Improve housekeeping and solid waste removal	2,280	21,050	1
Milk packaging and storage	Rationalize milk packaging and increase milk refrigeration efficiency	4,650	6,950	8
White cheese	Reuse whey	0	350	Immediate
Boiler house	Upgrade boiler and restore softening unit	350	3,290	<1
Garage	Collect used oil	90	440	<3
Milk receiving and pasteurization	Milk tank level controls	1,800	22,100	7
	Food quality valves	11,230		
TOTAL		20,400	54,180	<5

Cost/benefit analysis

Table 2.17 illustrates the breakdown of the cost/benefit analysis for the cleaner production techniques implemented in the factory.

The implementation of the above improvements has resulted in daily savings of 350 kilograms of milk. Additional benefits include:

- Reduced pollution loads.
- The elimination of floor spills.
- Improved hygiene and safety.
- Water consumption dropped by 6%.
- Mazot consumption has decreased by 10%.
- Solar consumption has decreased by 5%.
- Electricity consumption reduced by 9%.

Textile sector: Sulfur black dyeing in textile companies

Sulfur black dyes are widely used in the textile industry because of low cost and excellent washing and light fastness properties. Sulfur dyes are water insoluble and must be first converted to a water-soluble form, by adding a reducing agent, traditionally sodium sulfide, so that the fibre can absorb the dyes. After dyeing the fabric, the dye is converted back to insoluble form with the addition of an oxidizing agent, often acidified dichromate. This step prevents washing out of the dye from the fabric. However, both sodium sulfide and acidified dichromate are toxic and hazardous to handle. Their usage may leave harmful residues in the finished fabric and generate effluents that are difficult to treat and that damage the environment.

This project was implemented by the Egyptian Environmental Affairs Agency "EEAA" through the SEAM Program in three different factories – El-Nasr Spinning and Weaving, Dakahleya Spinning and Weaving and AmirTex – at no capital cost and resulted in cost savings of 2–16% on consumable materials (SEAM, 1999d). The measures implemented by the SEAM project demonstrate how sodium sulfide and dichromate can be safely substituted without a decline in fabric quality. As a result the advantage of sulfur black can be retained, whilst eliminating the adverse environmental and health impacts. Other benefits included the elimination of toxic and hazardous materials from the workplace and environment, reduced wastewater treatment costs, improved fabric quality, and increased productivity.

El-Nasr Spinning and Weaving is a large public sector factory built in 1963 and currently employing 7,000 staff. It processes an average of 8,000 tons of raw fabric per year, of which 20% is spun cotton yarns, 12% is polyester blend yarns, and 68% is gray fabric. The main products are cotton or blended yarns, white and dyed cotton and blended fabrics. Approximately 52.5 million meters of fabric were produced annually.

Dakahleya Spinning and Weaving is a public sector company with annual production of 11,400 tons of spun yarns and ready-made garments. The factory was built in 1965 and employs 4,000 staff. It comprises three spinning departments, an open end spinning unit and a tricot plant with a weaving unit, a dyehouse and a tailoring hall. The main products are cotton yarns, cotton knitted fabrics, polyester blended fabrics, and ready-made knitted cotton garments.

AmirTex is a privately owned company with around 100 employees. The factory was built in 1984 and comprises a weaving and knitting department, a printing unit and a dyehouse. The main products are cotton yarns, cotton knitted fabrics, polyester blended fabrics, and ready-made knitted cotton garments. Average annual production is 720 tons of cotton, polyester and blended fabrics.

Opportunity assessment
The following CP opportunities were identified in the sulfur black dyeing process through an industrial audit:

- High sulfur content in the dyehouse wastewater of 40–170 mg/l.
- Acidified dichromate resulted in concentrations of hexavalent chromium of 27 mg/l in the dyehouse wastewater.
- Bad odor resulting from the use of sodium sulfide in the dyeing process.
- Worker safety issues associated with the handling of toxic and hazardous chemicals.
- High steam, energy and water consumption was noted during the sulfur black dyeing process.
- Fabric losses resulting from the presence of excess sulfur on the material.

TABLE 2.18
Possible Chemical Substitutes

Substitute for sodium sulfide	$/ton
• Glucose (75%)	280
• Dextrose	700
• Dextrine	260
• Hydrol (glucose water)	0
Substitute for acidified dichromate	**$/ton**
• Hydrogen peroxide	480
• Sodium perborate	560
• Ammonium persulphate	1,400
• Sodium bromate	1,400+
• Potassium iodate	1,400+

CP techniques implemented

An evaluation was undertaken to assess the viability, costs, and quality of using various potential substitutes for sodium sulfide and acidified dichromates. A summary of possible substitutes and costs is shown in Table 2.18.

Laboratory trials were initially used to determine the optimum combination of reducing sugars and alkali. Pilot trials were then carried out to refine the preferred substitute:

- Substitution of sodium sulfide: Dextrine and hydrol shown in Table 2.18 were rejected as they gave a poor depth of shade. Glucose and dextrose both gave good depths of shade when used with sodium hydroxide. Glucose was therefore preferred as a substitute to sodium sulfide because of its lower cost.
- Substitution of dichromate: In El-Nasr Spinning and Weaving, sodium bromate and potassium iodate, shown in Table 2.18, were rejected as these chemicals are corrosive, unsafe to handle and expensive. Hydrogen peroxide was also rejected, as it is not suitable for use with woven fabrics. Sodium perborate and ammonium persulphate were both acceptable; however, sodium perborate was preferred due to its lower cost. In Dakahleya Spinning and Weaving and AmirTex, hydrogen peroxide was preferred as it is particularly suitable for processing knitted fabrics.

A summary of fabric quality, before and after, at Dakahleya Spinning and Weaving is shown in Table 2.19. During the trials it was noted that when glucose is added in stages the depth of shade was significantly improved.

TABLE 2.19
Fabric Quality

	Before (conventional process)	After (modified process)
Washing fastness	2–3	3
Dry rubbing fastness	1–2	2
Wet rubbing fastness	2–3	1–2
Depth of shade	Satisfactory	Satisfactory

TABLE 2.20
Savings in $ per Ton of Fabrics Processed

Factory Shade	El-Nasr		Dakahleya	AmirTex
	Gray	Black	Black	Black
Savings in:				
Chemical costs	2	(14)	(2)	10.5
Water use	1	1	10	3
Steam	2.50	2.50	24	21
Electricity	2.50	2.50	6	0.20
Labor	16	16	30	1
TOTAL ($)	24	8	68	35.70

Note: Figures in brackets indicate an increase in cost.

Cost/Benefit Analysis
Direct financial savings
No capital expenditure was necessary for implementation, as the benefits have been principally achieved through substitution and process optimization.

Despite some increases in chemical costs, overall savings of 2–16% for all consumable materials have been achieved in the three factories for each ton of fabric processed. A breakdown of the savings is given in Table 2.20.

El-Nasr Spinning and Weaving: Chemical costs were reduced for gray shades but were much higher for black shades as larger volumes of glucose were required to produce acceptable results. Savings in steam and electricity were 16% and 22% respectively. Savings of $24/ton and $8/ton were achieved for gray and black dyeing respectively.

Dakahleya Spinning and Weaving: A slight increase in the chemical costs was more than offset by significant savings from process optimization. This included eliminating two hot washes after dyeing and optimizing the number of cold and overflow washes carried out. As a result, steam, water, and electricity costs were reduced by 38–39%. Processing time was reduced from 13 hours to 8 hours thereby increasing production capacity.

Around 24 tons of fabric per annum are dyed with sulfur black dyes
Annual savings on current production: $1,632
Annual benefits from increased production capacity: $3,684

AmirTex: Savings of $35.7 per ton were achieved; however, where hydrogen peroxide was already used as the oxidant, the savings reduced to $8 per ton. AmirTex processes 21.6 tons per year using sulfur black dyes. Due to improved fabric quality AmirTex has increased production to 70 tons per year and reduced the use of more expensive reactive dyes, a saving of approximately $86 per ton.

Annual savings on existing production: $419
Annual savings by reducing use of reactive dyes: $4,160

Improved fabric quality
Elimination of free sulfur now avoids the past problem: residual sulfur on the fabric progressively oxidized to form sulfuric acid, which then attacks and may eventually destroy the fabric.
At El-Nasr factory, fabric strength was improved by 5% by using glucose instead of sodium sulfide.

Improved productivity
The modified process not only improved product quality but also reduced wastage. In all factories, the modified process was shorter than the conventional process, thereby saving on time and labor costs.

Environmental benefits
Concentrations of both hexavalent chromium and sulfur were significantly reduced in the effluent coming from the dyeing line at the three factories. BOD levels increased, as shown in Table 2.21 due to the use of glucose; however, it is not likely to significantly increase either wastewater treatment costs or the characteristics of the final effluent.

Savings on effluent treatment requirements
Elimination of sodium sulfide and acidified dichromate made the final effluent easier to treat, as it became less toxic and corrosive. At AmirTex savings on wastewater treatment requirements were estimated to be around $10 per ton of fabric processed.

Improved working conditions
Workers interviewed at the factory were very positive on the improvement to working conditions. The bad odors and inhalation of sulfur fumes have been removed and the substitute chemicals are much safer to handle.
A summary of benefits and achievements for the three factories is shown in Table 2.22.

TABLE 2.21
Effluent Characteristics Before and After Implementation*

Concentration (mg/1)	El-Nasr		Dakahleya	AmirTex
	Gray	Black	Black	Black
Sulfide				
Before	68	117	40	103
After	1	2.5	0.7	1.5
Hexavalent chromium				
Before	26	27	n/a	Nil
After	Nil	Nil	n/a	Nil
BOD				
Before	360	540	347	660
After	1,275	2,150	1,233	1,070
TDS				
Before	3,385	3,900	1,510	–
After	2,690	3,450	1,950	–

* From the end of the dyeing line.

TABLE 2.22
Benefits and Achievements

	El-Nasr	Dakahleya	AmirTex
Electricity consumption reduced by:	22%	38%	8%
Steam consumption reduced by:	16%	39%	21%
Chemical costs reduced (increased) by:	2%	(6%)	4%
Water consumption reduced by:	13%	39%	15%
Labor costs reduced by:	23%	57%	6%
Processing time reduced by:	22%	38%	6%
Other achievements include:			

- Improved productivity
- Improved working conditions
- Phasing out of toxic sodium sulfide and acidified dichromate

Textile sector: Combining preparatory processes in textile companies

Sizing is a very important process in the textile industry to increase the tensile strength of threads and reduce fiber breakages during the weaving process. This process can be done by coating the threads with a "size"/"starch". "Desizing", either with acid or enzymes, then removes size from the fabric, so that chemical penetration of the fabric in later stages is not inhibited. Desizing, scouring, and bleaching are always used in spinning and weaving wet processes.

Scouring is carried out to remove impurities that are present in cotton, both natural (e.g. waxes, fatty acids, proteins, etc.) and processing (e.g. residual size, dirt, and oil). This is usually done with sodium hydroxide and produces strongly alkaline effluents (around pH 12.5) with high organic loads.

Bleaching is used to whiten fabrics and yarns, using sodium hypochlorite or hydrogen peroxide. Many cotton processing factories in Egypt use sodium hypochlorite as it is cheaper than hydrogen peroxide. It is, however, toxic and is now not preferred or banned in many countries.

Desizing, scouring, and bleaching are frequently undertaken as three separate steps in the preparatory stages for textile wet processing. Through chemical substitution and process optimization, it is possible to combine two or three of the three preparatory processes, thereby reducing water and energy consumption as well as shortening the processing time (SEAM 1999). As combined scouring and bleaching requires that bleaching by sodium hypochlorite is replaced with hydrogen peroxide, the overall processing sequence generates effluents that are less harmful to the environment.

Combined desize and scour was undertaken at Misr Beida Dyers Company in Alexandria, Egypt, and combined scour and bleach was implemented at Giza Spinning and Weaving Company, Egypt. These modifications have reduced operating costs and increased production capacity through a shortened processing sequence. Recently, El-Hesn Company, located in the Egyptian industrial estate, 10th of Ramadan, Sharkia, Egypt, combined the three processes of desizing, scouring, and bleaching in one process.

Misr Beida Dyers Company is a public company in Kafr El Dawar, Alexandria. It was established in 1938, and occupies a 270 acre site. The factory pre-treats, dyes, prints, and finishes cotton fabrics and cotton/synthetic blends, processes yarns, and cotton. During 1997/8, Misr Beida Dyers pre-treated about 1,182 tons of woven fabric on jiggers, of which about half were pre-treated using separate processes (desizing, scouring, and half bleaching). The remaining fabric was pre-treated using a combined process consisting of desizing/scouring, using a proprietary chemical, followed by half bleaching. The factory processes and manufactures cotton and polyester/cotton garments and finished fabrics. In 1996/7, 1,440 tons of fabric were processed, of which 240 tons were full bleached and 1,200 tons half bleached and dyed. Full bleaching is carried out using hypochlorite and half bleaching using hydrogen peroxide only. About 90% of the fabric produced by Giza Spinning and Weaving is knitted, the remainder woven.

CP opportunity assessment

Through an industrial or cleaner production audit at the two factories, a number of environmental and productivity related concerns were identified in the desizing, scouring, and bleaching processes, as follows:

- Sodium hypochlorite, a toxic and hazardous chemical, was used in the bleaching process at Giza Spinning and Weaving.

TABLE 2.23
Recipe for Combined Desize and Scour: Woven Fabric – Misr Beida Dyers

Processing recipe (for 1 ton of fabric)	Combined desize and scour	Half bleaching
NaOH (38% Be) (l)	133	75
Espycon 1030 (kg)	4.8	1.5
Egyptol PLM (kg)	2.4	–
Ammonium persulfate	1.2	–
Na$_2$SiO$_3$ (kg)	–	16
H$_2$O$_2$ (35%) (kg)	–	13.3

- Strong odor of chlorine was noted, resulting from the use of sodium hypochlorite in the bleaching process.
- Worker safety issues were noticed, associated with the handling of toxic and hazardous chemicals such as sodium hypochlorite.
- High steam, energy and water consumption was noted during the desizing, scouring, and bleaching processes.
- Some re-processing of fabric was required due to insufficient fabric whiteness and uniformity of dyeing.

CP techniques implemented
Combined desize and scour – Misr Beida Dyers Company
Seven production trials were conducted to improve the efficiency of the combined desizing and scouring process. Four fabric samples of 250–313 kg size were treated on the normal wide jiggers and three samples of 750–850 kg size were processed on the existing Vald Henriksen jiggers.

In the trials, concentrations and rates at which chemicals were added were varied as well as temperature, number, and timing of washes. Initial problems of attaining acceptable fabric wettability and whiteness index were eventually overcome. The final acceptable processing recipe for desize/scour followed by half bleaching of 1 ton of fabric appears in Table 2.23.

Fabric quality after bleaching and dyeing with Cibachrome Orange®, showed improved wettability and whiteness and consistent color properties. A comparison of results is shown in Table 2.24.

In optimizing the modified combined process it was possible to eliminate two hot washes in the half bleaching process. More expensive chemicals were phased out and replaced with ammonium persulfate and Egyptol. Overall there was a reduction in water, energy, and steam consumption. Processing cycle time was shortened by 2 hours (18% reduction in processing time).

Combined scour and bleach – Giza Spinning and Weaving Company
Trials were undertaken to combine the scour and bleach processes more efficiently and to phase out the use of sodium hypochlorite in bleaching. Hydrogen

TABLE 2.24
Comparison Between Fabric Quality Before and After Using the Combined Process – Misr Beida Dyers

Fabric characteristics	Before (conventional 3 stage)	After (modified combined)
After bleaching:		
Handle	Harsh	Soft
Wettability (seconds)	6	2
Whiteness index	68.3	70.1
Tensile strength (kg/cm^2)	50	52
Residual strength (%)	69.4	72.2
After dyeing:		
Color uniformity	Uniform	Uniform
Color fastness		
• Perspiration – alkaline	4	4
• Perspiration – acidic	4	4
Washing fastness	4	4

TABLE 2.25
Recipe for Combined Scour and Bleach: Knitted Fabric – Giza Spinning and Weaving

Processing recipe (for 1 ton of fabric)	Combined scour and half bleach	Combined scour and full bleach
Nionil N (kg)	10	10
NaOH (47%) (l)	75	105
H$_2$O$_2$ (50%) (kg)	30	105
Organic stabilizer (kg)	10	30
Optical brightener – Uvitex 2B (kg)	–	5
Softening – Knit Soft (kg)	–	30
Softening – acetic acid (11%) (kg)	–	50

peroxide was used to substitute sodium hypochlorite in the full bleaching process.

Tests were initiated on a laboratory scale winch using four standard 9 kg samples of different fabric types (i.e. ribs, riblycra, single jersey, and interlock). Using and refining the preferred laboratory scale recipe, pilot scale trials (95 kg) and production scale trials (35–452 kg) were carried out on single jersey fabric. Tests were conducted for both half and full bleach fabric. Single jersey was used as it is the main processed fabric. Final recipes for pre-treating 1 ton of knitted fabric are summarized in Table 2.25.

Tests on fabric quality gave comparable results between the conventional and modified combined processes as shown in Table 2.26. Subsequent dyeing of the modified half bleached fabric also gave acceptable depth shades.

TABLE 2.26
Comparison Between Fabric Quality Before and After
Using the Combined Process – Giza Company

Fabric characteristics	Half bleach	Full bleach
Whiteness index		
Before	67.1	103
After	66.2	104
Wettability (seconds)		
Before	3.0	2.5
After	1.5	1.5

Adopting combined processing meant that a number of stages could be combined or eliminated entirely. Stages were optimized for greater productivity and financial savings as outlined below:

- Two hot washes and one cold wash in the half bleaching process were eliminated.
- One hot wash, two cold washes, and three flotation rinses in the full bleaching process were eliminated.
- Reuse of bleaching bath for the optical brightening step in full bleaching process.

One of the greatest benefits has been in the processing time for half bleaching which has been more than halved. This has increased the production capacity from a previous maximum of 4.5 to 9.0 tons per day, an equivalent of 480 tons per year.

Cost/benefit analysis
No capital expenditure was necessary for implementation as the benefits were achieved by chemical substitution and process optimization. The benefits were therefore immediate once the modifications were complete.

Misr Beida Dyers company produces 1,182 tons per annum of half bleaching. It is noted that the increase in chemical costs before and after was offset by savings in utilities and labor as indicated in Table 2.27.

A comparison of costs for Giza Spinning and Weaving before and after modification is summarized in Table 2.28. Processing time has been considerably shortened in the half bleaching process and enabled production to be lifted by 40%. In addition to the savings on the modified bleaching process, additional annual benefits for the increased production are realized based on a $23 net margin per ton.

Improved fabric quality
The whiteness and absorbency were improved by the modified process at both Misr Beida Dyers and Giza Spinning and Weaving.

TABLE 2.27
Savings in $ (per Ton of Fabric) in Half Bleaching: Woven Fabrics – Misr Beida Dyers

Savings in	Before	After	Savings
Chemicals	41.10	41.30	(0.20)
Water use	2.10	1.50	0.60
Steam use	10.80	7.90	2.90
Electricity use	3.10	2.50	0.60
Labor	8.40	6.80	1.60
TOTAL ($ per ton)	65.50	60.00	5.50
Annual savings	for 591 tons		$3,250.50
Processing time (h)	107	8.7	2
Increased capacity	107 ton/year		18%

Note: Figures in brackets indicate an increase in cost.

TABLE 2.28
Savings in $ (per Ton of Fabric) for Half and Full Bleaching: Knitted Fabric – Giza Spinning and Weaving

Savings in	Half bleach			Full bleach		
	Before	After	Before	After	Before	After
Chemicals	22.20	27.80	(5.60)	95.20	101.70	(6.50)
Water use	6.20	2.50	3.70	13.30	5.20	8.10
Steam use	19.20	11.50	7.70	33.40	28.40	5.00
Electricity use	1.00	0.50	0.50	2.50	1.90	0.60
Labor	6.50	3.10	3.40	16.20	11.80	4.40
TOTAL ($ per ton)	55.10	45.40	9.70	160.60	149	11.60
Annual savings	for 1,200 tons		$11,640.00	for 240 tons		$2,784.00
Processing time (h)	7.5	3.5	4.0	18.5	13.5	5.0
Increased capacity	480/y		40%	48 ton/y		40%
Annual benefits	for 480 tons		$4,656.00	for 48 tons		$556.00
TOTAL			$16,296.00	TOTAL		$3,340.00

Note: Figures in brackets indicate an increase in cost.

Environmental benefits and improved working conditions

Sodium hypochlorite, a toxic and hazardous chemical, has been phased out of the bleaching process at Giza Spinning and Weaving. As a result worker conditions and safety have improved and the amount of halogenated organic hydrocarbons in the final effluent has been minimized. Water and energy consumption has also been reduced.

A summary of economic benefits and achievements for both Giza Spinning and Weaving Company and Misr Beida Dyers is shown in Tables 2.29 and 2.30.

TABLE 2.29
Summary of Economic Benefits

Summary of net savings ($ per year)	
Savings on operating costs – Giza Spinning and Weaving	$14,459
Savings on operating costs – Misr Beida Dyers	$14,350
Net benefit on increased production capacity – Giza Spinning and Weaving	$16,176
TOTAL ANNUAL BENEFITS	$44,985

TABLE 2.30
Benefits and Achievements

Savings	Giza Spinning and Weaving		Misr Beida Dyers
	Half bleach	Full bleach	Half bleach
Cost of chemicals	(25%)	(14%)	(1%)
Water consumption	59%	61%	30%
Steam consumption	40%	15%	27%
Electricity consumption	53%	27%	19%
Cost of labor	53%	27%	19%
Processing time	4 hours	5 hours	2 hours

Other achievements include:
- Improved productivity
- Improved fabric quality
- Improved working conditions

Note: Figures in brackets indicate an increase in cost.

Textile sector: Conservation of water and energy in spinning and weaving company (SEAM, 1999f)

El-Nasr Company for Spinning and Weaving is one of the largest public sector textile factories in Egypt, with annual production of over 52 million meters of fabric and around 7,000 employees. The factory was built in 1963 and is located in Gharbiya Governorate on 90 acres in the center of El-Mahalla El-Kobra.

The main activities are spinning, weaving, and wet processing. The company possesses an average of 8,000 tons of raw fabric per year, of which 20% is spun cotton yarns, 12% is polyester blend yarns, and 68% is gray fabric (50% commissioned by clients), using about 1,000 tons of different processing ingredients (chemicals, dyes, auxiliaries, etc.) per year. The main products are cotton or blended yarns, white and dyed cotton and blended fabrics, as well as terry fabrics, upholstery fabrics, and bed covers and sheets. The main objective of this case study is the conservation of water and energy and other simple CP techniques with high return in El-Nasr Company for Spinning and Weaving.

Process description
There are six main processing departments within the factory:

- Weaving: Wrap yarns are wound, starch sized, dried, and woven with the weft yarns.
- Cone pre-treatment and dyeing: Cone yarns are half or full bleached, dyed with vat, reactive or naphthol dyes, soaped, and then softened in Thies Kires unit.
- Fabric bleaching: Continuous pre-treatment is carried out either in Brugman or Gaston County bleaching ranges (rope form for pure cotton), or in the Kyoto range (open width for blends), while semi-continuous pre-treatment is carried out in the Bennenger range (open width for bed sheets). Mercerization is carried out using either Bennenger or Textima ranges.
- Fabric dyeing and printing: Dyeing is performed by exhaustion or by padding. Naphthol dyeing is also carried out. Printing is conducted using reactive or pigment dyes; the fabric is printed with the printing paste, dried, and thermo-fixed.
- Fabric finishing: Performed on Artos stenters, usually as normal, resin, or silicon-elastomer finishing.
- Fabric packaging: Fabrics are cut, rolled, or folded, then covered and packaged.

Service units
Factory service units include boilers, cooling towers, a caustic recovery plant, water and wastewater treatment plants, laboratories, storage, and maintenance facilities.

Energy consumption
There are two main sources of energy: electricity and different types of fuel, such as heavy oil, diesel, natural gas, and Naphtha.

Water consumption
The factory uses 1.9 million m^3/y of water.

Wastewater generation
The factory generates around 1.5 million m^3/y of industrial wastewater from different factory departments. This is disposed of to the public sewerage network, normally in compliance with the legislative discharge according to local environmental regulations.

CP opportunity assessment
Considering cleaner production opportunities that were identified as a result of the industrial audit, the following are the various improvement areas:

- Dyestuffs are stored with the lids unsecured, such that hydrolysis sets in and shelf life is reduced.

- Final fabric products are not given adequate protection during storage, so that soilage can easily occur.
- Various parts of steam and hot water pipes are not insulated and a great amount of heat is lost.
- Steam condensate from all departments is put directly to the drain rather than re-circulated as feed water, causing an unnecessary wastage of water.
- Huge amounts of thermal energy are lost in the flue gases of the boiler which are exhausted to air.
- Considerable amounts of hot effluent from the different units in the pre-treatment and dyeing departments are directly discharged to the sewer, with great heat losses.
- Huge quantities of final washing water in the bleaching ranges are directly discharged without reuse.

The measures which have been identified for implementation are briefly outlined below. During the audit stage, particular attention was paid to those improvements that could be carried out at low or no cost to the factory. These are easy to implement and often entail significant savings.

CP techniques implemented

- Improve storage facilities – dyes and fabrics (good housekeeping): The storage conditions in the dyehouse store were investigated and recommendations made to prevent the hydrolysis and lumping of reactive dyestuffs, which was achieved by ensuring that the dye containers were tightly closed and the storage space was closed when not in use.
- Material substitution and optimization of chemical usage (input material change): The different process chemicals used in the factory were thoroughly examined, with the aim of identifying optimum process chemicals, taking into account process and fabric requirements, resource, and environmental constraints. The following alternatives were identified:

Application	Original	Substitute/Alternative
Neutralization	Acetic acid	Formic acid
Dyeing	Sulfur and naphthol dyes. Dyes which involve the use of heavy metals	Other dyes. Dyes which do not involve heavy metals
Printing	Ammonium phosphate	Ammonium sulfate
Desizing/scouring poly/ cotton blend	Leonil LB-ET	Ammonium persulfate

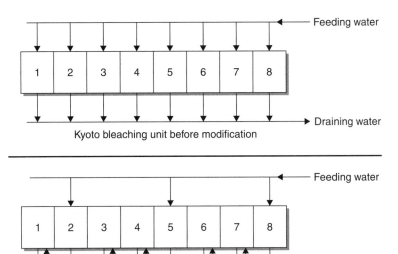

FIGURE 2.2 Bleaching unit before and after modification

- Water and energy conservation: The factory was inspected to determine the sources of water and energy losses. It was found that the pre-treatment and dyeing departments offered the greatest potential for savings and efforts were focused in these areas as follows:
  - Collection and reuse of steam condensate (equipment modification): Around 20% of the steam being used in the different processing departments is now being recovered in the form of condensate, which is stored in a water collection tank and re-circulated to the process water feed lines by means of pumps and piping network.
  - Upgrade insulation of steam and hot water network (better process control): In the pre-treatment department, the steam network was assessed to identify areas of heat losses. This was shown to result almost entirely from inadequate insulation. Once all necessary insulation was completed, energy recovery consumption was reduced, such that 6,576 million kcal of heat will be saved annually.
  - Counter current flow (process control): In the eight washers of the Kyoto bleaching unit range, the least contaminated water from the final wash is recirculated 1–2 times in the preceding wash, before being finally discharged. Using this approach, water is fed to three wash units instead of eight as shown in Figure 2.2. This required the installation of new pipelines, tanks, pumps, valves, and filters.
  - Installation of automatic shut-off valves in bleaching ranges (better process control): An evaluation of all bleaching ranges revealed that process water and steam were being consumed even

when the ranges were not in use. This excess consumption can be eliminated by the installation of shut-off valves at different control points to eliminate the waste of water, steam, and chemicals.

– Recycling of final washing water in the bleaching ranges (on-site recycling and process change): In the Brugman and Gaston County bleaching ranges, the final clean wash water coming from the bleaching stage can be reused in another, earlier washing stage for washing after scouring. In each of these two ranges, the final washing unit has been connected to the washing unit for the scouring process. This required the installation of pipelines equipped with pumps.

Cost/benefit analysis

Throughout industry, cleaner production and environmental protection measures can offer real financial benefits in terms of:

- Reduced raw materials consumption.
- Waste minimization.
- Reuse or recycling of in-plant materials.

Implementing these measures will also result in reducing environmental pollution load such that discharge consent was set within the limits of the law. The benefits achieved in this factory reflect the potential of the importance and power of CP techniques. A summary of CP techniques used in this factory are presented below:

- Good housekeeping, which resulted in the preservation of expensive dyestuffs.
- Input material change, which resulted in: (a) savings in process chemicals, (b) reducing rates of water consumption, (c) optimizing energy consumption, (d) reducing wastewater loads, and (e) improving operation and productivity.
- Better process control, which resulted in water savings, energy savings, and wastewater treatment savings.

The significance of these benefits manifests itself in the cost/benefit analysis shown in Table 2.31.

The benefits and achievements in this case study can be summarized as follows:

- Water consumption has dropped by 20%.
- Thermal energy consumption has dropped by 5%.
- Boiler fuel consumption has decreased by 5%.
- Wastewater volume has decreased by 20%.
- Chemicals and dyestuff costs have dropped by 5%.

TABLE 2.31
Cost/Benefit Analysis

Factory department	Action	Capital and operation cost ($)	Saving/y ($)	Payback period (months)
All	Improve storage facilities	0	4,000	Immediate
All	Optimize chemical usage	0	6,140	Immediate
All	Steam condensate recovery	7,895	23,700	<4
All	Upgrade insulation of steam and hot water networks	8,420	23,700	<5
Fabric pre-treatment	Counter current flow in Kyoto range	7,720	38,900	<3
Subtotal		24,035	96,440	<3
Additional measures for Improvement				
Fabric pre-treatment	Install shut-off valves	6,400	7,870	<10
Fabric pre-treatment	Recycling final wash water	5,260	24,780	<3
Yarn dyeing	Heat recovery from hot liquors	14,035	18,800	<9
Subtotal		25,695	51,450	<6
Overall cost benefits		49,730	147,890	4

Oil and soap sector: Oil and fats recovery at oil and soap company

Tanta Oil and Soap factory is located in Tanta-Gharbiya, Egypt, covering an area of 35 acres (147,000 m^2) and is one of three that are owned by the Tanta Oil and Soap Company chain. The plant was established in 1934. The Tanta factory produces an average of 18,000 ton/y of edible oil extracted from cotton seeds, sunflower seeds, and soy bean, and 12,000 ton/y of ghee using palm and other oil seeds. The factory also produces 9,000 ton/y of glycerin and 48,000 ton/y of animal fodder. The main objective of this case study is to recover oil and fat from the industrial wastewater effluent (SEAM 1999a).

Process description
Production processes in the factory are as follows:

- Seed receipt and preparation: Following receipt, the seeds are sent for mechanical preparation. This consists of mechanical screening and magnetic separation to remove any impurities which may be present. Following separation are the processes of crushing, flaking, and cooking.
- Oil Extraction: The prepared seeds are mixed with hexane in a continuous counter current system to produce a hexane–oil mixture (miscella) and seed cake. The seed cake is separated from the miscella, dried cooled, pressed, and used to produce animal feed. The hexane is recovered from the miscella under vacuum (using direct and indirect steam) and reused in the system. The remaining crude oil is then cooled and sent for refining.
- Oil refining packaging: Crude oil and ghee are processed using the following steps:
  - Degumming (for sunflower seeds or soybean) and neutralization – gums are removed in a batch process, using phosphoric acid. Neutralization is done by adding caustic soda to remove free fatty acids from crude oil to produce semi-refined oil.
  - Bleaching – color is removed from the oil using fuller's earth followed by filtration.
  - Deodorization – unpleasant odours are removed from oil by high temperature vacuum distillation.
  - Packaging – the reined, bleached, and deodorized (RBD) oil is bottled in automatic filling lines.
- Soap and glycerin production: Fats are saponified in a batch process, by mixing with caustic soda and heating with direct and indirect steam. After saponification, soap is separated from the lye solution to be dried, blended with additives, homogenized, cut, and pecked. Glycerin is separated from the lye solution and distilled.

Service units
Service units include boilers, cooling towers, pumping stations, power transformer stand-by diesel power generators, and storage and maintenance facilities.

Energy consumption
The two main sources of energy are mazot and electricity. Mazot and solar are used in the boilers to generate steam. Average annual consumption 15,000 tons of mazot and 600 tons of solar. Annual electricity consumption is around 10.5 million kWh.

Water consumption
The factory consumes an average of 16,800 m^3/day of water of which 1,800 m^3/day is process water, and 1,500 m^3/day cooling and vacuum water. This

water is taken entirely from group water boreholes within factory premises. Approximately 35 m^3 of drinking water is taken from the public network every day.

Wastewater generation
The factory generates about 16,000 m^3/day of industrial wastewater from different factory steams, including process effluents, boiler blow down, cooling water, vacuum water, and steam condensate. The wastewater is discharged to Akhnawy drain near the factory.

CP opportunity assessment
The main objective is to set cleaner production opportunities which were initially identified through an industrial audit of the factory (SEAM 1999a). The following improvement opportunities were identified as being of particular importance:

- Upgrade loading and unloading procedures for oil ghee and fats to minimize spillage.
- Improve housekeeping in the factory acids splitting unit.
- Improve oil recovery from processing unit effluents; especially oil refining and packaging unit.
- Improve handling of animal fodder ingredients to prevent losses.
- Develop water consumption control, by installing water meters for monitoring water consumption at various units, and installation of self-closing taps to reduce water consumption in service units.

CP techniques implemented/Cost/benefit analysis
1. Source reduction was implemented through good housekeeping and preventive maintenance through upgrading loading and unloading procedures. During the loading and unloading of oil, significant levels of leakage and spillage of ghee and fatty matter from batch reactors and separators were occurring. These losses were entirely eliminated as a result of issuing improved procedural instructions and by improving the supervision of transfer operations.

The cost/benefit analysis for using this CP technique revealed the following:

- The implementation cost is nil.
- The annual saving is $3,6190.
- The payback period is immediate, which is considered encouraging for investors.

2. Equipment modification was made through the installation of three gravity oil separators (GOS) manufactured by the factory. The GOSs were installed on the oil washing line, immediately after the point where water is recovered from the oil and ghee refining batch reactor. Recovered oil and ghee was discharged and lost to refinery effluent before the installation of the

GOS. Fatty matter is recovered from two sources: the mucilage produced during neutralization and fatty matter from refinery effluents in oil separators.

The cost/benefit analysis for using this CP technique revealed the following:

- The implementation cost is $48,000.
- The annual saving is $35,650.
- The payback period is 1.3 years, which is considered from the first category thus it is also encouraging for investors.

Moreover, using this technique has contributed to the improvement of the oil losses which prior to installation were 11.98% and after installation were 11.2% resulting in annual savings of $15,480.

Another equipment modification was implemented for the animal fodder production unit where a cyclone vacuum system was installed to collect the suspended matter and transfer it directly to the raw material intake system. This was employed because the animal fodder production unit was generating heavy dust emissions, especially during the loading and unloading of the raw material.

The cost/benefit analysis for using this CP technique revealed the following:

- The implementation cost is $22,385.
- The annual saving is $18,860.
- The payback period is 1.2 years, which is considered from the first category thus it is also encouraging for investors.

3. Process change was implemented by segregating the cooling water, vacuum water, and process water from one another in parallel with rehabilitation of the two existing cooling towers. This change required the installation of new pipelines, filling materials, valves, and connections in addition to some civil work.

The cost/benefit analysis for using this CP technique revealed the following:

- The implementation cost is $38,600.
- The annual saving is $21,050.
- The payback period is 1.8 years, which is considered from the first category thus it is also encouraging for investors.

The implementation of the cleaner production techniques has resulted in total savings of around $111,750 with an average payback period of around 1 year. It also resulted in:

- Water consumption reduced by 23%.
- Oil and grease concentrations in the final effluent reduced by 99%.

- Working conditions improved in the animal feed production units.
- Annual recovery of oil, ghee, fats, and animal fodder amounted to $90,705.
- Capital investment costs reduced by about $87,720.
- BOD loads in the final effluent reduced by 85%.

Final remarks
The following achievements were obtained as a result of applying the cleaner production technique:

- Conservation of natural resources.
- Avoidance of end-of-pipe treatment.
- Reduction of capital costs.

Tanta Company has applied the "win–win" strategy where the environment, the consumer, and the worker were protected while improving the industrial efficiency, profitability, and competitiveness.

Oil and soap sector: Waste minimization at edible oil company

This case study will present different steps for the oil extraction process at Sila Oil Company, located in Fayoum, Egypt, highlighting the different cleaner production techniques that could be implemented in some or all of these steps. The CP techniques that were utilized in this case study are good housekeeping, equipment modification, input material change, and on-site recycling. Finally, it was noticed that cleaner production techniques not only result in substantial financial savings due to reduction in energy and raw material consumptions, but also result in better quality products within a more environmentally friendly process (SEAM 1999c).

Process description
Extraction of oil from oil-bearing seeds consists mainly of four major steps:

- Pre-treatment of the raw material: During this step the oil-bearing seeds are cleaned, dried, and de-hulled. Then clean seeds are weighted and passed to a de-stoner and a magnet to remove stones and metals, then prepared for oil extraction and cooking.
- Extraction of oil: There are three main methods for extracting oil from seeds, which are: hydraulic pressing, expeller pressing and solvent extraction. In this process 50% of the crude oil content is extracted using expellers, while the seed cake containing 30% oil is sent to solvent extraction unit where the seed cake is mixed with hexane to produce a solvent oil mixture (miscella), and an extracted meal (2% oil content), which is sent to a desolventizing, toasting, drying, and cooling unit. Crude oil is extracted from miscella by a three stage evaporation

system. The evaporated hexane is recovered within the system and reused.

- Purification of oil: During this stage oil is purified by removing all impurities such as color, odor, fatty acids, gums, and residual soap. This step takes place in the following stages:
  - Degumming to remove about 0.15% gums.
  - Neutralization using caustic soda to remove fatty acids to generate soap stock 5%.
  - Washing and separation by centrifuge.
  - Drying and bleaching to remove the color.
  - Deodorization of the bleached oil by vacuum distillation.
- Packaging of the final product: A brine chiller is used to cool the refined, bleached, deodorized oil in a buffer tank. Then the final product is transferred to 0.75 liter or 2.0 liter bottles, capped, labeled and cased.

CP opportunity assessment

Cleaner production opportunities were identified during the industrial audit of the factory in order to achieve maximum conservation of raw materials and energy. The following are some of the identified cleaner production opportunities (SEAM 1999c):

- High amounts of steam losses due to the bad condition of the steam lines regarding insulations and valves. These losses were estimated to be 34 ton/day.
- In the seed receiving unit, broken seeds and hulls are treated as wastes, rather than being reused in the oil extraction process.
- Heavy oil leaks and spills occurring during delivery and transfer in the receiving area account for 2–5% of total fuel consumption.

CP techniques implemented

- Good housekeeping: This technique of cleaner production was utilized in this case study in the form of a preventive maintenance program with a total implementation cost of $2,630. The preventive maintenance program included in-factory servicing of the expeller and modification of the packing of the cooling tower, and it has resulted in annual savings of $4,386 and $880 respectively.
- Process modification: Process modification was implemented in this case study via modifying and redesigning the production process to allow for the more fresh seeds to be fed into the expeller with a total implementation cost of $1,750. This was achieved by converting the path of the recycled sunflower seed fines from the expeller (original design) to the extraction plant (modified design). As a result, the capacity of the expeller has been increased by 40 ton/day of sunflower leading to an annual saving of $21,050.

TABLE 2.32
Cost/Benefit Summary

Factory unit	Action	Capital/ operating costs ($)	Yearly savings ($)	Payback period (month)
Refinery	Use of liquid caustic soda	None	43,860	Immediate
Steam	Upgrade system network	5,260	97,010	<1
Receiving	Recovery of broken seeds	1,580	81,270	<1
Preparation	Reuse of fines	1,750	21,050	1
All	Preventive maintenance program	2,630	5,260	6

- Material substitution: The neutralization process that involves the removing of fatty acids to generate soap stock utilizes solid caustic soda as a processing material which leads to corrosion. This material has been changed to caustic soda solution which is less corrosive and with a lower cost of $175/ton against $368/ton. In addition to this substantial decrease in cost in which the daily neutralization cost dropped by 47%, soap stock quality has been improved and better working conditions have been achieved. Annual saving for this process was calculated to be $43,860.

Cost/benefit analysis

Cost/benefits analysis, shown in Table 2.32, is the process of weighing the total expected costs versus the total expected benefits of one or more actions in order to choose the best or most profitable option. While costs are either incurred once or may be on-going, benefits, on the other hand, are most often received over time. Therefore, the effect of time would be accounted for in the analysis by calculating a payback period or projected payback period. The payback period is the time it takes for the benefits of a change to repay its costs and it is calculated by dividing the cost over the annual benefit. Many companies look for payback over a specified period of time, for example three years.

Example of calculation of the payback period for the preventive maintenance program:

Payback period = cost/annual benefit

Payback period = 15,000/30,000 = 0.5 years × 12 = 6 months

Wood furniture sector: Waste and pollution minimization in wood furniture companies

Wood lumber is the raw material for wood manufacturing. It is usually in the form of logs of different lengths and thicknesses depending on the type of wood. Small workshops depend on wholesalers of wood for their need for raw material, which is generally not pre-treated (derosinated), and may contain above standard knots, compared to large furniture industries which may endure importing their own high quality woods. The most popular wood types used are beech, oak, MDF, arrow, white wood, Swedish wood, rosewood, mahogany, etc. Beech wood is characterized by being strong, heavy, resistant to shock, and easy to machine. It is preferred for carved furniture, dining room furniture, and luxury cabinets. Mahogany is easy to slice but difficult to machine; it is usually used for interior woodwork. Swedish wood is usually used for floor panels. White wood is one of the weakest types and is susceptible to deform over time; because of its low price it is used to manufacture cheap furniture.

A number of chemicals have been used in the wood industry such as fillers, sealers, solvents, thinners, paints, pigments, binders, resins, additives, stains, sealers, washout, topcoats, powder coating, acrylic lacquers, vinyl coatings, catalyzed topcoats, urethane coating, etc.

Process description

Drying

Wood standard water content is (10–12%) by weight. Imported wood usually has higher water content than the standard. Wood is a hygroscopic material, which tends to take up moisture and retain it, hence water content increases during shipment and storage. Wood must be dried before it is manipulated. Drying increases a wood's dimensional stability, compact ability and thus its overall strength. It also reduces wood staining, decay, and susceptibility to insect attack.

There are two methods for wood drying: air drying and kiln drying. Air drying may take from two to four months depending on temperature and humidity.

In this process wood timber is stacked in open air. As the timber tries to achieve moisture balance with its surroundings it loses some of its water content. Air drying is a cheap process but it takes time and it is hard to control the resultant water content.

On the other hand, kiln drying takes a much shorter time than air drying. By controlling the temperature of the kiln, specific water content can be achieved to suit the intended use of furniture. A large factory in Damietta uses a computerized kiln that can detect through X-rays the initial water content of timber and automatically adjust the temperature to achieve the aimed water content. It is capable of taking 9 meters of wood at a time and uses surface condensation to get rid of internal water content. It should be noted that in kiln drying the temperature should be controlled such that the wood loses water content without creating high moisture gradients to avoid cracking.

Sawing

Sawing is the cutting of timber usually as near as possible to required size. If wood is sawed across the grain it is called crosscutting, and if it is sawed parallel to the grain it is called ripping. Sawing can be done with a portable manual saw as in small workshops or with automated sawing machines in large factories and medium workshops. Sawing machines can be circular, band, scroll, or radial.

Planing

Planing is the process by which the faces or planes of timber are made flat, and usually perpendicular to each other. A planer is used for this purpose. In large factories a power planer or jointer is used which consists of a set of blades set on a rotating cutter head.

Sanding

Sanding is used to smooth the surface. In small workshops medium grit paper (100–120 grit) is used for the purpose. Sanding is made along the grain never against it. Some times the surface is wiped with a wet rag to cause the surface fibers to swell and raise. A diluted solution of glue or resin is applied with a rag to roughen raised grains. Then a finer grit paper is used to remove raised fibers (200–250 grit). For large-scale production entities an automated disk, belt, or roller machine is used. Sanding using different grades of grit paper is used along many stages of the manufacturing process.

Assembly

Assembly of wooden parts can be done before or after the surface finishing process. In small workshops furniture is usually assembled before finishing.

Assemblies in small workshops depend mainly on the use of glue and nails. In larger factories interlocks are used. This gives the furniture an antique essence. It is always recommended to use interlocks for high quality furniture. Since iron nails contract and expand more than wood due to different weather conditions climate differences will affect the nails more readily causing this type of assembly to weaken. Interlocks can be made out of wood waste logs using a lathe.

Veneer is a thin layer applied to the furniture using glue and pressure. Veneer is then sanded and sealed to pass through a staining process similar to that of the painting process described next. The only difference is that a polyester layer is not used in the staining process.

Finishing coating process

- Derosination: Because certain types of wood contain rosin (a naturally occurring resin) which can interfere with the effectiveness of certain finishes, a process known as derosination may be employed.

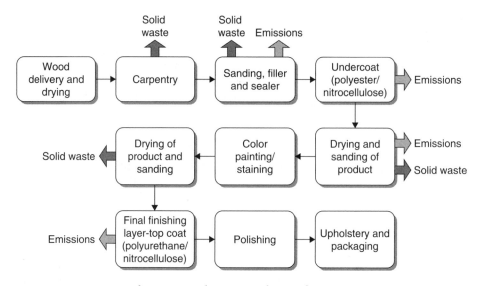

FIGURE 2.3 Waste and emissions during wood manufacturing process

Derosination is accomplished by applying a mixture of acetone and ammonia to the surface of the wood. Spent acetone and ammonia are the primary outputs from derosination.

- Opened pores are filled with calcium carbonate filler using a filler knife. As filler starts to dry, it becomes tacky. Excess filler is wiped or sanded.
- A nitrocellulose sealer is then used, followed by sanding with 200 grit sandpaper.
- Polyester pre-coat and sealing: Polyester powder dissolved in acetone or turpentine in a ratio of 1 to 1 is sprayed on to the surface. Three coatings are applied and each is followed by sanding with 220 grit sandpaper.
- Painting is performed utilizing water-based colors followed by fine sanding. Co-solvent turpentine is used as an additive to the color.
- A wash coat of nitrocellulose is used followed by fine sanding.
- Polyurethane topcoat is sprayed to protect the upper layer.
- Polishing sebedag and oil are applied to the surface and rubbed with cotton.

Waste and pollution generated

The furniture industry has mainly two areas with major environmental impacts: the management of solid wood waste and the emissions from the polyester and polyurethane spraying process.

TABLE 2.33
Summary of Process Materials Inputs and Pollution Outputs in the Wooden Furniture Industry

Process	Material input	Air emissions	Process waste	Other waste
Drying				
Drying kilns	Raw wood	Vapor, possible chemicals used in pre-treatment of raw wood		
Machining				
Sawing/Planing	Dried wood		Wood chips, sawdust	Wood chips, sawdust
Sanding	Wood	Wood flour	Wood flour	Wood flour
Assembly				
Gluing/Veneer application	Hot melts, polyvinyl acetate, solvent-based adhesives	Solvent emissions		Spent solvent-based adhesives
Sanding	Assembled furniture	Wood flour	Wood flour	Wood flour
Pre-finishing				
Watering/Sanding	Assembled furniture, water, adhesives, resins		Wood chips, sawdust, adhesive and resin particles	Wood chips, sawdust, adhesive and resin particles
Finishing				
Staining	Mineral spirits, alcohol, solvents, pigments	Solvent emissions		Pigment wastes
Wash coating	Nitrocellulose-based lacquers	Solvent emissions		Spent solvents and lacquers
Filling	Calcium carbonate, silica, pigment	Solvent emissions		Spent solvents, stains, drying oils, synthetic resins, thinners, pigments

(Continued)

TABLE 2.33 (Continued)

Process	Material input	Air emissions	Process waste	Other waste
Sealing	Nitrocellulose-based lacquers,	Solvent emissions		Spent solvents and lacquers
Priming (outdoor furniture or metals)	Fungicide, water repellent			
Painting	Toluene, pigments, epoxy-ester resins, aromatic hydrocarbons, glycol ether, halogenated hydrocarbons, vinyl acetate, acrylic	Solvent emissions		Spent solvents, pigments, resins, etc.
Topcoat application	Polyurethane, organic solvents	Solvent emissions		Spent denaturated alcohols, resins, shellac, etc.
Sanding (intermittently between each of the above finishing applications)	Finished piece of furniture	Particulates that include wood flour, adhesive, resin, lacquers, etc.	Particulates that include wood flour, adhesive, resin, lacquers, etc.	Particulates that include wood flour, adhesive, resin, lacquers, etc.
Rubbing/Polishing	Sebedag and oil			Spent lubricants, detergents, oils
Cleanup operations Brush cleaning, spray-gun cleaning	Solvents, mineral spirits, alcohols	Solvent emissions	Spent solvents, etc.	Spent solvents, mineral spirits, alcohols

Solid waste
The solid waste is mainly divided into four types according to its size:

- Wood chips: Chopped blocks of wood of different widths, thicknesses, and lengths.
- Wood shavings: Thin shredded wood from leveling operations.
- Sawdust: Fine wood powder or dust generated from sawing operations.
- Wood flour: Very fine wood dust generated from sanding.

The solid waste occupies a large space, exposing the workshop or factory to fire risks if a source of ignition is present, which would result in the loss of a valuable and expensive material. In addition, acute exposure to fine wood dust can cause health hazards to workers such as irritation of the eyes, skin, or respiratory tract, while chronic exposure may result in dermatitis, asthma, or respiratory system congestion.

Polyester and polyurethane spraying emissions
The painting workshop should be equipped with spraying booths. Polyester is always sprayed first and then sanded as discussed before. The top layer is polyurethane. There are many health hazards associated with polyester and polyurethane spraying. Their emissions contain acetone (from solvents), aerosol, and other harmful chemicals. Breathing moderate to high levels of acetone for short periods of time can cause nose, throat, lung, and eye irritations, headaches, light dizziness, increased pulse rates, nausea, vomiting, unconsciousness, and possibly coma. While breathing aerosol causes nearly the same health problems in addition to coughing, wheezing, chest tightness, and asthma.

CP opportunity assessment
Solid wastes
Before going through recycling of solid waste such as small pieces of wood, sawdust, wood shavings, etc. it is always recommended to install a dust collection system such as a bag filter to protect the working environment from any suspended solids in the air. As wood parts are machined and sanded, a substantial amount of sawdust is generated. Dust collection systems can provide safe, waste reduction benefits, but must be properly designed to be effective, safe and efficient. This improves worker health and safety by keeping the dust out of the breathing air and off the floor. It extends equipment life and decreases maintenance by keeping dust away from machinery, and collects and keeps the sawdust from becoming contaminated with dirt and other contaminants so that it can be recycled.
Wood shavings:

- Used as animal bedding for chickens, horses, cattle, etc.
- The bedding waste could be used as a soil additive.
- Could be used as a thermal insulation.

Saw dust and wood flour:

- Used as a fuel by converting them into briquettes.
- Used in particle board or MDF manufacturing.
- Added to flooring tiles to give thermal and resilient properties.

General precautions:

- Burning treated or coated wood can release regulated hazardous air pollutants.
- Waste streams should not be combined (i.e. mixing wood with glue and sawdust from rough mills) as it may inhibit optimal secondary use.

Cost/benefit analysis
Solid waste
- The estimated amount of wood input material loss in any factory is about 20–30%.
- By applying the above recommendations by the workshop/factory and recycling the wood waste or selling it at a suitable price as raw material to other industries; will reduce the loss percentage (5–10%), as well as bring a return for the furniture workshop/factory owner.
- By simple calculations the economic savings can be predicted for beech wood as an example:
  - $1 \, m^3$ beech ranges between $316 and $333.
  - Economic loss $= 0.25 \times 325 = \$81.25/m^3$.
  - By adapting cleaner production opportunities $81.25 - (.07 \times 325) = \$58.5$ is saved per m^3, and this is without adding the return from recycling that will certainly cover the initial cost of the wood raw material.
  - Economic savings $= 72\%$ (not including recycling return).

Spraying chemicals
- Raw material substitution or elimination is the replacement of existing raw materials with other materials that produce less waste, or act as a non-toxic waste.
  - Using high-solids coatings: Traditional sealers and topcoats are 20% solids or less, meaning that 80% or more of the coating evaporates and is wasted even if the transfer efficiency (TE) was 100%.
  - Replacing thermoset surface coatings with thermoplastic.
- Process or equipment modification is recommended in order to reduce the amount of waste/pollution generated, for example:
  - Manufacturers can change to a paint application technique that is more efficient than conventional spray guns, for example, high-volume/low-pressure (HVLP) guns. HVLP guns provide transfer efficiency (TE) as high as 40 to 65% compared to 20 to 40% for

conventional air spray guns and this will reduce overspray and hence will save raw material.

- Fixed painting table can be modified to a turning or rotating painting table to facilitate the painting process from all directions.

- Reducing air emissions is necessary when spraying polyester and polyurethane or any other chemical as these are very harmful to the working environment if not done in a controlled area. The emissions' impact can be reduced and controlled by adopting several measures:
  - A good ventilation system to collect the dust especially after polyester spraying where the white "polyester" dust is very dangerous to the worker force.
  - The chimney of the painting booth should be high enough from any surrounding buildings and could contain a filter. Biofiltration is recommended as a control technology, where contaminated exhaust air passes through a biofilter for contaminant removal. The biofilter consists of organic matter, such as tree bark and compost, the pores of which are filled with water. Biologically active micro-organisms are present, partly free-floating in the water and partly attached to the organic matter. Other recommended filters are catalytic oxidizers.
  - Always use the spraying booth for painting processes. This improves product quality and protects the surrounding environment as well as the working environment.

- In the area of materials reuse and recycling, waste separation is the first step for on-site or off-site recycling. Waste separation involves avoiding both the mixture of different types of waste and the mixture of hazardous wastes with non-hazardous wastes. This makes the recovery of hazardous wastes easier by minimizing the number of different hazardous constituents in a given waste stream. Also, it prevents the contamination of non-hazardous wastes. Specific examples include segregating spent solvents by solvent types, and segregating non-hazardous paint solids from hazardous paint solvents and thinners. Solvent recycling is the use or reuse of a waste as an ingredient or feedstock in the production process on-site.
  - Reuse of cleaning solvent for painting equipment as a reducer for the next batch of the same or darker color.
  - Distilling "dirty" solvent and reusing it for cleaning.
  - Inclined table to collect the drops of sealer/waste paint for reuse, instead of being lost on the floor or above the table.

- Loss prevention and housekeeping is the performance of preventive maintenance and equipment and materials management to minimize opportunities for leaks, spills, evaporative losses, and other releases of potentially toxic chemicals.
  - Spray guns can be cleaned by submerging only the front end of the gun in the cleaning solvent.

TABLE 2.34
Cost/Benefit Analysis

Pollution prevention process	Economic and environment savings and benefits	Payback period
Process modification: Implement alternatives to conventional spray gun systems	• Material consumption reduction: 15% • Waste volume from spray booth cleanup reduction: 50%	Payback period: 1 year
Reuse: Flush equipment first with dirty solvent before final cleaning with virgin solvent and use cleanup solvents in formulation of paint	• Waste savings/ reduction: 98% • Solvents reduction: 98.4%	Payback period: 1 year

- Routine maintenance of spray gun equipment can prevent equipment from breaking down and leaking.
- Allocate a designated area for chemicals. It is also recommended to have a simple measuring device to determine exactly the amount of chemicals to be mixed and assure a homogeneous mixing.

Table 2.34 represents a sample of the economic savings gained and the payback period from implementing cleaner production techniques.

Iron and steel sector: Wire rods factory

This case study revolves around a Chinese manufacturing company. The case involves a factory that specializes in wire rods. The factory is located in Beijing, China, and consists of 700 employees. The factory imported two wire mills with the capacity to produce 700,000 tons of wire rods annually. The mill was listed as one of the worst polluters in China, and a project was assigned for cleaner production in 1989. In order to comprehend the case study, one must be acquainted with the manufacturing steps. First of all, the raw materials are obtained in the form of billets where they are transported into a furnace for melting in order to shape the metal. After heating the billets to a high temperature the metal is then rolled into a coil shape of diameter 6.5–10 microns. The coil is then cooled and collected in order to be compacted and packaged for delivery.

An in-depth look at these manufacturing steps reveals the amount of energy used and the emissions from the heating process of the metal. The report from the factory showed that "450 tons of waste lubricates were generated from the cooling process of the coils, while 350 tons of waste hydrodynamic oils were produced from collecting and packing the coil" (Cleaner Production in

China 2006). These wastes were reprocessed for reuse by the factory. Moreover, other wastes occurred during the manufacturing process and these are emissions from the fuel combustion during the heating process by the furnace into the atmosphere and they are close to 451 million m³ of waste gas. Finally, huge amounts of water have been discharged at an annual rate of 1.36 million m³ of wastewater. The factory needs to be efficient in using the resources and also to cut down on the emissions and wastes into the environment.

CP opportunity assessment

Generally the factory had several options and they were divided into two main classes. Class A was a low cost option that was implemented directly for enhancing firm efficiency and protecting the environment. Class B involves investments. The firm had to take immediate action. The following Class A actions involved:

- Informing personnel in the factory of the importance and seriousness of cleaner production.
- Training programs for operators.
- Dealing with any leaks, drips or spills that may be found in the factory along the production line.
- Improving the management of the factory and the product in turn.
- Using the water efficiently and minimizing its use to where it is needed.
- Improving the rate of the waste oil recovery.
- Improving insulation of the furnace, so there will not be any heat dissipation.
- Reducing the iron oxides level.
- Improving the transportation of oil without any leaks or spills during the process.
- Reducing consumption of water and electricity.

CP techniques implemented

The above-mentioned CP opportunities had a low cost investment and could be applied directly; they are housekeeping cleaner production techniques.

Other improvements were made that had higher cost or investment – four were implemented while two were planned for the future, and they are as follows:

- Installing an oil fog system in the steamer controlled cooling line thus reducing oil consumption.
- Reforming the hydrodynamic system from the ration pump to the various pumps.
- Replacing the hydrodynamic system pipes to reduce oil loss.
- Installing a waste oil connector in the coil collection oil station to reduce oil discharges.

Planned for changes:

- Reforming the collecting coil system from the insert valve to the proportion valve thus reducing oil discharges.
- Replacing the burners with ones operating on oils or gases.

Cost/benefit analysis

The changes done to the factory revealed huge changes in the gas emissions as 50.48 million m^3 of the waste gas was reduced and there was a decrease of 45 tons in the discharge amount of wastewater. The changes that were implemented directly in the short term would yield a decrease in waste gas "by 43.52 million m^3 per year accounting for 9.53% of total waste gas emissions and decrease waste water by 126 tons a year accounting for 12.1% of total waste water discharges. If the mill would carry out the long-term options, the mill would decrease waste gas of 60 million m^3 per year accounting for 13.3% of annual total emissions and reduce waste water of 169 tons a year accounting for 11.44% of annual total discharges" (Cleaner Production in China 2006).

Questions

1. Discuss the main obstacles for implementing cleaner production techniques, and the methodology to remove such obstacles.
2. Does cleaner production always pay? Why? And how?
3. What is the difference between good housekeeping and better process control as two major techniques for cleaner production?
4. What is the role of the government to promote cleaner production?
5. Discuss different mechanisms to encourage/enforce the general manager/investor to implement CP within their facilities.
6. What is the difference between cleaner production and end-of-pipe treatment?
7. Does cleaner production require high technology? Explain.

Chapter 3

Sustainable Development and Industrial Ecology

3.1 Introduction

Concerning environmental issues, there is a common misconception that environmental protection comes at the expense of economic development or vice versa. This is clearly portrayed when communities faced with economic crises settle for alternatives that sacrifice environmental integrity such as incineration, treatment, or the construction of landfills even though these solutions in fact are extremely important but are not sustainable and economically expensive.

Sustainable development is formally defined by the World Commission on Environment and Development (WECD) as "development that meets the needs of the people today without compromising the ability of future generations to meet their own needs". Therefore, sustainable development refers to a shared commitment towards steady economic growth, given that this economic growth does not compromise the satisfactory management of available environmental resources. Resource allocation, financial investments, and social change are directed in a sound manner that guarantees their sustainability or continuation with time and thus they are made consistent with both future and present needs. Another notable definition for sustainable development by Sustainable Seattle is "economic and social changes that promote human prosperity and quality of life without causing ecological or social damage" (Redefining Progress 2002). Industries are therefore encouraged to flourish but also to realize their impacts on the environment and society around them. Thus, it can be concluded that sustainable development is a concept that is not only exclusive to policy makers and environmentalists, but should be a matter of concern to industries, the business community, and society.

The practice of sustainable development is not a new one. This is a concept which has been repeatedly used over time in an effort to sustain and/or preserve resources of any type. However, formal attention and labeling of this concept began during the 1970s. In 1972, the global community came together in Stockholm to discuss international environmental and development issues for the first time in "the United Nations Conference on Human Environment". This conference was the first significant link between business and environment to take responsibility for the environmental problem that uncontrolled industrial development was causing. The conference resulted in the creation of United Nations Environment Program (UNEP) to adopt a global action plan for protecting the environment. From its creation until now it tried to develop guidelines and tools for the above cause. In 1986, the "World Commission on Environment and Development WCED" was established. This commission's report is what first spread the term "sustainable development" and it became the benchmark for thinking about global environmental and development issues. The crest of global attention towards sustainability was during the United Nations Conference on Environment and Development in 1992 held in Rio de Janeiro. During this conference an action agenda was produced. This Agenda 21 was a comprehensive global plan of action for local, national, and global sustainable development. An equally comprehensive summit was the Johannesburg Summit in 2002 which was more focused on eradication of poverty as it also revived the commitment towards global sustainable development. These summits, augmented with vast global efforts, have aided in increasing awareness as well as multilateral agreements concerning various sustainability issues and critical environments.

The concept of sustainable development is a methodology that attempts to encompass social, technological, economic, and environmental aspects. Thus, focus is on the interactions and impacts of these four factors on each other rather than the fallacy that they are independent of one another. The elements of Environmental, technological, social, and economic growth are seen to reinforce each other thus attaining "win–win" solutions that do not compromise any aspect. In order to develop a methodology for sustainable development, a number of tools are required. The main tools necessary for implementing sustainable development are cleaner production (CP), environmental management system (EMS), 7Rs Golden Rule, industrial ecology (IE), environmental impact assessment (EIA), and information technologies (IT) as will be explained in detail in Chapter 4.

3.2 Industrial Ecology

Since the beginning of human history, industry has been an open system of materials flow. People transformed natural materials; plant, animal and minerals into tools, clothing and other products. When these materials were worn out they were discarded or dumped, and when the refuse buildup became a problem, the habitants changed their location, which was easy to do at that time due to the small number of habitants and the vast areas of land.

The goals of industry must be the preservation and improvement of the environment. With increasing industrial activity all over the world new ways have to be developed to make large improvements to industrial interactions with the environment.

An open industrial system – one that takes in materials and energy, creates products, and waste materials and then throws most of these away – will probably not continue indefinitely and will have to be replaced by a different system. This system would involve, among other things, paying more attention to where materials end up, and choosing materials and manufacturing processes to generate a more circular flow. Until quite recently, industrial societies have attempted to deal with pollution and other forms of waste largely through regulation. Although this strategy has been partially successful, it has not really gotten to the root of the problem. To do so will require a new paradigm for our industrial system – an industrial ecology whose processes resemble those of a natural ecosystem (Frosch, 1994).

Industrial ecology (IE) is the study of industrial systems that operate more like natural ecosystems. A natural ecosystem tends to evolve in such a way that any available source of useful material or energy will be used by some organism in the system. Animals and plants live on each other's waste matter. Materials and energy tend to circulate in a complex web of interactions: animal wastes and dead plant material are metabolized by microorganisms and turned into forms that are useful nutrients for plants. The plants in turn may be eaten by animals or die, decay and go around the cycle again. These systems do, of course, leave some waste materials; otherwise we would have no fossil fuels. But on the whole the system regulates itself and consumes what it produces (Frosch, 1994).

Industrial ecology is a new approach to the analysis and design of sustainable political economies (Frosch, 1995). Allenby (1999) calls industrial ecology the science of sustainability. Several other characteristics of stable ecosystems also suggest new norms to pursue in thinking about sustainability. Prigogine (1955) observed several very interesting features about steady state biological systems. One is that they are in a state of minimum entropy production, that is, the system is functioning with the least degree of dissipation of energy (and materials) thermodynamically possible in a real situation. These systems also exhibit a high degree of material loop closing. Materials are circulated through a web of interconnection with scavengers located at the bottom of the food web turning wastes into food. Even long-lived biological systems eventually succumb to environmental and internal stresses. They are not ideal models for a concept that implies flourishing forever. Ayres (1989) coined the term industrial metabolism as the web of flows of energy and material. When modeling an industrial economy consisting of an interconnected system of energy, material, and money flows such a system will supply an analytic means to repair the break in both the economic and environmental sciences. Daly (1977), and others have stressed the importance of including material flows in economic flows analysis, noting the fundamental connections of economics to natural resources. Daly (1977), and earlier Georgescu-Roegen (1971), developed a steady state framework for describing modern economic systems and for designing policy, invoking basic laws of thermodynamics and ecological systems behavior as part of the grounding. Expanding the typically sectoral or firm-level models used by policy analysts and corporate planners to material and energy flows during the entire life cycles of economic goods, in theory should reduce the probability of suboptimal solutions and of the appearance of unintended consequences. To convert part of these ideas into an industrial design context, a set of design rules has to be established for the innovation of more environment friendly and sustainable products and services. A few of these rules were developed by Ehrenfeld (1997):

- Close material loops.
- Use energy in a thermodynamically efficient manner; employ energy cascades.
- Avoid upsetting the system's metabolism; eliminate materials or wastes that upset living or inanimate components of the system.
- Dematerialize; deliver the function with fewer materials.

Industrial ecology as the "normal" science of sustainability (modifying slightly the phrase) as used by Allenby (1999) promises much in improving the efficiency of humans' use of the ecosystem. Technological improvements are not always better in the full sense of sustainability without taking the environment into consideration, where zero pollution is a must for industrial ecology. Cooperation and community are also important parts of the ecological metaphor of sustainability. Industrial ecology is the net resultant of interactions among zero pollution, cleaner production, and life cycle analysis according to the cradle-to-cradle concept.

3.3 Industrial Ecology Barriers

Even the industrial ecology concept has a lot of advantages from economical, environmental, and social points of view; there are still some barriers for implementation. The barriers to industrial ecology fall into five categories (Wernick and Ausubel, 1997) namely technical, market and information, business and financial, regulatory and regional strategies.

Technical barriers

Technical issues are one of the main challenges for industrial ecology to approach a cradle-to-cradle concept. It requires a lot of innovation to convert waste into money or prevent it at the source. Overcoming the technical barriers associated with recovering materials from waste streams is necessary but an insufficient step for stimulating the greater use of wastes in the economy.

Technology making recovery cheap (or expensive) and assuring high quality input streams must be followed by encouraging regulations and easy informational access. Finally a ready market must be present.

Market and informational barriers

These are inseparable from institutional and social strategies. Due to the absence of direct governmental interference, the markets for waste materials will ultimately rise or fall based on their economic vitality. Markets are sophisticated information processing machines whose strength resides in a large part on the richness of the informational feedback available. One option for waste markets is dedicated "waste exchanges" where brokers trade industrial wastes like other commodities. By using "Internet technology" to facilitate the flow of information, the need for centralized physical locations for either the waste itself or for the traders in waste may be minimal. Research is needed on waste information systems that could form the basis for waste exchanges. A stock market can be developed based on waste material. Systems would need to list available industrial wastes as well as the means for buyers and sellers to access the information and conduct transactions. The degree to which such arrangements would allow direct trading or rely on the brokers to mediate transactions presents a further question. As part of the market analysis for waste materials, research is needed to understand past trends regarding the effect of price disparities between virgin and recovered materials, and to assess the effect of other economic factors associated with waste markets, such as additional processing and transportation costs. A further matter for investigation concerns whether some threshold level of industrial agglomeration is necessary to make such markets economically viable. Progress is already being made on this front.

The Chicago Board of Trade (CBOT), working with several government agencies and trade associations, has begun a financial exchange for trading scrap materials. Other exchanges such as the National Materials Exchange Network (NMEN) and the Global Recycling Network (GRN) facilitate the exchange of both materials recovered from municipal waste streams and industrial wastes. Analysts might propose ideas for improving or facilitating the development of these exchanges. The value of such exchanges as a means of improving the flow of information depends on the deficiency of the current information flow, and how much this particular aspect of recycling plays in its success or failure. The CBOT is different from the other exchanges in that it is a financial market – starting as a cash exchange with hopes that it will evolve into a forward and/or futures market.

A simple waste exchange is premised on the notion that opportunities for exchange are going unrealized. A cash exchange has a related premise that there is a need for what economists call price discovery. Finally, a futures or forward market exists to allow the risk associated with price volatility to be traded independently of the commodity.

Business and financial barriers

Private firms are the basic economic units that generate innovative ideas which, among other goals, serve to enhance environmental quality. Corporations employ a spectrum of organizational approaches to handle environmental matters. In some cases the environment division of a corporation concerns itself exclusively with regulatory compliance and the avoidance of civil liability for environmental matters. For other firms the environment plays a more strategic role in corporate decision making. Decisions made at the executive level strongly determine whether or not companies adopt new technologies and practices that will affect their environmental performance. Also, the manner in which corporations integrate environmental costs into their accounting systems, for instance how to assign disposal costs, bears heavily on its ability to make both short- and long-term environmentally responsible decisions.

Research is needed to better understand the role of corporate organization and accounting practices in improving environmental performance and the incentives to which corporations respond for adopting new practices and technologies. Such studies would examine the learning process in corporate environments as well as investigate how corporate culture influences the ultimate adoption or rejection of environmentally innovative practices.

Regulatory barrier

Environmental regulation strongly induces companies to appreciate the environmental dimensions of their operations. Businesses must respond to local, national, and international regulatory structures established to protect environmental quality.

Although few question that regulations have helped to improve environmental quality, many argue that wiser, less commanding regulations would improve quality further at less cost. Agreements on hazardous waste tightly regulate the transport of these wastes across state and national boundaries, perhaps reducing opportunities for reuse and encouraging greater extraction of virgin stocks. Elements of the national regulatory apparatus for wastes, heavily control the storage and transport of wastes and dictate waste treatment methods that also serve to dissuade later efforts at materials recovery.

Regional strategies barrier

Often geographic regions may provide a sensible basis for implementing IE. Industries tend to form spatial clusters in specific geographic regions based on factors such as access to raw materials, convenient transportation, technical expertise, and markets. This is particularly true for "heavy" industries requiring large resource inputs and generating extensive waste quantities. Furthermore, the industries supporting large industrial complexes tend to be located within reasonable proximity to their principal customers. Due to the unique character of different regions this work could proceed in the form of case studies of regions containing a concentration of industries in a particular sector.

3.4 Eco-Industrial Parks

As noted in Chapter 2 from the definition of cleaner production, there are three stages during the life cycle of any item for consumption. Evans and Stevenson (2000) have illustrated these three stages as follows:

- Production processes: Conserving raw materials and energy, eliminating toxic raw materials, and reducing the quantity and toxicity of all emissions and wastes.
- Products: Reducing negative impacts along the life cycle of a product.
- Services: Incorporating environmental concerns into designing and delivering services.

Brewster (2001) has implied from the cleaner production definition that CP focuses only on individual activities or a single production process rather than focusing on the environmental impacts of the entire range of industrial activity. With the evolvement of CP, many decision makers, scientists, and engineers begin to break our dependence on single use of the finite natural sources that will lead to the ultimate depletion of these sources. As an alternative, biological ecosystems should be our model guidance to establish the industrial system with no "waste" but only residual materials that could be consumed by another process in the same industry or a different one. This preceding recognition is the main concept for the industrial ecology as explained before. Therefore industrial ecology seeks strategies to increase eco-efficiency and protect the environment by minimizing the environmental impacts to be within the allowable limits. In other words, industrial ecology seeks to move our industrial and economic systems toward a similar relationship with the Earth's natural systems or "artificial ecology". IE seeks to discover how industrial processes can become part of an essentially closed cycle of resource use and reuse in concert with the natural environmental systems in which we live. There are some similarities between IE and CP, but CP puts more emphasis on the sustainability of industrial practices over time and more frequently looks beyond individual firms and their existing processes, products, and services.

One of the most important goals of industrial ecology (Frosch, 1994) – making one industry's waste another's raw materials – can be accomplished in different ways. The most ideal way for IE is the eco-industrial park (EIP). These are industrial facilities clustered to minimize both energy and material wastes through the internal bartering and external sales of wastes. Robert Frosch – an executive at General Motors – put the question in 1989 "Why would not our industrial system behave like an ecosystem where the waste of a species may be the resource to another species?" (Wikipedia, 2006). One industrial park located in Kalundborg, Denmark, has established a prototype for efficient reuse of bulk materials and energy wastes among industrial facilities. The park houses a petroleum refinery, power plant, pharmaceutical plant, wallboard manufacturer, and fish farm that have established dedicated streams of processing wastes (including heat) between facilities in the park. The gypsum from neutralization ("scrubbing") of the sulfuric acid produced by a power plant is used by a wallboard manufacturer; spent fermentation mash from a biological plant is being used as a fertilizer, and so on. For more detail regarding Kalundborg, see case study on p. 100. The success of the EIP depends on the ability to innovate, access to talent, markets, and the ability to meet profit conditions or cost constraints and on achieving close cooperation between different companies and industrial facilities.

Nemerow (1995) defines EIP as "a selective collection of compatible industrial plants located together in one area (complex) to minimize both environmental impact and industrial production costs. These goals are accomplished by utilizing the waste materials of one plant as the raw materials for another with a minimum of transportation, storage and raw material preparation." There are a lot of definitions regarding EIP but all of them have taken into consideration the three main criteria for sustainable development namely, environmental, economic and social dimension and they emphasize the main role of eco-industrial parks as a tool for industrial ecology and for achieving the objectives of sustainable development.

From the above discussions, one can defend EIP as "a community of manufacturing and service businesses seeking conservation of natural and economic resources in order to reduce production cost and protect the environment as well as public and occupational health". The word community can be defined as a local community within the same facility or surrounding community within the industrial estate or nearby community or global community across a broader region. The global community is not yet realized because of distances. This could be done between two industrial estates where some wastes might match different industries in two different communities, especially if the industrial communities were not designed initially to act as an EIP.

EIP aims at achieving economic, environmental, social, and government benefits as follow:

- Economic: Reduce raw material and energy cost, waste management cost, treatment cost, and regulatory burden, and increase competitiveness in the world market as well as the image of the companies.
- Environmental: Reduce demand on finite resources and make natural resources renewable. Reduce waste and emissions to comply with environmental regulations. Make the environment and development sustainable.

- Social: Create new job opportunities through local utilization and management of natural resources. Develop business opportunities and increase cooperation and participation among different industries.
- Government: Reduce cost of environmental degradation, demand on natural resources, and demand on municipal infrastructure, and increase government tax revenue.

3.5 Recycling Economy/Circular Economy Initiatives

Developing an eco-industrial park is a complex process because it requires integration among information technologies, innovation, extended producer responsibility, design for environment, and decision making. Several models with slight differences are encountered in many countries. One of the models is the 1996 Act of the German Federal Government of the recycling economy (RE). After a short period of time, the Japanese Government established a program to achieve the RE concept by implementing a good product design and a comprehensive resources recovery. The circular economy (CE) initiative is then undertaken by the Chinese Government. The circular economy approach to resource-use efficiency integrates cleaner production and industrial ecology in a broader system encompassing industrial firms, networks or chains of firms, eco-industrial parks, and regional infrastructure to support resource optimization. Different initiatives were undertaken in other countries such as the USA and Canada.

Germany: The German Government passed a new Act in 1996 to move Germany toward a recycling economy using a closed loop economy law. This law traces the life cycle of production, consumption, and recovery or disposal in order to minimize the amount of waste generated in the manufacturing processes and encourage the product design that can be easily reused or recycled according to the principle of "extended producer and consumer responsibilities" discussed in Chapter 1 (Dietmar, 2003).

Upstream strategy for waste prevention and enhancing recyclables is the main key behind the German recycling economy legislation. Germany has started a number of EIPs and waste exchange projects to support implementation of the recycling economy law. The German institutes and consulting firms supporting implementation of the German legislation provided guidelines in the use of cleaner production, life cycle analysis and design for environment tools in the industrial sectors.

Japan: According to the Japan Environmental Agency (1998), Japan currently consumes 1,950 million ton/y of natural resources and imports 700 million ton/y from overseas. In the same time, a total of 450 million tons of waste (industrial and municipal) are generated per year. Over 60% of this waste is either incinerated or landfilled. Current estimates predict that remaining landfill capacity will be exhausted by 2007. As a result, Japan's government has created a comprehensive program for achieving a recycling economy through a series of laws such as the Basic Law for Promoting the Creation of a Recycling Oriented Society and the law for the Promotion of Effective Utilization of Recycled Resources. The foundation for the basic recycling law was a report by the Industrial Structure Council (July 1999). Other supporting legislations were issued to support the recycling law such as the Law Concerning Promotion of Separate Collection and Recycling of Containers and Packaging (2000), which accounts for more than 25% of general waste.

China: The Government of China introduced a circular economy as the outcome of over a decade's efforts to practice sustainable development by the international communities, and is the detailed approach towards sustainable development (Xiaofei, 2006). Sun Youhai, a law-drafting official with the Environment and Resources Committee of the Standing Committee of the National People's Congress, stated that "China has chosen a circular economy as the major means to combat environmental degradation and pursue sustainable development, but the development of the economy still lacks strong legal support and therefore it is necessary to draft such a law in time to help build a sound material-recycling society" (China Internet Information Center, 2005).

Since circular economy is a new concept in China, there are no clear mandates of each ministry on circular economy to date, and this problem may be solved by the end of 2006 after the issuance of the Guiding Principle of Promoting the Development of Circular Economy which the National Development and Reform Committee (NDRC) is now preparing for the State Council, yet Xiaofei presumes their mandates according to the existing ones (2006).

The circular economy (CE) initiative has been developed in China as a strategy for reducing the demand of its economy upon natural resources as well as the damage it causes to natural environments. The CE concept calls for very high eco-efficiency of using natural resources as a way of improving the quality of life within natural and economic constraints for sustainable development. China's planned economy is a top-down approach, which is similar to the economy model in most developing countries.

The circular economy model encompasses both the cleaner production techniques and industrial ecology strategy in a broad system that aims at industrial establishments, eco-industrial parks, and services in order to optimize the use of natural resources. The whole community has to play a role in achieving the CE concept, from the government to the firms all the way to the public. The fundamental actions that should be applied are stated below (Circular Economy, Xiaofei 2006).

- The first action within the individual firm, the managers must seek to apply the cleaner production hierarchy discussed in Chapter 2.

- The second action is to reuse and recycle resources within the established industrial parks. This will ensure that the resources will be circulated within the local industrial area.
- The third action is to integrate different production and consumption systems in such a way that resources circulate between industries and the community.

The circular economy initiative is meant to widen the opportunities and chances for local and foreign investments to take place. This is accomplished through the three actions stated above, and they all include development of resource recovery, cleaner production enterprises, and public facilities to achieve the concept of the circular economic industry. Another addition to the third regional level would be integrating management of material flow among suburban, urban, and rural areas (Circular Economy Xiaofei, 2006).

The circular economy concept joins the two basic approaches of cleaner production and industrial ecology with its application as eco-industrial development. The central strategies should revolve around establishing eco-industrial parks and networks within the borders of the country. On the other hand, eco-industrial parks confine themselves to the basic idea that one company utilizes the wastes of another, which goes against the understanding of how such established parks can meet the goal of a circular economy in a region. Countries must focus on meeting the demands of a circular economy and not just the establishment of eco-industrial parks by the actions stated earlier (Circular Economy Xiaofei, 2006).

Circular economy plans in some regions in China have been able to link eco-industrial development with cleaner production as the main strategies for applying a circular economy. This proves that eco-industrial parks form a link to the success of a circular economy; however, the parks should not be perceived as companies that use each other's wastes, but to look at the broader view.

United States: Reviewing the experience of the United States shows that a variety of stakeholders have important roles to play in promoting the development of eco-industrial parks and networks. Yet due to the stringent health, safety, and environmental regulations (Martin *et al.*, 1996), combinations of top-down and bottom-up approaches are required.

The US has been actively involved in the promotion of EIPs since 1994 when a major contract was granted to Research Triangle Institute in South Carolina and Indigo Development in California to assess the potential application of industrial ecology to economic development with the Environmental Protection Agency (EPA) being the primary federal agency interested in the concept at the time, which resulted in new White House initiatives to "reinvent regulation" and promote community economic development (Peck *et al.*, 1997). Ms Suzanne Giannini-Spohn, US-EPA, Office of Policy, Planning and

Evaluation, stated that the policies and programs that are needed to promote EIP development are those that promote resource efficiency, recycling, pollution prevention, and environmental management systems (Peck *et al.*, 1997). In 1995, EPA took up the challenge, looking at ways to improve the efficiency of its regulatory programs and reduce the burden on all concerned through cutting red tape, partnerships, flexibility, and facilitating compliance, which resulted in some projects and initiatives to promote the concept of EIP.

Also, research conducted by the US-EPA on EIPs has resulted in the identification of a number of regulatory strategies for encouraging the development of industrial ecosystems and the implementation of pollution prevention in the context of EIPs that could be summarized as follows (Peck *et al.*, 1997):

- Modifying existing regulations.
- Reforming existing permitting and reporting processes.
- Moving toward performance-based regulations.
- Promoting the use of facility-wide permitting.
- Promoting the use of multimedia permitting.
- Utilizing market-based approaches, such as emissions trading.
- Utilizing voluntary agreements such as covenants.
- Implementing manufacture extended waste liability regulations, which will impact on the design, production, use, reuse, and recycling of products.
- Promoting technology diffusion within and between industrial sectors.
- Providing opportunities for technology development and commercialization.
- Providing technology development grants specific to industrial ecology applications.

University and research institutes have also supported the development of EIP extensively in cooperation with local and federal governments. It is very important to develop a partnership among different stakeholders with universities and research institutes to develop innovative techniques and guidelines for industrial parks to follow. EIP concept, methodology, and strategies should also be included with the educational system within colleges and universities.

Canada: The strategic and policy framework exists at the federal and provincial levels to support activities that promote EIPs in Canada as discussed by Peck *et al.* (1997) and falls under the following five main guidelines:

- Securing Our Future Together Plan
- Business Plan and Sustainable Development Strategy
- Sustainable Development Strategy
- Renewed Canadian Environmental Protection Act and Toxics Management
- Pollution Prevention Federal Strategy

The outcome of main guidelines developed by Canada to move toward sustainable development was reflected in several programs and action plans with no programs specifically designed to support EIP development in Canada, yet programs that may be applicable to EIP have been identified by Peck *et al.* (1997) with their incentives, objectives, and outcome mechanisms.

Several federal strategies, programs, initiatives, and tools have been developed to help implement EIPs. Although these programs are not intended specifically for EIP, they are generally directed to material and energy saving, and innovation which fit the criteria of EIP. Legislation and incentive mechanisms to encourage EIP in Canada were developed such as:

- Overcome traditional fragmentation by collaboration among public agencies, design professions, project contractors, and companies.
- Developers may need to make a strong case for banks to finance a project with a longer payback period.
- Get contracts with major companies to locate in EIP; this will help prove the concept to financiers.
- Public development authorities may be better prepared to bear the possible increase in development costs than private developers. Or the public sector may fund some aspects of the development of an EIP with strong public benefits.
- Companies using each other's residual products as inputs face the risk of losing a critical supply or market if a plant closes down. To some extent, this can be managed as with any supplier or customer relationship (i.e. keeping alternatives in mind and writing contracts that ensure reliability of supply).
- Exchange of byproducts could lock in continued reliance on toxic materials. The cleaner production solutions of materials substitution or process redesign should take priority over trading toxics within an EIP site.

3.6 Eco-Industrial Park Case Studies

Based on previous discussion, a well-designed eco-industrial park (EIP) will close the material flows and energy cascading within an industrial community. In other words, eco-industrial parks can be considered an "Industrial community of manufacturing and service companies to enhance their eco-efficiency through improving their economic and environmental performance by collaboration among each other in the management of the natural resources." EIPs proved that it is the most valuable approach in the industrial zone from the economical, social, and environmental points of view. One famous model for eco-industrial parks is Kalundborg industrial estate located in Kalundborg town (harbor town) at the north west of Denmark, 75 km west of Copenhagen. It will be discussed in detail in the following case study to demonstrate the benefits of and approach to EIP. New EIPs have been designed and engineered by researchers, companies, and developers in different parts of the world such as the Netherlands, Austria, Canada, USA, Denmark, Spain, Costa Rica, Australia, Finland, etc. The world marches to adopting EIP concept, and thus more and more EIP is introduced all over the world every day such as:

- Crewe Business Park, UK
- Knowsley Park, UK
- Londonderry Eco-Industrial Park, UK
- Trafford Park, UK
- Emscher, Germany
- Value Park, Germany
- Environmental Park, Italy
- Hartberg Ecopark, Austria
- Styria, Austria
- Herning-Ikast Industrial Park, Denmark
- Vreten, Solna, Sweden
- Sphere Ecoindustrie D'alsace, France
- Parc Industriel Plaine de l'Ain (Pipa), France
- Burnside Industrial Park in Dartmouth, Canada
- New Eco-Industrial Park in Hinton, Alberta, Canada
- Brownsville, Texas, USA
- Chattanooga, Tennessee, USA
- Baltimore, Maryland, USA
- Cape Charles, Virginia, USA, etc

The concept of Eco-Industrial Parks (EIP) was first developed by Indigo Development (Indigo Development 2006). In the early 1990s, innovators at Dalhousie University (Nova Scotia, Canada) and Cornell University (Ithaca, New York, USA) conceived related frameworks for industrial park development (Côté and Cohen-Rosenthal, 1998). Indigo introduced this concept to staff at the US-EPA in 1993. The EPA then included an EIP project in an Environmental Technology Initiative and recommended that the President's Council on Sustainable Development adopts EIPs as demonstration projects in 1995. From 1994 to 1995 Indigo collaborated with Research Triangle Institute in a major US-EPA cooperative research grant focused on EIPs.

An eco-industrial park or estate is a community of manufacturing and service businesses located together on a common property. Member businesses seek enhanced environmental, economic, and social performance through collaboration in managing environmental and resource issues. By working together, the community of businesses seeks a collective benefit that is greater than the sum of individual benefits each company would realize by only optimizing its individual performance.

TABLE 3.1
EIP in USA and Developer Organization (Gibbs and Deutz, 2004)

EIP	Developer
Devens Planned Community, MA	Public Agency
Philips Eco Enterprise Center, MN	Community non-profit
Port of Cape Charles Sustainable Technology Park, VA	Public agency
Gulf-Coast By-product Synergy Project, Freeport, TX	Private companies
Londonderry Eco-Industrial Park, NH	Private sector
Redhills Ecoplex, MS	Public agency
Dallas Eco-Industrial Park, TX	Local authority
Ecolibrium, Computer and Electronic Disposition, Austin, TX	Public sector consortium
Front Royal Eco-Office Park, VA	Public agency
Basset Creek, MN	Consultants/Local Authority

The main goal of an EIP is to improve the economic performance of the participating companies while minimizing their environmental impacts and complying with environmental regulations. Components of this approach include environmentally friendly design of park infrastructure and plants (new or retrofitted), cleaner production, pollution prevention; energy efficiency; and intercompany partnering. An EIP also seeks benefits for neighboring communities to assure that the net impact of its development is positive.

There is no one single model or methodology to follow for developing EIP and accordingly there cannot be a single Act or strategy to be addressed. It seems that in Canada the development of EIP is still more of a top-bottom approach, unlike the US where many private companies are taking the lead as shown in Table 3.1. A combination of the Canadian system and American system might be the best: to develop strategy and awareness programs and seek benefits and compliance with environmental regulations according to a bottom-up system. The development of a circular economy still lacks strong legal support and therefore it is necessary to draft such a law in order to achieve the desired objectives.

These experiences showed some important support tools that have been developed in the process of EIP development. These tools include a mechanism to make the park work, financial tools, and model codes among the members of EIP. An information management system to facilitate the interconnectedness is identified as a major tool (Peck, 1998; Peck et al., 1998; El Haggar, 2005) with more research needed in this area.

The following case studies will illustrate some ideas and history behind each EIP to help readers or researchers to develop their own methodology to approach EIP in their country/industrial estate. The methodology might change from one industrial estate to another and from one country to another according to the awareness level, type of industry, and industrial estate and accordingly type of raw material/waste generated, culture, available information, etc. Some case studies will be handled in some detail and others will be handled very briefly for the sake of demonstrating different ideas, methodologies, concepts, criteria, etc.

Kalundborg Eco-Industrial Park, Denmark (Indigo Development 2006)

Kalundborg is a typical example to demonstrate the benefits and methodology of implementing industrial ecology within an existing industrial estate toward an eco-industrial park (EIP). The example of Kalundborg is often quoted because it is simple enough to demonstrate the idea of an industrial ecosystem. Moreover, it helps to visualize the benefits of industrial ecology and the main criteria for the implementation of such ecosystems as well as the way the methodology of industrial ecology was structured.

As discussed in this chapter, industrial ecology can be defined as "a manmade artificial ecosystem that follows the natural ecosystem". The main point behind this definition is to eliminate the words waste/pollution from manmade systems, and replace them with raw material or byproduct. If this is applied, then this definition will be more accurate, leading to an overall increase in efficiency – "eco-efficiency".

Kalundborg was not planned to be an eco-industrial network in the first place. Gradually, a network of industries has slowly developed over 30 years resulting in the formation of an industrial ecosystem or symbiosis. The case of Kalundborg started in 1961 with an idea to reduce the limited supplies of underground water usage and replace it with surface water from Lake Tisso. The first company to start this idea was an oil refinery called Statoil. The first partnership was between the city of Kalundborg which took the responsibility of constructing the pipelines and the oil refinery which financed the project. Building on this first project were other projects where the parties concerned started to realize the benefits from this partnership. Nowadays, Kalundborg is one of the largest complexes in which a mutual exchange of input and output is applied between different industrial organizations. The main partners in this industrial complex are the following (Saikkuu, 2006):

- Asnaes Power Station, Denmark's largest coal fired power plant, producing electricity with a capacity of 1,500 MW – it produces heat for the town of Kalundborg (4,500 households) and other industries in the estate.
- Statoil refinery, Denmark's largest oil refinery, with a capacity of 3.2 million tons/yr and expanded to 4.8 million tons/yr recently.
- Gyproc, a Swedish company producing 14 million square meters of gypsum wallboard (plasterboard) annually.

- Novo Nordisk, a multinational biotechnological company producing pharmaceuticals including 40% of the world's supply of insulin and industrial enzymes with annual sales over $2 billion.
- The Municipality of Kalundborg supplies district heating to the 20,000 residents, as well as water to the homes and industries.

Other partners joined the symbiosis based on the waste of the core partners as a byproduct. They ended up with the following added partners as will be explained later in detail:

- Bioteknisk Jordrens Soilrem – a soil remediation company that joined the symbiosis in 1998.
- Fish farm – consists of 57 fish ponds.
- Fertilizer farm.
- Cement company – Aalborg Portland and road paving.
- Kemira – a sulfuric acid producer.

The history of structuring Kalundborg

The history of structuring Kalundborg as an eco-industrial park was based mainly on individual and independent agreements according to the following time history and as shown in Figures 3.1, 3.2 and 3.3:

- In 1959 the Asnaes power station commissioned.
- In 1961 the Statoil refinery commissioned. The oil refinery started to utilize water from Lake Tisso through a pipeline that was financed by the refinery and built by Kalundborg city.
  - Saving the underground water resources.
  - *Driving force*: Enforcing community regulation/requirement.

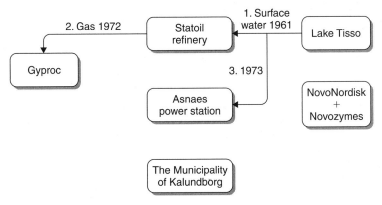

FIGURE 3.1 Kalundborg industrial symbiosis in 1975 (Saikkuu, 2006)

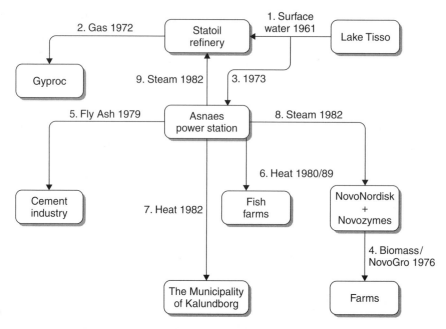

FIGURE 3.2 Kalundborg industrial symbiosis in 1985 (Saikkuu, 2006)

- In 1972, Gyproc was established. Gas for Gyproc plant was piped from the refinery after the refinery removed the sulfur from the gas.
  - Statoil gas is cheaper for Gyproc plant than using oil.
  - *Driving force*: Economic benefits.
- In 1973, the Asnaes power station expanded drawing water from Lake Tisso.
  - Saving the underground water resources.
  - *Driving force*: Enforcing community regulation/requirement.
- In 1976, regulation placed significant restriction on the discharge of organics into the sea. Since Novo Nordisk used to mix the industrial sludge with wastewater and discharge it to the sea, it found that the most cost effective way for sludge disposal/utilization was to give it for free to farmers as fertilizer. This was done by pipeline and trucks.
  - *Driving force*: Enforcing regulation as well as economic benefits.
- In 1979, the Asnaes power station started to sell fly ash to cement factories. Asnaes built an ash silo with an unloading facility to accomplish this duty.
  - Saving money by not landfilling fly ash.
  - Making money out of selling fly ash.
  - Less cost for cement producers.
  - *Driving force*: Economic benefits.

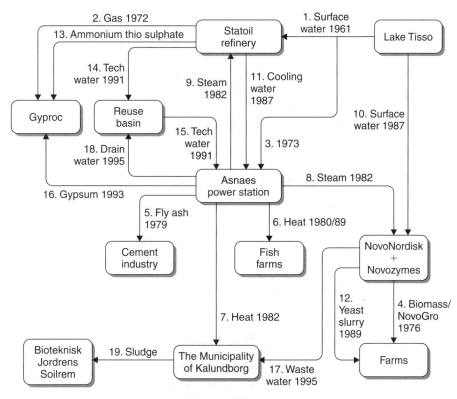

FIGURE 3.3 Kalundborg industrial symbiosis in 2000 (Saikkuu, 2006)

- In 1981, Asnaes started to supply heat to the Kalundborg community. There was an oil crisis in the late 1970s due to the 6th of October war in the Middle East. Oil prices increased and heating security was in question. The Kalundborg city was mostly heated by oil. There was a need to restrict the use of the oil heating system. Asnaes took care of this by supplying steam/hot water to the city.
  - *Driving force*: Security and economic reasons.
- In 1982, Asnaes delivered steam to Statoil and Novo Nordisk. Novo need to renovate and upgrade their boilers to get the required heat. The construction of a pipeline was more cost effective than an upgrade. For Statoil, it was much cheaper to get steam from Asnaes.
  - *Driving force*: Economic benefits.
- In 1987, Statoil piped cooling water to Asnaes power station. Considering the rare water resources there, and knowing that thermal pollution was being criticized at Asnaes, the most cost-effective solution was to utilize the waste cooling water from Statoil in Asnaes.
  - *Driving force*: Economic benefits.

- In 1989, Novo Nordisk switched from Lake Tisso to well water.
- In 1989 Asnaes power station started fish farming to solve the problem of thermal pollution. The sea water used to cool the condenser of the thermal power plant was utilized to develop an artificial fish farm that makes the fish grow faster in such a warm temperature.
  - *Driving force*: Economic benefits.
- In 1990, Statoil began selling molten sulfur to Kemira in Jutland. Excess gas from the operations at the Statoil refinery was treated to remove sulfur, which was sold as raw material for the manufacturing of sulfuric acid at Kemira, and the clean gas was then supplied to Asnaes power station and to Gyproc as an energy source.
  - *Driving force*: Economic benefits.
- In 1991, Statoil sent treated wastewater to Asnaes for utility use. Due to community pressure as well as related regulations, Statoil invested in a wastewater biological treatment plant to supply clean water to Asnaes.
  - *Driving force*: Enforcing regulation and economic benefits.
- In 1992, Statoil sent desulfurized waste gas to Asnaes.
- In 1993, Asnaes supplied gypsum to Gyproc. Asnaes power station installed a desulfurization unit to remove sulfur from its flue gases, which allowed it to produce calcium sulfate (gypsum). This is the main raw material in the manufacture of plasterboard at Gyproc. By purchasing synthetic "waste" gypsum from Asnaes power station, Gyproc were able to replace the natural gypsum imported from Spain.
  - *Driving force*: Enforcing regulation and economic benefits.

In summary, materials and energy are being exchanged in the city of Kalundborg among different companies and with the community in a closed loop such that the waste or byproduct of one company is taken as the raw material for another. It all started in 1961 with the need to use the surface water from Lake Tisso for a new oil refinery (Statoil) to save groundwater. Originally, there was no planning of the overall network; it just evolved as a collection of one-on-one deals between different industries that resulted in economic benefits for both partners in each deal. Figures 3.1 to 3.3 illustrate the network of companies in the symbiosis, showing the extent of the material and energy exchanges.

Material and energy flow analysis
Material and energy exchanges and savings started in Kalundborg in 1961 as the Statoil refinery began using water from Lake Tisso instead of groundwater, saving around 2 million cubic meters of water per year. Then Gyproc located its facility in Kalundborg to take advantage of the fuel gas available from Statoil. By the early 1970s, Statoil refinery agreed to provide its excess gas (byproduct) to Gyproc instead of burning it, which has been considered by Gyproc to be a source of low cost fuel. Later on as Statoil supplied both its

purified wastewater as well as its cooling water to the Asnaes power station, it thereby saved a total of 3 million cubic meters of water per year (instead of 2 million) as the same water was being "used twice". In 1976, the Novo Nordisk plant started materials flows by supplying sludge from its processes as well as from the fish farm's water treatment plant to be used as a fertilizer for a nearby farm. This sludge exchange totalled over 1 million tons per year. In addition, surplus yeast from the produced insulin was sent to farmers as animal food.

Enzyme production is based on fermentation of raw materials such as potato flour and cornstarch. The fermentation process generates about 150,000 cubic meters of solid biomass as well as 90,000 cubic meters of liquid biomass. Through proper repositioning of this waste, farmers have been using it as fertilizer, thus reducing the consumption of commercial fertilizers.

Another waste transformation is the yeast which is used in the production of insulin. Through the addition of sugar water and lactic acid it is converted into animal food. The insulin production builds on a fermentation process in which some of the main ingredients are sugar and salt, which are converted into insulin by adding yeast. After a heating process, the yeast, a residual product in this production, is converted into a much appreciated feed: yeast slurry. Sugar water and lactic acid bacteria are added to the yeast, making the product more attractive to animals (800,000 pigs).

The Asnaes power station is coal fired and operates at about 40% thermal efficiency producing huge amounts of energy. It uses salty seawater for its cooling needs saving the Lake Tisso water, and at the same time supplies the heated seawater to the 57 nearby fish ponds producing 200 tons of trout and salmon on a yearly basis.

In 1981, Asnaes began to supply the districts with steam for heating which replaced about 3,500 oil furnaces and significantly reduced air pollution. In addition, it provided steam to both Novo Nordisk and Statoil for their heating processes. After Statoil treated its excess gas by removing sulfur to comply with regulations on sulfur emission, it became possible to use the gas at the Asnaes power plant. Statoil's desulfurization plant reduces the sulfur content of the refinery gas whereby SO_2 emissions are reduced significantly. The byproduct is ammonium thiosulphate, which is used in the production of approximately 20,000 tons of liquid fertilizer roughly corresponding to the annual Danish consumption.

In 1992, the Asnaes power plant began using the treated gas from Statoil in place of coal. Statoil also supplies gas to Gyproc as its source of energy. In addition, the removed sulfur is sold as a raw material for the manufacture of sulfuric acid at Kamira. In 1993, the Asnaes power station added a desulfurization unit that removes sulfur from its gases and produces calcium sulfate as waste which is known as synthetic gypsum. The desulfurized fly ash is used by a cement company while gypsum is supplied to Gyproc as the main raw material for the manufacture of plasterboard instead of importing natural gypsum from Spain. In 1998, approximately 190,000 tons per year of synthetic gypsum were available from the power station.

Other types of wastes were also generated such as 13,000 tons of newspaper/cardboard which after a quality check are sold to cardboard and paper consuming industries in Denmark, Sweden, and Germany producing new paper, new cardboard, egg boxes and trays. Another 7,000 tons of rubble and concrete were used for different surfaces after crushing and sorting, and 15,000 tons of garden/park refuse were delivered as soil amelioration in the area as well as 4,000 tons of bio-waste from households and company canteens. The bio-waste is used in compost and biogas production. Four thousand tons of iron and metal were resold after cleaning for recycling as well as 1,800 tons of glass and bottles sold to producers of new glass.

In ecological terms, Kalundborg exhibits the characteristics of a simple food web: organisms consume each other's waste materials and energy, thereby becoming interdependent with each other. The exchange of reused and recycled materials and energy from the industries' byproducts resulted in large amounts of profit and cost savings. Through 1993, the $60 million investment in infrastructure (to transport energy and materials) has produced $120 million in revenues and cost savings. In 1998, the capital cost for this project was around $75 million. The savings were estimated to be $160 millions for a payback period less than 5 years. At the same time, tens of thousands of tons of water, fuel and other products are saved annually. The reductions in consumption of natural resources are as follows: 45,000 tons of oil/y, 15,000 tons of coal/y, and 600,000 m^3 of water/y. The amounts of reduced wastes and pollution are also significant: 175,000 tons of carbon dioxide/yr; 10,200 tons of sulfur dioxide/y; 4,500 tons of sulfur/y; 90,000 tons of calcium sulfate (gypsum)/y; and 130,000 tons of fly ash/y.

Although the industrial symbiosis of Kalundborg was developed due to business interactions between companies seeking to make economic use of their byproducts and waste material, it ended up with both economic as well as environmental benefits. Materials are being exchanged in a closed loop, companies are gaining profits, and the environment is protected by the reduced air, water and land pollution.

The Kalundborg case study proved that sustainability through industrial ecology can be profitable. The main obstacle to implementing industrial ecology methodology is the absence of an industrial ecology leadership. Industrial ecology leadership is a must to initiate and maintain the methodology. Industrial ecology leadership should be done on a voluntary basis. Socializing the top management of the industrial activities will enhance the communication skill and will develop mutual trust between partners and help people talk to each other and initiate eco-industrial systems.

Criteria for symbiosis

From the above analysis, one can develop the following criteria for successful implementing of an eco-industrial park:

- Database for wastes quantity and quality (analysis).
- Environmental and economical awareness.

- Possibility of mutual benefits between companies.
- Proximity between the locations of the companies forming the network.
- Communication between companies.
- Developing well-structured legislation and incentive mechanism to encourage such networking.
- Providing technical know-how through CP/IE experts.
- Providing facilities for partnership.
- Developing confidence between different establishments of the network.
- Developing trust between stakeholders including companies and government.

Administration

The Kalundborg center for industrial symbiosis was established and financed by symbiosis partners with the following objectives:

- Follow-up on the current Kalundborg symbiosis.
- Developing a database for Kalundborg symbiosis and other successful case studies.
- Developing an internal and external communication system to disseminate information and experience.
- Coordinating studies on the industrial symbiosis according to demand and need.
- Organizing of visits and study tours on the symbiosis.
- Consultation on new symbiosis projects.
- Contributing to forming new symbiosis projects.

Brownsville Eco-Industrial Park (Martin *et al.*, 1996)

This case study is a prototype developed by a group of researchers for the existing Brownsville eco-industrial park located in Texas, USA. The main reason for selecting this case study in particular is that the group of researchers who studied this EIP has developed five different scenarios based on the level of interaction and the extent to which wastes and resources are exchanged between members of the park. Not all of the EIP members participate in each scenario. In scenario 1, very few of the companies are working together. As we move from one scenario to the next, the level of cooperation and interactions among members of the park increases. In scenario 5, each EIP member interacts with the others in some way. This means that in one case study we will be able to trace the development of EIP and how this will be reflected on the economical and environmental benefits.

To develop the EIP prototype and its different scenarios, the researchers have collected data regarding the inputs and outputs of each company operating in the area of Brownsville, their willingness for using recycled material instead of virgin material that is currently used, and the potential for selling their byproducts instead of disposing of them as wastes. Finally, they summarized this data and prepared a flow diagram illustrating the inputs and outputs of each company. From this diagram, they identified several opportunities for symbiotic byproduct exchange.

EIP members

The prototype EIP contains 12 members; some of them are within the area of Brownsville, while others are located at remote sites. Therefore, this eco-industrial park is referred to in many literatures as a "virtual" eco-industrial park.

EIP port members

- Refinery: The refinery produces three products: naphtha, diesel, and residual oil. It expects to be producing approximately 8,300 barrels per day of each of these products. Its main input materials are light crude oil and energy.
- Stone company: The stone company brings limestone into the port and distributes it to companies in the area. At baseline, it sells stone to the asphalt company.
- Asphalt company: The asphalt company uses limestone from the stone company and residual oil from the refinery to produce asphalt for use on roads in the area.
- Tank farms: Clusters of tanks belonging to a variety of companies offload a variety of fluids brought into the port by ship and store them until they are delivered to their destinations by tanker trucks. The tanks sit in the port and frequently contain materials that must be kept warm to remain fluid. At baseline, they burn natural gas to generate the steam required to keep the materials warm.

Remote partners

- Discrete parts manufacturer: This company produces plastic and metal parts using screw machines, automated roll feed punch presses, and injection molding. At baseline, this company gives away used oil (about 100 gallons per month) to a recycler; it also landfills about 75% of its scrap plastics.
- Textile plant: This company assembles garments. It uses a small amount of solvents to wash parts. An outside party treats and disposes of compressor oil waste. A large quantity of high density polyethylene is landfilled.
- Auto parts manufacturer: This company uses plastic injection molding, metal stamping, and powdered metal forming to make small parts for assembly at a facility. A distant recycler buys the company's plastic scrap. The company also pays for disposal of several types of oil.

- Plastic recycler: This recycler accepts 12 types of plastic, grinds it, and sells the grind overseas. The company also manufactures plastic pellets from scrap.
- Seafood processor and cold storage warehouse: This company processes seafood and acts as a cold storage warehouse. It uses a great deal of water and electricity.
- Chemical plant: This plant manufactures anhydrous hydrogen fluoride. The major byproduct is $CaSO_4$ (gypsum). The company currently gives away gypsum to the Mexican Department of Transportation for use as road base. The gypsum is very pure and probably could be used in other applications (e.g. wallboard, concrete, tiles).
- Manufacturer of magnetic ballasts: This company produces electronic and magnetic ballasts. It currently landfills about 332 tons of waste asphalt per year.
- Gypsum wallboard company: This EIP member, located in Houston, is the only member not located in the Brownsville area. This wallboard producer relies exclusively on synthetic gypsum as an input to its wallboard production process.

EIP scenarios

In this section each of the five scenarios will be elaborated. The level of interactions between companies will increase gradually as we go from one scenario to the other until reaching a maximum level of interaction at scenario 5.

Scenario 1

At baseline, very few symbiotic relationships exist between these companies as shown in Figure 3.4 and as follows:

- The refinery sells its residual oil to the asphalt company.
- The stone company sells limestone to the asphalt company.

Scenario 2

In this scenario, the existing EIP members implemented pollution prevention activities independently from one another. This scenario is useful because it helps in revealing the benefits and limitations of individual pollution prevention efforts compared with the gains achievable by looking outside the plant boundaries for waste reduction opportunities.

This scenario describes some pollution prevention and recycling opportunities that can provide economic and environmental benefits to the companies acting independently. Experts in cleaner production techniques identified some of these opportunities during brief site visits to the companies.

- The discrete parts manufacturer introduces an aqueous cleaning system and an oil/water separation system.

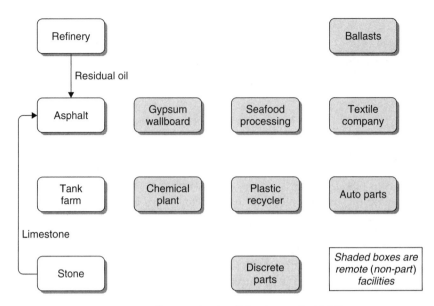

FIGURE 3.4 Scenario 1 – Baseline activities (Martin *et al.*, 1996)

- The textiles company recycles cutting room clippings.
- The automobile parts manufacturer purchases a ringer system for absorbent socks and rags.
- The seafood processor uses brown water for non-critical cleaning processes.

Scenario 3
This scenario represents the first development stage of the EIP as shown in Figure 3.5. This stage takes advantage of potential exchange opportunities that can take place with little or no additional investment.

- The discrete parts manufacturer sells scrap plastic, which is currently landfilled, to the recycler. He also purchases plastic pellets from the plastic recycler instead of from a remote source. The benefits arise from conducting both transactions with a local broker.
- The textile company sells plastic, which is currently landfilled, to the plastic recycler.
- The auto parts manufacturer begins selling scrap plastic to the local recycler, rather than the current recycler he uses in Chicago.
- The ballast manufacturer sells scrap asphalt to the asphalt company for mixing with its virgin materials.

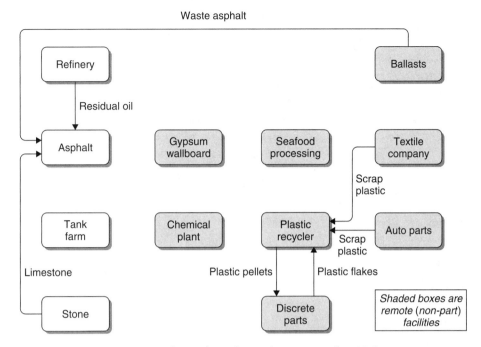

FIGURE 3.5 Scenario 3 – industrial symbiosis (Martin *et al.*, 1996)

Scenario 4
In this stage, the environmental and economic benefits of creating new businesses within the EIP will be demonstrated as shown in Figure 3.6.

- A power plant burning Orimulsion™, a heavy bitumen emulsified with water equipped with a steam pipeline to distribute process steam to other EIP members.
- A remotely located gypsum wallboard company.

These projects will require investment but will result in the following set of symbiotic relationships:

- The power plant delivers waste steam, through the pipeline, to the refinery and the tank farm. Once the energy in the steam is spent, the condensate is returned to the power plant and recycled to make more steam.
- The stone company delivers stone to the power plant for use in the scrubbers in the power plant's air pollution control system.
- The wallboard company receives waste gypsum from the power plant.

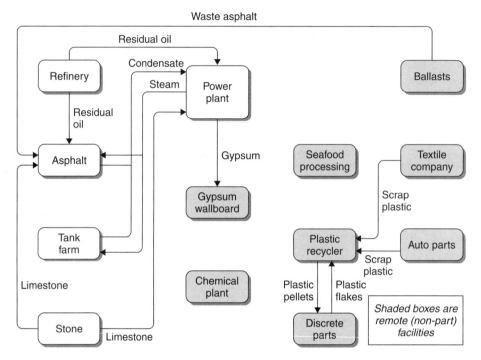

FIGURE 3.6 Scenario 4 – new EIP members (Martin *et al.*, 1996)

Scenario 5

In this stage, the researches assume that the remote partners are co-located with the remainder of the EIP members to study the additional benefits that could be derived from co-location. They also analyzed the provision of several joint services, which we assume the port can provide once the EIP has enough members to make these activities economically feasible. These joint services include a solvent recycler, an oil recycling operation, and a water pre-treatment plant. These changes produce the following opportunities:

- Each of the exchanges described in scenario 3 take place with lower transportation costs.
- The water pre-treatment plant provides clean water to the power plant.
- The solvent and waste oil recyclers are used by the discrete parts manufacturer, ballast manufacturer, auto parts manufacturer, and textiles company.

Comments and analysis

This study has demonstrated the development of different levels of interactions between Brownsville park's members starting from minimum interactions taking place at scenario 1 and increasing gradually until reaching

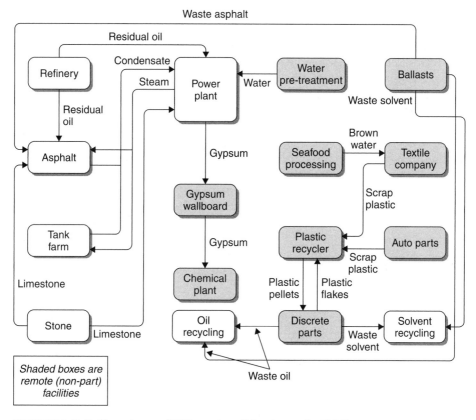

FIGURE 3.7 Collocation and EIP services (Martin *et al.*, 1996)

TABLE 3.2
Economic Indicator for Each EIP Scenario

Economic indicator	Scenario 3	Scenario 4	Scenario 5
Net annual economic benefit	$107,384	$4,658,786	$8,180,869
Return on investment	359%*	38%	59%
Payback period (year)	0.28	2.64	1.69

* This figure reflects only changes in net revenue from asphalt, since the plastics exchange required no initial investment.

maximum level of interaction and co-location at scenario 5. This breakdown in the level of interactions between the companies has allowed the researchers to trace the development and extent of the economic benefits resulting from each scenario alone. Thus, allowing them to know different circumstances at which industrial symbiosis might be more beneficial. Table 3.2 illustrates some economic indicators for each EIP scenario.

The researchers have concluded that the benefits of an EIP expand when companies are engaged in greater levels of cooperation as well as when they are located at a closer proximity to one another. Meanwhile, the opportunities to improve economic and environmental performance expand when an effective communication is established between members of the park so that companies are informed about how they might work together to improve the "industrial ecosystem" in their community.

Also, the researchers have noticed that in scenario 2, companies can achieve economic and environmental benefits through pollution prevention techniques that can be implemented with little or no investment. These opportunities require no cooperation or dependency between companies and they can be done at the unit level. In fact, this finding supports the vision that the industrial ecology approach is the net result of interaction among zero pollution, cleaner production, life cycle analysis, and eco-industrial parks (Peck, 1998; El Haggar, 2005).

From the previous analysis, some elements that play an important role in the success of any EIP project have been identified by the group of researchers and these elements are:

- The first and most essential input to the EIP is information about members' operations.
- The success of the EIP requires that members are open to depending on each other.
- To achieve the greatest economic benefits, the EIP will require substantial investment in infrastructure.
- The economic and environmental benefits to the EIP and the community are greater if the potential symbiosis opportunities are recognized during the planning stages of a park or plant.

Naroda Eco-Industrial Park, Ahmedabad, India

This case study is presented by UNEP. Naroda industrial estate is located in Ahmedabad in the northwest of India. Ahmedabad is the largest city in the state of Gujarat and has played an important industrial role in the estate because of its important textile industries. However, in the 1980s many of the textile industries in Ahmedabad closed. The city promoted other industries such as chemical, plastics, engineering, and pesticides industries. In the 1980s lots of textile chemical dyestuffs manufacturing was transferred from Europe and North America to India and other countries in Asia. Today almost 60% of dyestuffs exports from India are manufactured in Gujarat with approximately half coming from three industrial estates in Ahmedabad: Vatwa, Odhav, and Naroda.

Naroda industrial estate was established in 1964 by the Gujarat Industrial Development Corporation. Today there are approximately 900 industries located in Naroda industrial estate. They employ nearly 30,000 employees

TABLE 3.3
Industrial Sectors in Naroda Industrial Estate

Industrial sector	% of industries
Chemicals: dyestuffs and dye-intermediates	26
Engineering	24
Trading companies	9
Plastics	5
Textiles	5
Pharmaceuticals	3
Pesticides	1
Other industries	27

while 40,000 people depend indirectly on the industrial estate for their livelihood. Table 3.3 shows the significant industrial sectors in the Naroda industrial estate.

The estate provides services like water, power, and communications, in addition to infrastructure such as roads, schools, hospitals, post offices, banks, and a police station.

Administration/developer
EIP in Naroda was administered by Naroda Industries Association (NIA). NIA is an association made up of the owners of companies located within the estate. It achieved a number of projects in the areas of infrastructure, services, and environment. For example, NIA established Naroda Environmental Projects Ltd (NEPL) as a separate company to operate a landfill for the estate's hazardous wastes starting in 1997. NEPL also was responsible for the construction and operation of a common effluent treatment plant (CETP) which started in 1999. The CETP treats the wastewater for more than 200 companies. NIA also provides opportunities for its members to share information and learn about environmental approaches. The involvement of NIA in the estate activities is important for promoting environmental awareness for a large number of firms.

The eco-industrial network
The firms in the estate needed to enhance their environmental performance beyond mere compliance to regulations. They investigated more proactive approaches such as cleaner production to be promoted within the estate. The establishment of the CEPT led to a better understanding of the waste material flows within the estate thereby providing information on possible links between processes. Finally, the environmental and economic pressures had led firms in Naroda to improve their processes in order to improve their resource efficiency and their profitability. They started to achieve this improvement through applying cleaner production techniques. This helped to enhance

individual environmental performance. After that they wanted to enlarge the scope of their activities and to cooperate with different companies to look for resource recovery opportunities.

The eco-industrial networking project described in this study was carried out by NIA and the University of Kaiserslautern (Germany). The project started as a sponsored "workshop on industry and environment" held at the Indian Institute of Management in Ahmedabad in 1999. The first step was to understand the main types and quantities of wastes generated by the firms in the estate. A survey of 500 companies was carried out. As a result of the survey, the most important waste materials were:

- Chemical gypsum
- Biodegradable waste
- Mild steel scrap
- Waste acids in particular sulfuric and hydrochloric acids
- Chemical iron sludge

Chemical gypsum is generated by 19 chemical industries as a result of neutralization of their acidic wastewater with lime. This gypsum can be used by cement manufacturing companies provided that it meets certain specifications. An analysis for the process of recovering the gypsum as a raw material confirmed that it is economically viable.

Biodegradable waste is produced from a total amount of 10,000 kg of solid material and 90,000 liters of liquid wastes per year. This type of waste could be used to generate biogas as an energy resource for the industrial estate or for a housing development located nearby. An economic analysis showed that this energy recovery process is extremely favorable.

Sulfuric acid and mild steel scrap can be used as raw materials to make ferrous sulfate, a chemical used in primary wastewater treatment at the CETP.

Several of the partnerships described above are now being put in place with the support of NIA. Other possible partnerships that have been identified in the industrial estate are:

- Using sulfuric acid in the manufacture of phosphate for fertilizer.
- Using iron sludge to prepare synthetic red iron oxide.
- An alternative application for chemical gypsum in the production of plaster board.
- Reduction in raw material and energy use in the ceramic sector.

In conclusion, the use of a resource recovery project in the Naroda industrial estate has enhanced the interest of the firms in environmental management activities and has encouraged the industries in the estate to focus not only on their individual environmental aspects but also on the effect resulting from the large number of companies concentrated within the estate.

Analysis and comments

- Establishing an EIP takes time and happens in stages. Starting in 1964 Naroda began by increasing the awareness of business leaders by presenting them with past experiences of the industries that joined the industrial complex. Today, there are 900 plants in the Naroda industrial estate.
- The Naroda Industries Association (NIA) played a major role in establishing the EIP. It acted as a steering organization and managed the project. It keeps on looking for opportunities for partnerships, and creates opportunities for its members to exchange information and learn about various environmental approaches. The involvement of NIA is helping to sustain progress in the Naroda industrial estate.
- The firms in the estate were willing to establish the eco-industrial network as a result of environmental and economic pressures. They wanted to enlarge the scope of their activities beyond just complying with the regulations. They wanted to recover their resources to increase their profitability. This mature vision evolved from an awareness of the meaning of EIP.

Burnside Eco-Industrial Park, Canada

The Burnside park is located in Halifax Regional Municipality, Nova Scotia, Canada, in an area of 1,200 hectares of which three-quarters is currently developed, and it is operated by a municipality. The main aim is to protect water, air, and land because this was the demand of the Canadian Government. There are approximately 1,300 companies and 17,000 employees in those businesses. The park is one of the five largest in Canada (UNEP 2001). It has a solid waste management system which is considered the most highly sophisticated in Canada.

The success behind this eco-industrial park lies in the fact that municipalities in Canada such as Halifax Regional Municipality have "implemented a solid waste management system, which is widely viewed as the most sophisticated in Canada" (UNEP 2001). This system includes diversion of glass, some plastics, paper, and cardboard and aluminum, organic wastes, and construction and demolition debris from the landfill. The municipality has adopted a sewer use by-law which will limit discharges of certain materials into the sewers and a pesticide use by-law which will, within four years, ban the use of pesticides for aesthetic purposes within the city (UNEP 2001).

Burnside is designated primarily for light manufacturing, distribution, and commercial activities as shown in Table 3.4. One section of the park is designated as a business park and attracts computer, health, and technology companies. Although not specifically designated as such, another section has attracted many large trucking companies and their maintenance facilities. There is no "worker housing" at this time but there are two hotels in

TABLE 3.4
Sectors Represented in Burnside Industrial Park (UNEP 2001)

Accommodations	Distribution
Adhesives	Door and window manufacturing
Air conditioning	Electrical equipment
Automotive repair	Environmental services
Beverage products	Furniture manufacturing
Building materials	Food equipment
Business centers	Industrial equipment
Business firms	Steel fabrication
Carpeting and flooring	Machine shops
Chemicals processing	Medical equipment
Commercial cleaners	Paint recycling
Clothing manufacturers	Paper/Cardboard products
Communications equipment	Printing
Computer assembly and repair	Metal plating
Construction	Refrigeration
Containers and packaging	Transportation
Dairy products	Warehousing

the park and a third is being considered. Housing is an option that is under consideration for park expansion.

Businesses in the park must satisfy the federal, provincial, and municipal regulations that apply to them. These include requirements to prevent the discharge of pollutants that may be "hazardous to fish, restrictions of specified toxic chemicals such as PCBs and ozone depleting substances, and separation of solid wastes that can be composted or recycled for diversion from the landfill". These requirements are enforced through regulation or encouraged through fees such as tipping fees and sewer use fees. All of these requirements, whether regulation or economic instruments, along with the increasing environmental awareness of people in Canada, have created an atmosphere which is conducive to enhanced networking and the application of new environmental management strategies based in industrial ecology (UNEP).

All of the functions that were identified have been put into place. The Eco-Efficiency Centre is primarily an information clearinghouse and networking mechanism. According to UNEP:

The Centre conducts environmental reviews and encourages companies to join an Eco-Business program adopting an environmental code or policy, setting objectives and targets and, competing for reduction or conservation awards. Recently, the Centre has begun testing the Efficient Entrepreneur calendar and assistant developed jointly by the Wuppertal Institute and UNEP-DTIE to encourage companies to track their performance.

Moreover UNEP stated that

Both Dalhousie University and the Eco-Efficiency Centre act as educators in this endeavor. For the past seven years, we have written The Burnside Ecosystem column in the monthly park newspaper. The Centre prepares and publishes a series of fact sheets on various generic and sector specific topics which are distributed to businesses in the park. In addition, we have just concluded an agreement with the Metropolitan Chamber of Commerce which will result in the publication of 3 or 4 articles per year in their monthly magazine. Under the supervision of their professors, students undertake projects with companies in the park. For example, during the past five years, papers on the implications of environmental management systems with a gap analysis have been written and presented to 25 businesses in the park. The Centre encourages materials exchange and symbiotic relationships between companies and supports Clean Nova Scotia in the province-wide waste materials exchange. There is a possibility that the Centre will be taking this function over in the next few months. Finally, the project and the Centre encourage professors and students to collaborate with companies or sectors on applied research projects. One such study involves the integration of an environmental management system and the Natural Step into a furniture manufacturer owned by a multi-national corporation. Another project is a demonstration of an engineered wetland for the treatment of landfill leachate and runoff.

Analysis and comments
According to Mr Raymond Cote, the contact person in the park, what constitutes success in the park is as follows (Smart Growth 2000):

- Commitment from park owners;
- Flexibility in implementation of environmental regulations;
- Participation by capital owners;
- Appropriate economic instruments;
- Active information, education and interpretation; and
- A technical extension service.

With the following linkages:

- System Definition [Surveys and Database]; Industrial, social, brainpower inventories;
- Food Web [Materials Flow Database];
- Energy Conservation [Energy Audits];
- Resource Conservation [Waste Audits and Materials Exchange];
- Scavengers and Decomposers [Recyclers and Waste Managers];
- Information Exchange [Cleaner Production Centre Business Leaders Forum, Newspaper Column].

The park key features are:

- Six year multi-disciplinary, multi-institutional study of requirements;
- Cooperative partnership among academic, 3 levels of government, owners, developers and tenants;
- Phases/retrofitting and planning.

Burnside Eco-Industrial Park reflects an example of what an eco-industrial park should be like. The idea of having such parks requires a lot of study and analysis through university professors and students in cooperation with municipalities and owners.

The Bruce Energy Centre in Tiverton, Canada

Developing an eco-industrial park focused mainly around energy cascading is the objective of this case study (The Cardinal Group 2005). In this park, six companies are organized around Ontario Hydro's Bruce Nuclear Power Development (BNPD) to take advantage of its waste heat and steam generation capacity (steam serves as a potential source of heat energy for a broad range of industrial and agricultural processes such as dehydration, concentration, distillation, hydrolysis, and space heating). The main industries currently located in the park include:

- Bruce Tropical Produce Tomato: Which grows 2.3 million pounds of tomatoes each year in a hydroponic greenhouse; an amount equivalent to a 100 acre field. Steam from the BNPD is used to heat the greenhouse. The steam is transported in hot water coils and the condensate is then returned to the BNPD for reuse.
- Bruce Agra: Foods that process fruits and vegetables into concentrates, sauces, and purees. The food processing facilitators use steam energy from the BNPD to concentrate 84,000 gallons of raw products per day.
- Bruce Agra: Dehydrates locally grown crops to produce nutrient rich feeds for livestock and horses. The facility uses steam to run its dehydrators. This firm produces 90,000 tons of feed cubes per annum.
- Commercial Alcohols: The largest manufacturer and distributor of alcohol in Canada, and currently produces 23 million liters of industrial and fuel alcohol from 58,000 tons of locally grown corn. Steam from the BNPD is used in the distillation and ethanol processing of the alcohol.
- BI-AX International: A specialized company that manufactures a special polypropylene film for domestic and international markets. The polypropylene is heated in steam-driven ovens.
- St Lawrence Technologies: A research and development facility that specializes in finding ways to convert renewable resources to develop a wide range of products.

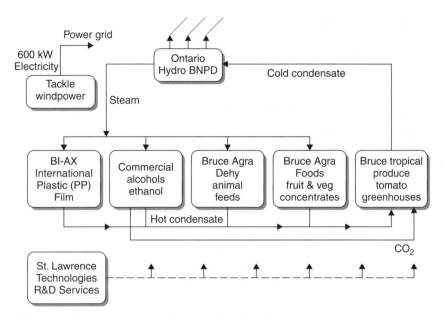

FIGURE 3.8 The Bruce energy linkages (Cardina Group)

The positioning of and interaction occurring between these industries enables the participants in the park to exchange recovered material so that byproducts of one firm serve as a raw material in another. Any residues generated at the Bruce Agra Foods facility, for example, are used either for animal feed or as an input for ethanol by Commercial Alcohols. Another established waste linkage in the park involves the use of carbon dioxide from the fermentation plant by Bruce Tropical Produce in their agricultural process. Figure 3.8 illustrates the types of linkages that exist between the firms in The Bruce Energy Centre. This project has been championed by Integrated Energy Development Corporation, a local industrial firm. It has resulted in substantial savings for the firms involved, an increase in local jobs, and environmental benefits.

Fairfield Ecological Industrial Park, USA

The Fairfield park is located to the Southeast of Baltimore, Maryland, USA. The park deals mainly with heavy Industries such as petroleum firms, chemical plants, trucking depots, asphalt manufacturing facilities, and others. During the last couple of years a lot of effort has been put in by the City of Baltimore, Baltimore Development Cooperation, and Cornwell University, for example, to come up with an outline for an efficient ecological industrial park (Smart Growth 2000).

The Fairfield mission statement is:

The Fairfield Ecological Industrial Park is an interdependent, partnering, environmentally conscious business/residential district. This district encourages recycling of products and services, and is committed to empowerment through improved employment, profits, education, health, and quality of life. This district cycles resources with minimum export of waste products and maximum export of value added products.

The Fairfield EIP has over 1,300 heavy industrial zoned areas and about 60 operating companies located in the South Baltimore Empowerment Zone. Transportation of raw materials and waste streams will be facilitated by port and rail. All available resources and other newly initiated ones will be used to achieve a closed loop for the production process (Smart Growth 2000).

The current economy of Fairfield is based on carbon-based industries. The major facilities in the area include:

- Oil company marketing sites.
- Asphalt manufacturing and distribution facilities.
- Chemical plants.
- Homologation companies which customize automobiles for export and import.

Other companies were developed to connect primary facilities to Fairfield. For transportation trucks, rail, and ports are used. Environmental companies control tank trucks, storage tanks, and cleanup operations. For manufacturing and distribution tire treading, box making, and materials handling machinery exist. Table 3.5 provides percentages for the current economies in the Fairfield EIP.

The Fairfield EIP is trying to achieve the following:

- Expand cleaner production programs.
- Integrate innovative environmental technologies.
- Expand its business networks.
- Implement an extensive master plan and fiscal impact analyses.

TABLE 3.5
Percentages for the Current Economies in the Fairfield EIP

Industrial classification	Percentage
Carbon	77%
Metal manufacturers	10%
Inorganic products	4%
Services	9%

Achievements

- Making plans and getting funds for demolition and site preparation.
- A state sponsored cleaner production monitoring program.
- Implementation of a land swap and ordinance process.
- Construction of new rail and road.
- Reuse of some unused sites by building new buildings and industrial projects.

Main goals

- Fairfield Housing Project: The goal of this project is to reduce the amount of waste that comes from the demolition process of the buildings and especially lead that has hazardous effects as a waste material.
- Hazardous waste: Recycling of waste that will help in using it as potential raw material for different industries.
- To avoid paperwork and duplicated reports.
- Voluntary inspection and maintenance: To ensure operation is at highest levels of efficiency.

Expected results

- Higher rates of solid wastes and hazardous wastes recycling.
- Eliminate illegal dumps.
- Decrease air emissions.
- Improve the quality of the air.
- Improve contaminant of non-point source run-off.

Analysis and comments

Of course there is still a great deal to be done for the completion of the closed loop of the Fairfield EIP but it is still considered as one of the more successful emerging examples. Still new ideas and creativity are being developed to improve the current situation. "By using the creativity of businesses, the desires of local residents, the experience of employees, the knowledge of educators, and the regulatory flexibility, the EIP will become a model of economic and environmental performance" (EPA).

In a study (Heeres *et al.*, 2004) of three designed EIPs from USA and three from the Netherlands, the Dutch parks (INES, RiVu, Moerjdik) were found more successful than the ones in the USA (Fairfield, Brownsville, Cape Charles). The participation in the Dutch companies was more active than the US companies. The presence of an anchor person or champion was found to be very important for the success of an EIP. The champions in Netherlands were local Dutch entrepreneurs. EIP's projects were mainly led by companies receiving local and regional financial and advisory support.

In the US cases, the most crucial point against success was a lack of companies' interest in the project. The US EIPs were initiated by local government and the companies were not interested in the project. In Fairfield and Brownsville, the majority of companies did not want to invest in the project because they saw it as financially risky and because they did not trust the local government. In addition, politicians in those areas saw the project merely as a job creation opportunity, not as economically or environmentally beneficial.

Questions

1. Make a comparison between natural ecosystem and artificial ecology – "industrial ecology".
2. What is the difference between industrial ecology and an eco-industrial park?
3. Make a comparison between industrial ecology and cleaner production.
4. Discuss the top-down approach versus bottom-up approach for administering EIP. Which approach do you prefer for your community? Why?
5. The recycling economy legislation approach will improve the utilization of natural resources. How can you structure such an initiative in your country? What are the benefits of a recycling economy?
6. Develop a strategic framework to initiate EIP in your community.
7. How does EIP relate to a cradle-to-cradle approach? What are the necessary tools to implement a cradle-to-cradle concept in the nearest industrial estate?

Chapter 4

Sustainable Development and Environmental Reform

4.1 Introduction

Ever since the Earth Summit of Rio de Janeiro in 1992, much has been written and continues to be written about "sustainability". Meanwhile, we have lost the ability to measure sustainability because we need to think first how we can develop sustainability and then develop indicator(s) to measure it (or a percentage of it). Our real environmental and economical problem in this century is that the development of science and technology has increased human capacity to extract resources from nature, and then process them and, use them, but it has not returned them to the environment for regeneration – the "cradle-to-grave" concept. A renewable natural resources concept is a must for sustainability, i.e. the word "disposal" should be removed from our daily dictionary. Unsustainable human activities are creating an open loop "cradle-to-grave" system that cannot continue and has one day to reach a conclusion. Closing the loop for renewable resources is the role of changing the "cradle-to-grave" concept to a "cradle-to-cradle" concept as discussed in Chapter 1.

The tool for sustainable human economic systems is the life cycle assessment according to the "cradle-to-cradle" concept and not the "cradle-to-grave" concept. A new hierarchy for waste management to approach the "cradle-to-cradle" concept was developed at The American University in Cairo in 2001 and upgraded in 2003, and is called the "7Rs rule or 7Rs cradle-to-cradle approach". The concept starts by developing Regulations to Reduction at the source, Reuse, Recycle, Recovery by sustainable treatment for possible material recovery (not waste-to-energy recovery). The last two Rs are Rethinking and Renovation where people should rethink about their waste (qualitatively and quantitatively) before taking action for treatment or disposal and develop a renovative/innovative technique to solve the waste problem. This approach is based on the concept of adapting the best practicable environmental option (not only from technical aspects but also from economical and social aspects) for individual waste streams and dealing with waste as a byproduct. This 7Rs cradle-to-cradle rule (Regulation, Reduce, Reuse, Recycle, Recovery, Rethinking and Renovation) is the basis for sustainability. This rule will ban the disposal and treatment facilities and develop renewable material resources.

Natural resources are becoming a very crucial issue for sustainable development because finding new sources of raw material is becoming costly and difficult. Concurrently, the cost of treatment and safe disposal of waste is exponentially increasing and locating waste disposal sites is becoming more and more difficult. Also, the impact of waste disposal on the environment is significant since it might contaminate air, soil, and water. In order to make waste management more sustainable, it should be moved from the traditional life cycle analysis following a cradle-to-grave system to a new system without disposal facilities following a cradle-to-cradle system.

4.2 Sustainable Development Proposed Framework

The sustainable development proposed framework is shown in Figure 4.1. The elements in the proposed framework are regulations, environmental management systems (EMS)/industrial audit (IA), cleaner production (CP), industrial ecology (IE), and compliance with regulations. The initial procedure in the proposed framework is to develop a set of environmental regulations. Strict enforcement of environmental regulations would force investors to carry an industrial audit (IA) or implement the environmental management system (EMS) (ISO14001: 2004) within the organization's policy and decision-making strategies to be able to identify the wastes and pollution in order to comply with environmental regulations. The information gathered from industrial auditing is used to perform a gap analysis to determine whether the goals have

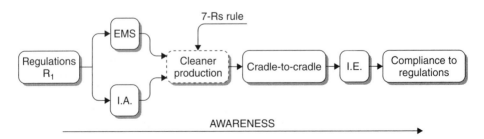

FIGURE 4.1 Sustained development proposed framework

been met. The main goal is compliance with regulations as well as conservation of natural resources, which will finally lead to sustainable development.

Regulations

Regulations are the basis for achieving sustainable development; these regulations are set by the government to provide organizations and projects with the policies they must abide by in order to reach sustainability. According to these regulations, it is necessary to develop a fine/incentive mechanism to encourage those who approach 100% sustainability by giving money/benefits and discourage those who ignore sustainability by taking money/fining according to a sustainability formula, which will be discussed later in this chapter.

Industrial audit/environmental management system

To attain sustainable development many elements have been outlined and researched. One of the elements that is widely discussed is integrating environmental management systems (EMS) or ISO 14001within industries. An EMS consists of a systematic process that allows an organization to "assess, manage, and reduce environmental hazards". Thus the continuous monitoring of environmental impacts concerning that organization is integrated into the actual management system guaranteeing its continuation as well as commitment to its success.

During the operation of the project an industrial audit (IA), sometimes called environmental audit (EA), must be implemented to evaluate its actual environmental performance. Industrial audit is considered a management tool that encompasses a set of environmental management techniques needed to ensure that the operation of the project complies with the environmental requirements and regulations, or the development of an Environmental Management System (EMS) within the organization will be implemented for continuously managing its environmental performance based on the ISO14001: 2004 standards. An EMS incorporates the environmental concerns into the organization's operation and enables it to reduce its environmental impacts as well as increase its operating efficiency.

An EMS is a part of the overall management system of an organization, which consists of organizational structure, planning, activities, responsibilities, practices, procedures, process, and resources for developing, implementing, achieving, reviewing, and maintaining the environmental policy.

In general an EMS should include the following elements, also shown in Figure 4.2:

- Management commitment and environmental policy.
- Planning for the environmental policy.
- Implementing the environmental planning.
- Evaluation and corrective/preventive actions.
- Management review.

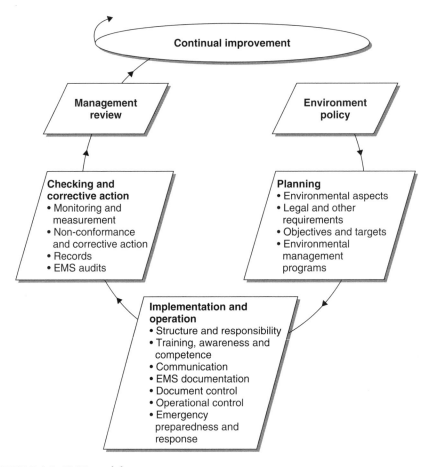

FIGURE 4.2 EMS model

The EMS provides several benefits through continual implementation and development which include:

- Financial benefits through cost savings as well as increasing local and international market competitiveness.
- Improving company's performance and image.
- Reducing business risks.
- Compliance with environmental regulations.

Periodic EMS audits must be carried out to check that the EMS is effectively implemented and maintained. At the same time, an EMS is considered a necessary tool for adopting the strategies of cleaner production (CP), which is the next step towards sustainable development. Cleaner production focuses on the prevention of waste generation at the source which is achieved by adopting the CP techniques discussed in Chapter 2 to enhance processes, products, or services that will lead to energy, raw material, and cost savings as well as protection of the environment and natural resources. A cleaner production audit is necessary to systematically identify and evaluate cleaner production opportunities.

Environmental management system/cleaner production

Environmental management systems (EMS) and cleaner production (CP) are located at the top of sustainable development tools. Huge efforts in spreading these concepts worldwide are dedicated especially to developing countries due to the immediate environmental and financial benefits they generate if properly applied as explained in Chapter 2.

By carrying EMS for existing activities, in addition to adopting the 7Rs rule, the organization will be in compliance with environmental regulations, which will facilitate the implementation of cleaner production techniques. Therefore, the 7Rs rule can now be called the hierarchy for cleaner production in order to approach cradle-to-cradle. However, the adoption of the 7Rs does not rely solely on investors; research institutes and universities should develop solutions to existing environmental problems and promote the concept of sustainable development. On the other hand, new investors should be encouraged to cooperate and establish a recycling unit to reuse/recycle the waste and produce raw materials/products that can be sold.

EMS can be implemented using cleaner production techniques or pollution control systems. The key difference between cleaner production and other methods like pollution control is the choice of timing, cost, and sustainability. Pollution control follows a "react and treat" approach, while cleaner production adopts a "prevent better than cure" approach as previously discussed. Cleaner production therefore focuses on before-the-event techniques that can be categorized as discussed in Chapter 2 and are as follows:

- Source reduction:
  - Good housekeeping
  - Process changes:
    - Better process control
    - Equipment modification
    - Technology change
    - Input material change.
- Recycling:
  - On-site recycling
  - Useful byproducts through off-site recycling.
- Product modification.

Cleaner production can reduce operating costs, improve profitability and worker safety, and reduce the environmental impact of the business. Companies are frequently surprised at the cost reductions achievable through the adoption of cleaner production techniques. Frequently, minimal or no capital expenditure is required to achieve worthwhile gains, with fast payback periods. Waste handling and charges, raw material usage, and insurance premiums can often be cut, along with potential risks. It is obvious that cleaner production techniques are good business for industry because they:

- Reduce waste disposal cost.
- Reduce raw material cost.
- Reduce health, safety and environment (HSE) damage cost.
- Improve public relations/image.
- Improve company's performance.
- Improve local and international market competitiveness.
- Help comply with environmental protection regulations.

On a broader scale, cleaner production can help alleviate the serious and increasing problems of air and water pollution, ozone depletion, global warming, landscape degradation, solid and liquid wastes, resource depletion, acidification of the natural and built environment, visual pollution, and reduced bio-diversity.

The EMS can provide a company with a decision-making structure and action plan to bring cleaner production into the company's strategy, management, and day-to-day operations as shown in Figure 4.3. As a result, EMS will provide a tool for cleaner production implementation and pave the road toward it. So, integrating cleaner production techniques with EMS as shown in Figure 4.3 will help the system to approach zero pollution and maximize the benefits where both CP benefits and EMS benefits will be integrated together.

Integrating CP strategies within the EMS (El-Haggar, 2003a; El-Haggar and Sakr, 2006) promotes their implementation and compliance with environmental regulations. The EMS provides a decision-making structure and action plan to incorporate cleaner production strategies into the company's management – plan strategy and day-to-day operations, therefore approaching minimum pollution levels and combining CP and EMS benefits. Cleaner production can be incorporated into the environmental policy of the organization as a commitment from the top management to encourage the organization to look after CP techniques everywhere as a solution to any environmental problem. During the planning phase of EMS, CP should be the main tool to achieve the objectives and targets.

Life cycle assessment/cradle-to-cradle

In the proposed framework structure for sustainable development shown in Figure 4.1, the current life cycle assessment (LCA) techniques will be modified to evaluate industrial activities adopting cradle-to-cradle concepts as

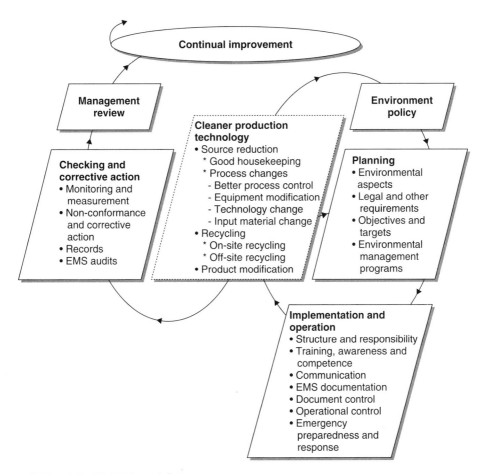

FIGURE 4.3 CP-EMS model

discussed in Chapter 1. Ideally, there should be no impact on the environment. Therefore, the lower the impacts the more efficient the cradle-to-cradle concept and the closer the organization is to resembling a natural ecosystem (industrial ecology), hence complying with regulations and realizing sustainable development.

Life cycle analysis (LCA) is a methodology for the environmental impacts associated with any product starting from the initial gathering of raw materials from the earth to the point at which all residuals are returned – to the earth a concept known as the "disposal" or cradle-to-grave approach. Unfortunately most manufacturing processes since the industrial revolution are based on a one-way flow according to a cradle-to-grave flow of materials. Based on industrial experience the cradle-to-grave flow of materials has proven to be inefficient due to depleting natural resources. Therefore, the cradle-to-cradle concept promotes sustainable development in a wider approach. It is a system of thinking based on the belief that human endeavors can emulate nature's elegant system of safe and regenerative productivity, by transforming industries to sustainable enterprises and eliminating the concept of waste.

Industrial ecology (IE)

Industrial ecology can be considered as the science of sustainability and promises much to improve the efficiency of the ecosystem as discussed above in Chapter 3. Technological improvements are not always better when considering sustainability without taking the environment into consideration, where zero pollution is a must for industrial ecology. Cooperation and community are also important parts of the ecological metaphor of sustainability.

The main tool for industrial ecology is cleaner production, which provides a framework for identifying the impacts of industries on the environment and for implementing strategies to reduce these impacts as it involves studying the interactions and relations between industrial systems and ecological systems. The ultimate goal of IE is to achieve sustainable development that will eventually lead to achieving compliance with the environmental regulations aimed at protecting the environment.

Industrial ecology aims at transforming industries to resemble natural ecosystems where any available source of material or energy is consumed by some organism. The managerial approach of IE involves analyzing the interaction between industry and the environment, through the use of tools such as life cycle analysis (LCA). The technical approach, on the other hand, involves implementing new process and product design techniques such as cleaner production and eco-industrial parks. The interaction of cleaner production, eco-industrial parks and life cycle analysis finally leads to industrial ecology. A set of design rules that promote industrial ecology has been established, the following rules were developed by Ehrenfeld (1997):

- Close material loops.
- Use energy in a thermodynamically efficient manner; employ energy cascades.
- Avoid upsetting the system's metabolism; eliminate materials or wastes that upset living or inanimate components of the system.
- Dematerialize; deliver the function with fewer materials.

Industrial ecology barriers

Even the industrial ecology concept has a lot of advantages from economical, environmental, and social points of view, but there are still some barriers to overcome.

The barriers to industrial ecology fall into six categories, namely technical, market and information, business and financial, regulatory, legal, and regional strategies as discussed in detail in Chapter 3.

Technical issues require a lot of innovation to convert waste into money or prevent it at the source. The markets for waste materials will ultimately rise or fall based on their economic vitality and can be enhanced through information technology tools. One option for waste markets are dedicated "waste exchanges" where brokers trade industrial wastes like other commodities. Developing business plans and providing financial support will help promote the concept of industrial ecology. The private firm is the basic economic unit and collectively constitutes the mechanism for producing inventions and innovations to practice. The government should regulate planning for industrial ecology and sustainable development. Industries tend to form spatial clusters in specific geographic regions based on factors such as access to raw materials, convenient transportation, technical expertise, and markets. This requires regional strategies provided by local governorates and federal government.

4.3 Sustainable Development Tools, Indicator, and Formula

Sustainable development and environmental protection cannot be achieved without establishing the concept of industrial ecology. The main tools necessary for establishing sustainable development and/or industrial ecology are cleaner production, 7Rs rule, environmental management system and life cycle assessment according to the cradle-to-cradle concept. The concept of industrial ecology will help the industrial system to be managed and operated more or less like a natural ecosystem according to the cradle-to-cradle concept, hence causing the least damage as possible to the surrounding environment. The 7Rs Golden Rule encompasses Regulation, Reducing, Reusing, Recycling, Recovering, Rethinking, and Renovation and is the basic tool for industrial ecology.

Industrial ecology is a generic concept that leads to a methodology which consists of more specialized tools that collectively transform an industry to resemble a natural ecosystem. In addition, industrial ecology requires an indicator to evaluate the performance of industries in implementing the IE tools. The cradle-to-cradle is the main indicator of industrial ecology. The degree to which cradle-to-cradle is achieved will give an indication of how close industry is emulating nature's ecosystems. Achieving industrial ecology will lead to sustainability.

Therefore, the main indicator for sustainability as a result of previous discussion is LCA based on a "cradle-to-cradle" concept rather than a "cradle-to-grave" concept. In other words, "cradle-to-cradle" is the indicator for sustainability. The degree to which cradle-to-cradle is achieved will give an indication of how close the industry or any manmade activity is to sustainability. This approach is similar to the "power factor" approach used in electricity, where power factors range from 0.6 to 1.0 worldwide. In Japan and some developed countries the power factor is 1.0 but in some other countries the power factor is less than 1.0. The deviation of power factor from 1.0 represents energy loss. Similarly, the difference between cradle-to-cradle and cradle-to-grave represents a loss in natural resources because of the "disposal" step. The sustainability formula might change from one country to another according to the level of environmental awareness and culture. The formula should have the same trend as the power factor formula used in electricity to calculate the penalty/bonus according to the standard power factor in the country and the energy cost.

4.4 Sustainable Development Facilitators

Environmental awareness and information technology can be considered as facilitators for sustainable development that will ignite the participatory process and assist people in spelling out their ideas, perceptions, attitudes, knowledge, etc. so as to attain self-development and compliance as well as conservation of natural resources.

Environmental awareness

As shown in Figure 4.1, the catalyst of the proposed sustainable development framework is awareness. Environmental awareness is an essential nutrient in each and every phase of the framework. Awareness will motivate individuals to carry out the various tasks and draw their attention to the benefits of preserving the environment; especially investors, and the damages that could otherwise take place in the long run. Eco-efficiency can lead to major economic gains and increased efficiency. The main problem is that in the short run activities that harm the environment are more attractive, either due to their low costs or ease of implementation. However, in the long run the cost of this damage can be significant and irreversible. Environmental awareness can be enhanced through information technology tools and principles.

Information technology (IT)

Since the 1980s, the computer-based information technologies have become the focus of the world. The environmental situation is continually changing as a result of human activities, so it is very important to obtain accurately and timely information on various environmental changes, as well as provide an excellent information source of global knowledge (UNEP, 1997).

One of the main tools for information technology is the Internet. The Internet is a tool that will improve interaction between people and institutions, give better access to government, and provide easy access to information.

IT is also faced with challenges, especially in the developing countries, such as:

- If new technology may not be available for all, in return it may create new divisions between the haves and have nots, i.e. developing countries get equal access to a new world of possibilities through technology.
- It can face barriers not directly related to the technology itself, such as:
  - Illiteracy in general
  - Computer illiteracy
  - Differing levels of technical and educational abilities of users.

Information technologies can help promote the concept of Environmental Awareness (EA) that might increase people's interest in the environment. People's response to improving their environment depends on the depth of their perception on environmental problems, and their willingness to act in favor of that. Participatory environmental awareness, such as bottom-up approaches, helps people to identify problems that concern them, understand how to solve these problems, and encourage them to be involved in the planning and implementation stages. To make participatory EA a success, an efficient information process is required as a tool. Information technology is the tool that matches the needs of modern times; however, it is faced with several challenges in developing countries.

Public environmental awareness develops gradually. The government cannot take the major role in this process, as it is controlled and directed mainly by the public. The required role of the government, NGOs, and educated individuals is to find efficient and creative means to reach the public through information technology.

4.5 Environmental Reform

Environmental degradation is the exhaustion of the world's natural resources: land, air, water, soil, etc. It occurs due to crimes committed by humans against nature. Individuals are disposing of wastes that pollute the environment at rates exceeding the wastes' rate of decomposition or dissipation and are overusing the renewable resources such as agricultural soils, forest trees, ocean fisheries, etc. at rates exceeding their natural abilities to renew themselves. Therefore, the environment's capacity to withstand the negative impacts due to human activities has diminished and environmental degradation has become a threatening issue.

To most investors overexploitation of natural resources is more profitable in the short run, due to cheap means of disposing of wastes, avoiding the costs of waste treatment and the excluding of social losses in cost calculations. However, in the long run natural resources will be depleted and the losses will be irreversible. Due to the severity of environmental degradation all over the world, the World Bank and other environmental institutions have conducted studies to present a cost assessment of environmental degradation (Sarraf *et al.*, 2004a, b).

Human resources are the most crucial element to reform. Developing these resources is the first step of reform. According to Abraham Maslow's theory of human needs, without satisfying the basic physiological needs one cannot expect the individual to be motivated further and take positive actions to society's improvement. We cannot discuss reform without providing healthy food for eating, clean air for breathing, clean water for drinking. This fulfillment is the first attachment and loyalty to one's country – creating patriots. Patriotism is the key issue to reform.

Maslow's hierarchy of needs reveals that individuals tend to fulfill certain needs before others. The most fundamental needs are the physiological needs: oxygen, food, water, etc. anything that they need to survive. In order of importance to individuals the other needs are safety and security, love and belonging, self-esteem, and finally at the peak of the hierarchy self-actualization (Sarma and van der Hoek, 2004).

Environmental degradation prevents individuals from attaining their two most basic needs: physiological safety and security. Due to depletion of natural resources such as air, soil, water, etc. people are less likely to have clean food, clean water, and/or good quality air. In addition, the wide spread of disease and disruption of natural ecosystems does not provide a safe environment for people to live due to high risk of disease breakout or natural disasters; hence safety and security needs are also unfulfilled. These two basic needs are deficiency needs; if a deficiency occurs in any of them individuals would directly try to eliminate it. Therefore, individuals will be reluctant to undergo any effort towards political, economical, social, or cultural reform unless their basic needs are fulfilled and sustained.

Environmental reform is a key issue behind development of human resources due to the need for natural resources as well as the severity of environmental degradation in most developing countries and some developed countries. Environmental reform is a constituent element in the main reform aspects such as political reform, economic reform, social and cultural reform. Thus this assumed reform plan cannot be complete without the inclusion of environmental issues because environmental degradation will obstruct the reform movement. Members of society cannot sustain endeavors of reform in the presence of a degrading environment. Environmental reform can be incorporated within each of the aspects of political, economical, social, or cultural reform.

Political reform

Regulations relating to preserving the environment and recycling economy should exist in any country seeking reform (political or environmental).

However, the reform plan should develop new and effective methodologies for enforcing these regulations. One of the major contributors to the current level of environmental degradation is lack of regulations or the poor enforcement of regulations.

Economic reform

Environmental and economic policies should be integrated and incorporated within the economic reform plan to achieve the environmental goals at the lowest cost and to determine the effects of other policy measures on the environment. Moreover, concepts from environmental economics should be adopted to conduct economic analyses of policies, analyze problems related to environment and natural resource management, perform cost/benefit analysis, set criteria for the evaluation of projects and public policies, understand economic structures of industries from the standpoints of efficiency, activity, etc.

Social and cultural reform

The fuel for environmental reform is awareness and interaction as discussed before. Members of society should interact and cooperate together to achieve environmental reform; similarly, developed and developing countries should cooperate with one another to eliminate environmental degradation in the world. Joint programs and research projects should be initiated to develop more environmentally friendly technologies and innovative techniques for resource and waste management. The exchange of knowledge and resources will strengthen the bond between different societies and promote social and cultural reform. Environmental reform is realized by promoting the concept of sustainable development as explained above.

Although the developing countries are progressing and there is rising economic growth, there is little attention given to sustainable development; hence environmental degradation is also rising. The recent preliminary study by the World Bank and METAP on cost assessment of environmental degradation in Middle East countries in 2004 revealed that pollution levels have not decreased since 1999; however, the final figures have not yet been released (Sarraf *et al*, 2004).

4.6 Environmental Reform Proposed Structure

A proposed structure that highlights the elements of environmental reform that should be adopted to be able to comply with environmental regulations is presented in Figure 4.4. The hierarchy of the elements in the proposed reform structure is regulations, environmental impact assessment (EIA), environmental management systems (EMS), cleaner production (CP), industrial ecology (IE), and compliance with regulations. Each element is succeeded with its own auditing procedure. The information gathered from auditing is used to perform a gap analysis to determine whether the goals have been met. The main goal is compliance with regulations, which will finally lead to sustainable development. Figure 4.4 illustrates the proposed environmental reform framework that integrates all the necessary elements to realize sustainable development and approach environmental compliance.

Regulations are the base for achieving environmental reform; these regulations are set by the government to provide organizations and projects with the policies they must abide by in order to protect the environment. According to these regulations, it is necessary to conduct an environmental impact assessment (EIA) at the early stage of planning a new project or modifying an existing one. An EIA is the primary process in the environmental reform structure that is necessary for identifying the environmental consequences of any proposed project and ensuring that the potential problems are addressed in a way that conforms to protecting the environment. During the operation of the project an environmental audit must be implemented to evaluate its actual environmental performance. Environmental audit is

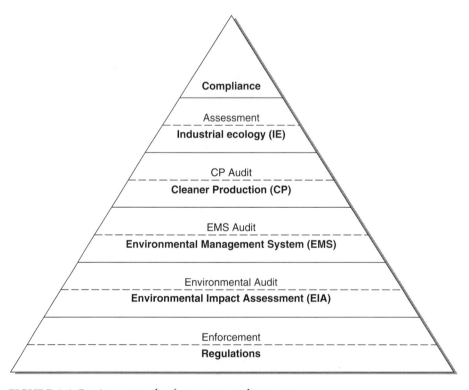

FIGURE 4.4 Environmental reform proposed structure

considered a management tool that encompasses a set of environmental management techniques needed to ensure that the operation of the project complies with the environmental requirements and regulations. The following process involves the development of an environmental management system (EMS) within the organization for continuously managing its environmental performance based on the ISO 14001 standards. An EMS incorporates the environmental concerns into the organization's operation and enables it to reduce its environmental impacts as well as increase its operating efficiency. Periodic audits must be carried out to check that the environmental management system is effectively implemented and maintained. At the same time, an EMS is considered a necessary tool for adopting the strategies of cleaner production (CP) which is the next step towards environmental reform. Cleaner production focuses on the prevention of waste generation at the source which is achieved by adopting CP techniques to enhance processes, products, or services that will lead to energy, raw material, and cost savings as well as protection of the environment. A CP audit is necessary to systematically identify and evaluate cleaner production opportunities. Cleaner production is one of the main tools of industrial ecology (IE) which provides a framework for identifying the impacts of industries on the environment and for implementing strategies to reduce these impacts as it involves studying the interactions and relations between industrial systems and ecological systems. The ultimate goal of IE is to achieve sustainable development that will eventually lead to achieving environmental reform and compliance with the environmental regulations aimed at protecting the environment and conserving the natural resources.

Regulations

Regulations are the base of the environmental reform structure since they provide the policies to all activities influencing the environment. Environmental regulation is a must to be able to perform environmental compliance. The law will necessitate the establishment of an environmental protection agency, whose responsibilities are to set regulations that investors have to follow, set the inspection procedures that ensure compliance, develop the standards that guarantee preserving the environment, prepare periodical reports on the state of the environment, etc.

Governments should adopt two methodologies or mechanisms for implementing the environmental regulations; providing financial incentives to encourage the community to engage in activities and projects that aim at protecting the environment and imposing fines and legal punishments in case of deviation from regulations.

Environmental Impact Assessment (EIA)

In 1969, the United States enacted the National Environmental Policy Act (NEPA). This Act requires that people think about environmental consequences before taking any action. Any developer who intends to establish a new project or extend/upgrade an existing one should perform an environmental impact assessment for his project to study the possible environmental consequences. If the consequences include negative effects then NEPA needs to implement mitigating measures or develop alternatives to the action that would produce similar end results but less damage to the environment (Bregman, 1999). This 1969 Act encourages the people to change their attitude from "build now and don't worry about its impacts" to "build now and worry about it later" to "plan for your impacts before build".

The main purpose of EIA is to ensure environmental protection and conservation of the natural resources including public and occupational health aspects against uncontrolled development. EIA is defined by UNEP as "a tool to identify the environmental and social aspects of a project prior to decision making. It aims to predict environmental impacts at an early stage in project planning and design. Find ways and means to reduce adverse impacts, shape projects to suit the local environment and present the predictions and options to decision makers." Another definition for environmental impact assessment is "the systematic examination of unintended consequences of a development project or program, with the view to reduce or mitigate negative impacts and maximize on positive ones" (El-Haggar and El-Azizy, 2003; EEAA, 1996). Through EIA the negative impacts of the project on the environment are studied and the alternative solutions for mitigating measures are presented with complete analysis from technical, economical, environmental, and social aspects to ensure sustainability of the project. Accordingly, proposed projects with allowable environmental impacts and proper mitigating measures are authorized for initiation from the Competent Administrative Authority (CAA). Proper enforcement of regulations necessitates conducting EIA studies to guarantee full control of environmental impacts for new projects or an existing project that needs expansion or renovation. Different mechanisms for implementing EIA for different countries according to their culture and environmental awareness level will be discussed in the next section.

Environmental management system (EMS)

EIA is a preliminary procedure for a project; it should be accompanied by an environmental management system to sustain the implementation of environmental policies. An EMS is a systematic process that allows an organization to assess, manage, and reduce the environmental hazards due to its operations. The monitoring of the organization's environmental impact is integrated into the actual management system to guarantee its continuation as well as commitment to success. The details of implementing EMS were discussed above.

Cleaner production (CP)

Adopting the cleaner production strategy is the key element for having acceptable results from EIA and EMS. Applying CP techniques to processes involves conservation of raw materials, and energy, elimination of toxic raw materials, and reduction of the quantity and toxicity of all emissions and wastes prior to their disposal. On the other hand, applying CP techniques to products involves reducing environmental impacts along the entire life cycle of the product starting from raw material extraction to final disposal. Adoption of CP techniques requires applying know-how, improving current applied technologies, and increasing people's awareness.

A major incentive for adopting a CP strategy is the reduction in production costs due to improved process efficiencies. On the other hand, minimal or no capital expenditure is required to achieve worthwhile gains with fast payback periods; in addition, waste handling and charges, raw material usage and insurance premiums can often be cut, along with potential risks.

General adoption of the cleaner production strategy can alleviate the serious problems of air and water pollution, ozone depletion, global warning, landscape degradation, solid and liquid waste accumulation, resource depletion, acidification of the natural and built environment, visual pollution, and reduced bio-diversity.

4.7 Mechanisms for Environmental Impact Assessment

EIA must be performed for new establishments or projects and for expansion or renovation of existing establishments. EIA studies the effect of the surrounding environment on the project as well as the effect of the project on the surrounding environment. It also looks at the different processes involved in product production, including inputs and outputs. EIA tries also to find ways of minimizing the environmental impacts of the project. This study if implemented properly will ensure sustainability for the project especially now that it has become necessary to provide this assessment before starting any project.

EIA was included within the environmental reform as a major component for proper planning. Proper planning will sustain the environmental reform. Because of the importance of EIA for sustainable development and environmental reform, this section will discuss the EIA processes with two different mechanisms applied worldwide. The first mechanism is the screening list approach which is suitable for developing countries, different organizations as well as World Bank financed projects. The second approach is the environmental impact assessment and environmental impact statement approach which has been implemented in the United States and other developed countries.

The EIA contents

The general description of the EIA contents can be demonstrated through the following seven steps:

- Project description: A brief description for the basic details of the project is required such as:
  - Process flow diagram(s) showing clearly all inputs (raw materials, water, and energy), output products, wastes, and emissions.
  - Construction phase, operation phase, maintenance activities, estimating staffing, plants facilities, and services.
- Surrounding environment: Description of the surrounding environment showing all details with the necessary maps to describe the neighbors and their impacts on the project and vice versa. Baseline environmental conditions are very important at this stage, the baseline data that will be collected such as water, soil, and air quality as well as weather conditions. The appropriate data of the surrounding environment for the EIA should be carefully reviewed. Project location with a general layout and the necessary maps of the surrounding environment with wind direction are very important for EIA.
- Screening: This is the step of deciding whether an EIA is required for a project. This might be left in a generic format as will be discussed later or will be left to the country to develop guidelines to identify the projects that must have an EIA or are exempted from EIA. Other countries might develop a screening list (A, B, or C) to classify the projects into three classes where each class reflects the severity of the potential environmental impact such as World Bank, Egypt, and some developing countries according to the following:
  - White list: Those likely to have minor environmental impact.
  - Gray list: Those that may result in significant environmental impact.
  - Black list: Those projects that require complete EIA due to their potential impact.
- Scoping: The stage of the identification of the issues that should be covered in the EIA and the geographic boundary. At that stage a site visit and consultation with relevant regulatory authorities must be done. Consultants must identify the characteristics of the development that might give rise to impacts. Also consultants should identify all regional and local issues that might be relevant to the EIA.
- Prediction of impacts: The impact prediction must encompass both construction and operation of the development. The impacts should be quantified, if possible, or described. The following impacts should be considered:
  - Magnitude and duration of impact.
  - Whether impacts are temporary or permanent.

- Direct and indirect effects, beneficial as well as adverse physical, sociocultural, biological aspects must be assessed.
- Evaluation of impacts: Legislative standards should be identified and followed to evaluate the impacts and their severity. Evaluation of the impacts should take into consideration the magnitude, duration of the impacts, and its indirect effects if any. Of course, all predictions and evaluations of predictions have an element of uncertainty associated with them. The consultant should try as best as he can to quantify the level of his uncertainty.
- Mitigating measures: These are the most important steps in EIA to mitigate the negative impacts. It is always recommended to use cleaner production techniques as an alternative to pollution control. Mitigation should always be incorporated in the early stages of the project during the planning and design phases by communication between the consultant and the developer. The mitigation strategy should set the environmental management principles that should be followed in the planning, design, construction, and operation phases of the development.

EIA screening list approach

The EIA evaluation process according to the screening list approach is shown in Figure 4.5. The Governmental Environmental Affairs Agency (GEAA) might accept, reject, or request a scoped EIA if needed with proper justifications. In the case of a scoped EIA, the developer is required to conduct a scoped EIA study for a certain impact or a process. Scoped EIA on a certain process means that the developer gathers more information on that process and explains it in greater depth so that the GEAA could see all its aspects to make sure that it does not represent severe negative environmental impact. In the case of rejection, the decision taken by the authority regarding the assessment or the measures to be implemented can be appealed to the Permanent Appeals Committee by a developer after receiving such a decision. From the list approach explained above, it is noticeable that it is a simple system to be used for the developing countries and organizations working with developing countries for the following reasons:

- The EIA process in the developing countries is considered to be a new concept. So most of the business leaders are not completely aware of it. The screening system makes it easy for them to locate their projects whether in A, B, or C lists with the help of screening list that covers approximately all the possible industries and activities and classifies them into white, gray or black lists. Meanwhile, A and B could have an EIA form as shown in Figure 4.5 covering the main aspects of EIA discussed before such as: project description during construction and operation, surrounding environment, impact analysis, and mitigating measures.

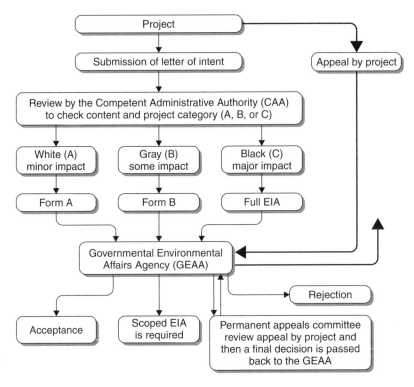

FIGURE 4.5 EIA process in Egypt

- The number of projects subjected to the EIA system is large and will form a heavy burden on the administrative authorities and the GEAA. The list approach will ease the management of those projects.
- The screening list approach uses limited economic and technical resources of the administrative authority, GEAA, and the developer in the best possible way

EIA–EIS process approach

According to the US National Environmental Policy Act (NEPA) whenever the US Federal Government takes a "major Federal action significantly affecting the quality of the human environment" it must first consider the environmental impact in a document called an environmental impact statement (EIS) as shown in Figure 4.6. An EIS typically should have the following sections:

- A statement regarding the purpose and need of the proposed action.
- A description of the environment impacts.
- Mitigating measures for the environmental impacts.
- A range of alternatives to the proposed actions including mitigating measures. Alternatives are considered the "heart" of the EIS.

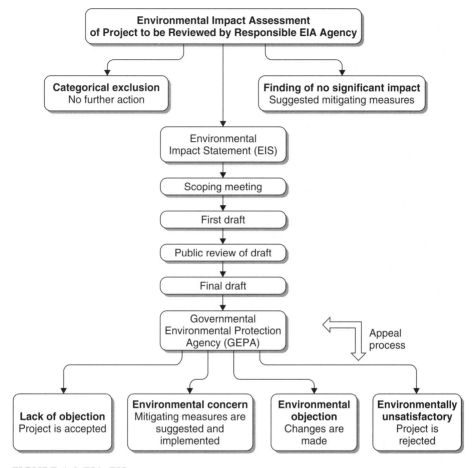

FIGURE 4.6 EIA–EIS process

- An analysis of the environmental impacts of each of the possible alternatives.
- Selecting the best alternative based on economical, technical, and social aspects.

Not all federal actions require a full EIS. If the action is not likely to cause a significant impact the agency may prepare a smaller, shorter document called an environmental assessment (EA). However, EAs are only appropriate if there will be "no significant impact". An EIA should be performed before establishing any development. It discusses the consequences of the project and the mitigation measures that can be built into the action that would produce similar end results but less damage to the environment. The EIA–EIS process as shown in Figure 4.6 will be discussed in the following section (Bregman, 1999). At the beginning of the process, the EIA is reviewed by the responsible agency that determines whether:

- No further action is required in case of categorical exclusion: Many federal actions are of a routine nature and generate no significant impacts on the environment. For example, routine procurement of goods and services. These actions have minimal or no cumulative effect on environmental quality. That is why the responsible agency decides that no further action is required for those types of activities.
- Finding no significant impact: This is for the activities that have minimal environmental impacts. Mitigating measures are suggested to lessen the negative environmental impacts. No further EIS is required for those activities.
- An environmental impact statement (EIS) is required: EIS is the document that examines the consequences of a project or an establishment. It is done for every major federal action that may have significant impact on the environment.

The EIS process

The EIS process, as described by Bregman (1999) in his book *Environmental Impact Statements*, is as follows:

Task 1 Scoping meeting: The purpose of the scoping meeting is to determine the scope of the draft EIS and to identify the major project issues to be addressed in it. Individuals and firms that may have an interest in the project impact are invited to participate in that meeting. Also the public, represented in groups or individuals, can participate.

Task 2 First draft: The product of the scoping meeting is an EIS first draft report. It describes the existing environmental conditions and evaluates the project alternatives. A brief discussion of the scoping process and the comment received from the public are also included in the report. The potential impact of each alternative is assessed as well as the "do nothing" alternative. The report includes the potential short-term impacts such as those associated with the construction phase like noise and dust. Long-term includes air, water pollution, wildlife displacement, and overloading of infrastructure. Mitigating measures are included with their costs and benefits quantified if possible.

Task 3 Public review of draft: Public participation in the EIA process is very important because it has a potential to lead to a better project. The public often have good suggestions for items to be incorporated or given more emphasis in the EIS report. The public are usually concerned for social and environmental aspects of the project and how it affects the natural resources, wildlife, and historical monuments if any. The contribution of the public in

reviewing the EIS report makes the public a partner in the project. They know the actual facts about the proposed project.

Task 4 Final EIS draft: After the responsible agency and the public have reviewed the EIS first draft and suggested changes, a final EIS draft is prepared. Administrative or policy questions are answered by the agency and given to the EIS preparer. He develops answers to the technical comments. The final draft is then submitted to the EPA for review and approval. Eventually, EPA takes one of the following decisions:

- Lack of objection: Project is accepted.
- Environmental concern: Mitigating measures are suggested and implemented.
- Environmental objection: Changes should be made.
- Environmentally unsatisfactory: Project is rejected.

4.8 Sustainable Development Road Map

By investigating the elements of the proposed environmental reform structure shown in Figure 4.4, one can decide whether or not the country/community follows a good reform and how far we are from the good environmental reform and what are the missing elements in the current practice. One of the main reasons is the absence of a comprehensive framework that provides a strategy towards sustainable development. Such a framework can be used by members of society to identify their possible roles. In addition, it can be used by the government to design an efficient and strict monitoring system to evaluate compliance to regulations. The reason for the absence of such a framework is the lack of awareness. Therefore, a framework has been proposed in Figure 4.4 that integrates all the essential elements of environmental reform.

The sustainable development road map can be developed by the environmental reform proposed structure covered in this chapter which reveals the different elements of environmental reform. The last two sections of this chapter can be considered as a conclusion of the first four chapters of this book. It is clear that the poor–moderate–good implementation of most elements is primarily due to lack of awareness. Therefore, a methodology for integrating the different elements of environmental reform was illustrated by the sustainable development proposed framework section. Using this framework the members of society can determine the different endeavors that need to be adopted to comply with regulations and finally realize sustainable development.

The catalyst of this framework is environmental awareness, which is the main barrier towards sustainable development. The benefits arising from adopting eco-efficiency are numerous and can lead to increased efficiency of operation and economic gains. However, most investors are reluctant to adopt eco-efficiency due to their lack of awareness of its benefits and the dangers of the sustained level of environmental degradation. Unfortunately, the benefits arising from environmentally polluting activities are more obvious since they take place in the short run but the damages can be irreversible. Without strong efforts towards increasing awareness, people's selfish behavior and short sightedness will continue to obscure their devastating activities.

Questions

1. Discuss the LCA based on cradle-to-grave versus cradle-to-cradle. Can a cradle-to-cradle concept be practically implemented.
2. Develop a formula for sustainability in your country.
3. What are the facilitators for sustainable development?
4. Develop a detailed road map for sustainable development in your community.
5. Compare between the proposed framework for sustainable development in this chapter and the sustainable development framework in your country.
6. How does environmental reform relate to political, economic, social, and cultural reforms?
7. How does environmental degradation relate to environmental reform? Compare between the proposed environmental reform discussed in this chapter and the environmental action plan in your country using a gap analysis approach.

Chapter 5

Sustainability of Municipal Solid Waste Management

5.1 Introduction

Municipal solid waste is considered one of the major global environmental problems, especially in developing countries. Solid waste in general can be classified into industrial or manufacturing, agricultural and forestry, municipal, and other types such as sludge, water canal floating weeds, construction and demolition waste, clinical waste, etc. as shown in Figure 5.1. Municipal solid waste (MSW) is considered the most important solid waste because of

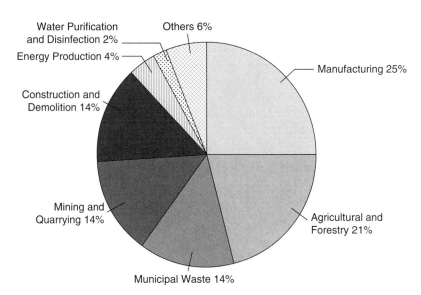

FIGURE 5.1 Composition of total waste generated in OECD region [OECD, 2001]

149

its nature and impact on our community and consists of hazardous and non-hazardous waste. It is a fact that solid waste composition differs from one community to another according to their culture and socioeconomic level. However, solving the problem of MSW in general is very challenging because of its heterogeneous nature. On the other hand, solving the problem in rural and developing countries is more challenging because of two factors: the low socioeconomic level of the majority of the population and their lack of awareness of the size of the problem as well as the lack of a suitable technology platform needed to face the problem.

These problems can be classified as social, economic, technical, and environmental. Social problems, include environmental awareness and visual pollution caused by the piles of garbage in the streets which in turn causes psychological problems proven to affect individual work efficiency. In addition to this, there are economic problems concerning the cost of MSW management including cost of collection, sorting, incineration, and landfill.

Another economic problem arises from leaving MSW without recycling, which is loss of resources and energy. Moving to the technical problem, the lack of suitable technologies to convert the heterogeneous MSW into products, even after sorting, represent a very challenging approach for sustainable development in order to save the natural resources for future generations. On the other hand, the unavailability of a technological platform in developing countries makes it a must to import the technology needed, which might be infeasible because of the cost and the need for adaptation of the foreign technologies to suit developing countries' environments. Concerning the environmental problem, bad odors of the waste attract all kinds of flies and mosquitoes which carry diseases to humans causing health problems in addition to pollution. Therefore, the main objective of this chapter is to develop a cradle-to-cradle approach for MSW in order to achieve a sustainable municipal solid waste management for conservation of natural resources.

The most general type of waste produced by humans on earth is MSW. In order to properly evaluate the rating of MSW within the whole waste production sectors, an overall view of the amounts of waste produced is needed. According to the Organization for Economic Cooperation and Development (OECD) Environmental Outlook regarding waste, the following represent the composition of the waste sectors in OECD countries: the major waste constituent results from manufacturing at 25% (industrial waste) followed by agricultural at 21%, and then MSW 14%, mining 14%, construction 14% and other minor sources (OECD, 2001) as shown in Figure 5.1.

However, the waste composition in developing countries will not be the same as developed countries; the typical waste composition in Egypt (El-Haggar, 2000b) shows that agricultural waste is the biggest constituent, while MSW represents 21.5% of the total waste produced. The composition of total waste generated in Egypt is presented in Table 5.1.

Global resource consumption is growing and accordingly the waste generated is also growing but at a much higher rate. The rate of waste generated

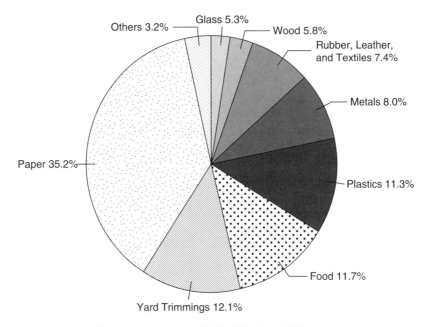

FIGURE 5.2 US MSW composition in 2003 (USEPA, 2005)

TABLE 5.1
Types and Amounts of Solid Waste in Egypt (El-Haggar, 2000)

Type	Average annual amount (ton)	Percentage (%)
1. Agricultural waste	23 million	35
2. Water canals and drain cleaning waste	20 million	30.3
3. Municipal waste	14–15 million	21.5
4. Industrial waste	4–5 million	6.25
5. Construction and demolition waste	3–4 million	4.5
6. Sludge	1.5–2 million	2.3
7. Hospital waste	100–120 thousand	0.15
TOTAL	65.6–69.1 million tons	100%

seems to be proportional to the social standards, i.e. the more advanced and wealthy societies (individuals) produce more waste. The amount of municipal waste estimated in 2004 according to the Global Waste Management Market Report is about 1.82 billion tons with a 7% increase compared to 2003. The

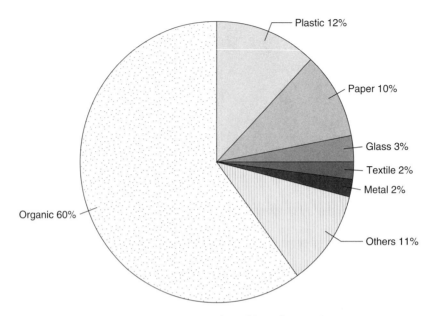

FIGURE 5.3   MSW compositions in Egypt (World Bank, 2005)

amount of waste is estimated to rise by 31% in 2008 to reach nearly 2.5 billion tons. This huge amount would cover the total continent of Australia at a thickness of 1 mm (Research and Markets, 2004).

As previously mentioned, consumption patterns differ within one community, and within one country even within seasons. Accordingly the waste produced would differ in composition. For example, MSW composition in the United States where organic materials are the largest components, is as follows: Paper and paperboard products account for 35% of the waste stream, with yard trimmings and food scraps together accounting for about 24%; plastics comprise 11%; metals make up 8%; and rubber, leather, and textiles account for about 7%. Wood follows at 6% and glass at 5%. Finally, other miscellaneous wastes make up approximately 3% of the MSW generated in 2003 in the US. A breakdown, by weight, of the MSW materials generated in 2003 is provided in (USEPA 2005).

MSW composition in Egypt – a developing country – compared to that of the United States – a developed country – would be very much different; the typical waste composition in a wealthy country was illustrated in the values presented for the US. Paper constitutes the majority of the waste where wrapping or packaging is used a lot. On the other hand, the MSW composition in Egypt would reveal high organic waste content constituting the greatest portion of MSW at 60%, followed by plastic at 12% and paper at 10%, glass, textiles, and metals at 3%, 2% and 2% respectively, and others at 11%; a breakdown by weight percentage is illustrated in World Bank (2005).

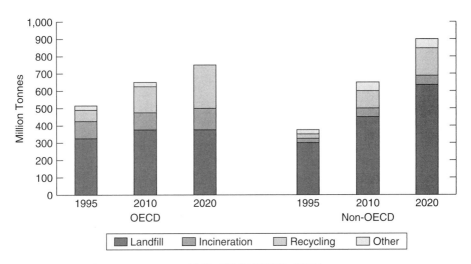

FIGURE 5.4 MSW management 1995–2020 (OECD, 2001)

The MSW composition in Egypt is typical of mid to low income countries, where consumption patterns are similar.

In the mid-1990s, over 50% of municipal waste incinerators in OECD regions were equipped with energy recovery systems. Standards for landfilling and incineration have been strengthened in a number of OECD countries. According to the OECD Environmental Outlook report, the MSW system is expected to change considerably in the future. The reference scenario projects a significant increase in the share of waste that is diverted to recycling and a decrease in that which is landfilled. In 2020, about 50% of municipal waste is likely to be landfilled, 33% recycled and 17% incinerated in OECD regions – compared to 64% landfilled, 18% recycled, and 18% incinerated in the mid-1990s (OECD, 2001). The report highlights the possibility that non-OECD regions are also projected to show significant changes in waste treatment methods, thus landfill is expected to decrease from about 80% in 1995 to about 70% in 2020, and recycling to increase from about 10% in 1995 to about 20% in 2020. Figure 5.4 illustrates the MSW in OECD and non-OECD countries.

Municipal solid waste is a concern for it means not only an increase in depletion of natural resources, but it also contributes to environmental pollution and health problems. MSW adversely impacts the environment as a result of mismanagement or behavior in the following manner (EPA, 2005).

- Airborne pollution through exhaust from open burning in dump sites or incineration.
- Health risk impact due to enhancing the growth of bacteria, flies, and rodents in dump sites.

TABLE 5.2
Generation and Recovery of Materials in MSW (EPA, 2005)

	Weight generated (million tons)	*Weight recovered (million tons)*	*Recovery as percent of generation (%)*
Paper and paperboard	83.1	40	48.1
Glass	12.5	2.35	18.8
Metals			
Steel	14	5.09	36.4
Aluminum	3.23	0.69	21.4
Other non-ferrous metals	1.59	1.06	66.7
Plastics	26.7	1.39	5.2
Rubber and leather	6.82	1.1	16.1
Textiles	10.6	1.52	14.3
Wood	13.6	1.28	9.4
Other materials	4.32	0.98	22.7
Other wastes			
Food, other	27.6	0.75	2.7
Yard trimmings	28.6	16.1	56.3
Miscellaneous inorganic wastes	3.62	—	less than 0.05
TOTAL MSW	236.28	72.31	30.6

- Release of greenhouse gases into the atmosphere through methane release or carbon dioxide from emissions.
- Groundwater contamination from uncontrolled leachate in dump sites.

In addition, landfilling is not the best utilization practice for the land use plans. Accordingly nations are attempting to enhance solid waste management in sustainable systems and increase recycling options, which is the main objective of this book.

According to US-EPA in 2005: "In 1999, recycling and composting prevented over 60 million tons of materials from ending up in landfills. Today, the U.S. recycles over 28% of all its wastes, a number which has doubled in the past fifteen years. 42% of all paper, 40% of all plastic bottles, 55% of all drink cans, 57% of all steel packaging and 52% of all major appliances are now recycled" (EPA, 2005). Table 5.2 shows the types of materials in MSW, their generations, recovery amounts, and percentages as per the EPA 2005 report.

As a result of previous discussion, the main objective of this chapter is to approach sustainability to MSW, as defined in Chapter 4, by applying the 7Rs Golden Rule and developing a sustainable municipal solid waste management system in order to approach 100% recycling of non-hazardous MSW

FIGURE 5.5 Conveyor belts

using the new hierarchy of "cradle-to-cradle" discussed in Chapter 1. This system will utilize the non-renewable resources, eliminate the problems caused by MSW, decrease the cost of MSW management, and approach the cradle-to-cradle concept without any use of incineration or landfill.

5.2   Transfer Stations

Transfer stations are used for very crowded communities, such as those with narrow streets, as well as remote communities, for example rural communities with low quantity of waste. The transfer station is always recommended to transport waste directly from the collection points to the recycling center, which is usually situated some distance away from the generation point. Transfer stations, which can be strategically located to accept waste from collection trucks, can represent a suitable and more economic solution (El-Haggar, 2004c).

The transfer station can be divided into several workstations. The first workstation would include a conveyor belt as shown in Figure 5.5 where trucks are allowed to enter leaving all the waste to be sorted manually on the conveyor. A second workstation would include a glass crusher as shown in Figure 5.6, where glass is crushed into small pieces – "cullets" – ready for recycling. The third workstation would include a hydraulic press as shown in Figure 5.7 to compact the sorted solid waste such as plastics, metals, paper, textiles, etc. The hydraulic press is used to decrease their volume for easy storage, handling, and transport to other companies for recycling.

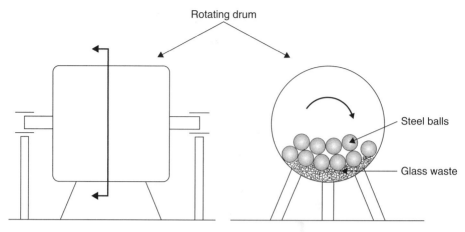

FIGURE 5.6 Glass crusher (Ball Mill)

FIGURE 5.7 Hydraulic press

An important consideration when designing the transfer station is to take into account the method of energy used and capital investment such as automatic sorting, manual sorting, magnetic separator, etc.

Recycling is the first process to be considered for solid waste management. The process of recycling is used to recover and reuse materials from spent products. That is why recycling is used to recover as much as possible

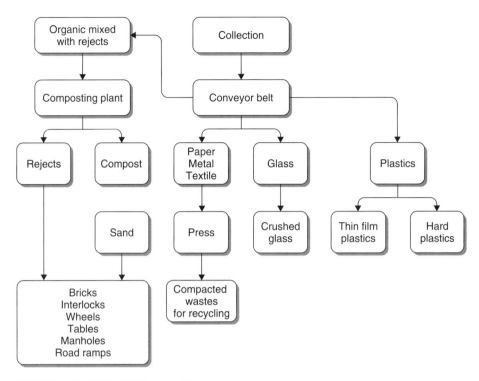

FIGURE 5.8 MSW 100% recycling system

from the waste and other treatment processes handle the remaining waste. Recycling will reduce waste and the limited resources will be conserved for future use since the scale of consumption of raw material is very large. Recyclable materials are plastics, papers, metals, bones, glass, and food wastes as well as unrecyclable wastes called "rejects". Figure 5.8 illustrates the system used to reach 100% recycling of non-hazardous MSW including recycling of unrecyclable wastes as will be explained in detail through this chapter.

Recycling (material recovery) can provide a cost-effective waste management approach otherwise it will be a burden in our society at large. This technique can help reduce costs for raw materials and waste disposal and provide income from a salable waste as well as protect the environment. The types of wastes that are separated and can be recycled easily with high benefits include paper and cardboard, aluminum cans and tin cans, plastics, textiles, bones, and glass. Organic waste or food waste recycling should be treated with special attention because it contains some rejects such as contaminated plastic bags and small pieces of glass, for example, as a result of unsorting of MSW at the source, which is the case in 60–70% of the world. There are many ways to recycle organic waste and convert it into soil conditioner (fertilizer) such as aerobic fermentation (composting), anaerobic fermentation (biogas), vermi composting, and co-composting processes. Composting is the most commonly used method to recycle organic wastes from technical, economical, and environmental points of view as will be explained later in this chapter.

5.3   Recycling of Waste Paper

The earth has limited natural resources, which must be conserved for future generations – "sustainable development". The scale of consumption of raw materials from forest and crop planting is very large in paper manufacturing. Hence one of the solutions to this problem is to recycle existing products' waste. Recycling may enable the recovery of primary material for reuse in the paper manufacturing process.

Paper is a "pliable material used for writing, packaging, and a variety of specialist purposes" (Biermann, 1993). Paper materials are classified as paper or paperboard. Paper includes newsprint, stationery, tissue, and many others while paperboard includes linerboard, corrugating media, milk cartons, etc. Kraft paperboards are any of the heavyweight papers generally above $134 \, g/m^2$, used in packaging. Bleached paperboards are made from bleached kraft pulp and are used in folded milk cartons, cups, and plates. Unbleached paperboards are used in linerboard and corrugating media for the production of corrugated boxes for example. Recycled fiber is being used in larger amounts in the production of unbleached paperboards (Biermann, 1993).

There are four main categories of paper: high-grade printing and writing paper, newsprint, corrugated/paperboard (including packaging), and tissue/towel products. The overall recycled content in each category varies, with tissue/towel containing the highest percentage of recycled material and printing and writing paper the lowest. So far, all pulp used for tissue/towel products as well as newspaper is imported in most developing countries. These products could be produced in developing countries very easily through simple recycling technologies with high return because they do not require high technologies or good quality (El-Haggar *et al.*, 2001; El-Haggar, 2001a).

Pulping processes

Paper consists of a web of pulp fibers normally made from wood or other lignocellulosic materials that is produced by physical or chemical means for separating the fibers to be used for paper production. In general, there are three processes for pulping, these are: mechanical, semi-chemical, and chemical pulping.

Mechanical pulping

Mechanical pulping is a process used for separating fibers without the addition of any chemicals. The process gives a pulp that is generally characterized by high yield that ranges from about 92 to 96%, and has high bulk, stiffness, opacity, and softness. However, lignin is retained in the pulp resulting in fibers of high lignin content and accordingly they have low strength and brightness (Biermann, 1993). This pulp is generally used for the making of newspapers, books, and magazines. The use of mechanical pulping is increasing worldwide due to the high yield of the process as well as the increasing competition for fiber resources.

Kraft (chemical) pulping

The word "kraft" means strong. Kraft pulping is one of a number of different chemical pulping methods; it uses sodium hydroxide and sodium sulfide at certain temperatures and pressures to dissolve the lignin of wood fibers and produce pulp. It is useful for any wood species and gives a high strength pulp characterized by long and strong fibers. This method is used to produce bags, wrappings, linerboards, and bleached pulps for white papers.

Paper making and recycling

Paper can be produced from wood pulp or from non-wood fibers. Most developing countries do not have enough wood for producing pulp to be processed for paper manufacturing and accordingly they use non-wood raw material from agricultural residues for pulping. Examples of non-wood raw materials are wheat straw, rice straw, and bagasse. Caustic pulping is used for non-wood raw materials in order to dissolve the non-cellulose organic fraction such as lignin, and leave behind the fibrous residual as pulp for paper production. The process steps for rice straw pulping start by shredding the rice straw material and cleaning it with water. The shredded fibers are then cooked in a rotatory digester for dissolving the lignin content while leaving the cellulose fibers. Vacuum filters are then used to separate the fibers from the cooking liquor. Fibers are then thickened by centrifuges and finally bleached by chlorination producing pulp ready to be used for paper manufacturing (Nour, 2002).

In Egypt, recycling paper is very important as it is estimated that Cairo produces 8,000 ton/day of municipal solid waste, 10% of which is paper and paperboard. The main objective of the recycling process is to reduce the amount of solid waste present as well as to recover some of the primary materials. Most of the papers that are used for newspapers or for educational text books are being imported to all developing countries including Egypt, which explains the marked increase in the price of press and publication in general. This crisis occurred during 1995 when the imported newspaper prices increased in Egypt and other developing countries by 40%. This will not only affect the press and publication costs but also the educational system where most of the government text books depend on this type of imported paper. This clearly shows the importance of implementing recycling in Egypt and other developing countries. It is also argued that by applying

TABLE 5.3
Worldwide Usage of Waste Paper (McKinney, 1991)

Year	Pulp and paper production (m ton)	Wastepaper consumption (m ton)	Apparent utilization rate (%)
1986	202	63	31
1990	237	85	36
1991	239	91	38
1992	246	96	39
2000	307	138	45

TABLE 5.4
World Recovered Paper Utilization (Kilby, 2001; CEPI, 1999)

Country	Recovery (000 ton)	Utilization rate (%)
EU	34,988	44
USA	32,943	38
Canada	4,810	26
Japan	16,378	55
Brazil	2,295	35
Mexico	3,395	93
Australia	1,463	58
Others	9,759	

recycling a huge reduction in both the volume of waste and the greenhouse effect would result, in addition to saving water and energy, which will help in having a better environment.

The process of paper making starts with fiber slurry preparation. First, fibers are mechanically and/or chemically pulped, and then a digester is used for cooking the cellulosic fibers. The resulting pulp is screened to remove undigested fibers, impurities, and any undesired particles. Pulp is then introduced through washers for separating pulping chemicals. Bleaching or treating the fibers with chemical agents is necessary to increase the brightness of the pulp. Pulp refining or beating is then carried out to increase the strength of the fiber to fiber bonds and accordingly increase the strength properties of the produced paper. The resulting pulp slurry is then applied to a fine screen for forming; water is allowed to drain by means of a force such as gravity or pressure difference developed by water column. Pressing after draining is then necessary for further dewatering by squeezing water from the sheet and finally drying the sheet by air or by passing it over a hot surface.

Two hundred and eighty years ago the paper industry introduced the concept of paper recycling because recycling is considered to be more cost effective than incineration or landfilling. This demonstrates the importance of paper recycling in the paper manufacturing process. Worldwide paper recycling has been improving since 1986 as shown in Tables 5.3 and 5.4.

According to the Leading Technical Association for the Worldwide Pulp, Paper and Converting Industry (TAPPI, 2005), the process of paper recycling involves repulping, screening, cleaning, deinking, refining, bleaching, and color stripping, and finally paper making as described below:

Repulping: A pulper, which is like a big container, is used for repulping paper. Paper, water, and chemicals are added and the pulper mechanically, chemically, and thermally breaks down the paper into fibers producing pulp slurry.

Screening: The pulp is forced through screens containing holes and slots of different shapes and sizes for removing small undesirable contaminants.

Cleaning: Pulp is cleaned and contaminants are further removed by spinning around the pulp in large cone-shaped cylinders.

Deinking: Pulp deinking is necessary for the removal of printing ink as well as other sticky materials like glue residue and adhesives. This process is divided into two stages: washing and flotation. In the washing process, small ink particles are rinsed from the pulp using water; larger ink particles and sticky materials are removed through the flotation process. During flotation deinking, pulp is fed into a flotation cell where air and chemicals called surfactants are injected into the pulp. The surfactants cause ink and sticky materials to loosen from the pulp and stick to the air bubbles. Air bubbles carrying the ink particles float to the top of the mixture creating foam. The foam is removed leaving the clean pulp behind.

Refining: Pulp is beaten or refined to separate any large bundles of fibers into individual ones. It also causes swelling of fibers and makes them more suitable for papermaking.

Color stripping and bleaching: If colored wastepaper is being recycled, then chemicals are used to remove the dyes, while if white recycled paper is required then the pulp is bleached with hydrogen peroxide, chlorine dioxide, or oxygen to make it whiter and brighter.

Papermaking: The recycled pulp fiber finally enters the paper machine for producing recycled paper sheets. Recycled fiber can be used alone, or blended with virgin fiber to give it extra strength or smoothness.

Recycled paper products

Due to technological developments, there is an increasing variety of new applications for recovered paper both within and outside the paper and board industry such as newsprint as well as printing and writing paper (Hyvärinen, 2001). Recycled paper and cardboards are widely used in Egypt in the manufacture of local craft and cardboard as well as board egg trays (El-Haggar, 2001a).

In 1994 the American University in Cairo (AUC) started a paper recycling subprogram within a program called "Industrial/Municipal Waste Management Program, IMWMP". A model of a paper recycling machine was

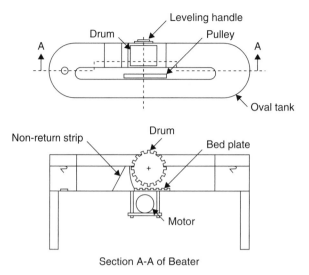

Section A-A of Beater

FIGURE 5.9 Schematic drawing of paper pulping (beater) machine

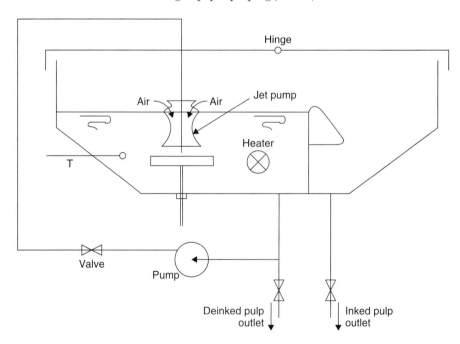

FIGURE 5.10 Schematic drawing of air injection flotation cell

designed and manufactured at AUC to test different factors affecting the recycling process and quality of produced paper as shown in Figure 5.9. A deinking system was also incorporated into the design to remove the ink mechanically from the recycled paper pulp as shown in Figure 5.10. Overall, the system proved to be highly effective in producing quality paper and the focus of the research has turned to optimize the paper recycling process. Different raw materials were tested to optimize the mixing ratio for better product quality.

5.4 Recycling of Plastic Waste

"Plastic recycling is still a relatively new and developing field of recycling. The post consumer items made from PET and HDPE resins have found reliable markets within the US and in ASIA" (Connecticut Metal Industries Inc., 2005). Applications for recycled plastics are growing every day. Plastics can be blended with virgin plastic to reduce cost without sacrificing properties. Recycled plastics are used to make polymeric timbers for use in picnic tables, fences, outdoor toys, etc., thus saving natural resources.

The problem of plastics wastes has increased tremendously since the use of plastics increased in most industrial, commercial, and residential applications. Households and industry produce huge amounts of plastic waste. Plastic waste causes severe environmental problems when incinerated or open burned on roadsides or in illegal dumpsites. Also plastic bags are a major source of littering residential areas, parks, and even protected areas.

About 50% of the total volume of plastic wastes consists of household plastic refuse, which are mainly in the form of packaging wastes. Once rejected, plastic packages get contaminated and when reusing them a more serious problem appears, which is the so-called commingled plastics, affecting in return the properties of the new recycled products (Wogrolly *et al.*, 1995).

The recycling of thermoplastics, or plastics, can be accomplished easily with high revenue. Each type of plastic must go through a different process before being recycled. Hundreds of different types of plastic exist, but 80–90% of the plastics used in consumer products are (1) PET (polyethylene terephthalate), (2) HDPE (high-density polyethylene), (3) V (vinyl), (4) LDPE (low-density polyethylene), (5) PP (polypropylene), (6) PS (polystyrene), and (7) PVC (poly-vinyl chloride). The most common items produced from post-consumer HDPE are milk and detergent bottles and motor oil containers. Soda, mineral water, and cooking oil bottles are made of PET.

Mechanical recycling involves cleaning, sorting, cutting, shredding, agglomeration, pelletizing, and finally reprocessing by injection molding, blowing or extrusion according to the required products. A simplified schematic diagram for the plastic recycling process is shown in Figure 5.11.

Recycled PET has many uses and there are well-established markets for this useful resin. By far the largest usage is in textiles. Carpet manufacturing companies can often use 100% recycled resin to manufacture polyester carpets in a variety of colors and textures. PET is also spun like cotton candy to make fiber filling for pillows, quilts, and jackets. PET can also be rolled into clear sheets or ribbon for audio cassettes. In addition a substantial quantity

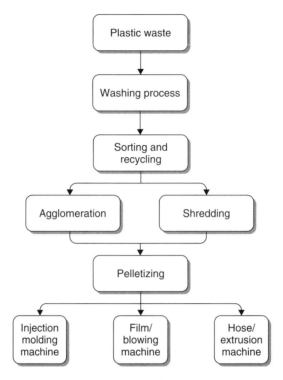

FIGURE 5.11 Plastic recycling process flow diagram

goes back into the bottle market. China is currently using it for the manu-facture of fiber optics (Connecticut Metal Industries, 2005).

Some common end uses for recycled HDPE are plastic pipes, lumber, flower pots, and trash cans, or it is reformed into non-food application bot-tles. Some end uses for recycled LDPE are plastic trash bags and grocery sacks, plastic tubing, agricultural film, and plastic lumber.

To recycle plastics there are mainly three stages needed. The initial stage is where the wastes are collected, sorted, separated, and cleaned; this stage is mainly labor intensive requiring little capital investment and relatively no technical skills. The second stage is where the collected wastes are being pre-pared for reprocessing: in this stage, the wastes are reduced in size by undergo-ing cutting, shredding, and agglomeration. The final stage is the reprocessing stage, where mixing, extrusion, blowing, injection, and product manufacturing takes place (Lardinos and van de Klundert, 1995) as shown in Figure 5.11.

Collection, sorting, separation, and cleaning stage

The first step in the initial stage is collection, which is usually carried out in developing countries by the informal sector and small individual enterprises and in developed countries by private companies. This stage is labor intensive and requires no technical experience. Collection is usually made direct to the final dumping site or via a transfer station, since these are the most practical methods in developing countries.

The next step in the initial stage is sorting of plastics, which depends on the manufacturing demands. It could be sorted based on color or type. Sorting is usually carried out by women and children in developing countries because of low wages and no technical skills are required. The working conditions at this stage are not hygienic because generally the sorting stage falls between collection and cleaning, which leaves the workers exposed to contaminated plastics dirt, and chemicals (Lardinos and van de Klundert, 1995).

In sorting there are basic guidelines that should be followed. Sorting separates plastic films from rigid plastics, each undergoing a further separation based on color and type. The waste plastics are first sorted and classified into bottles, transparent plastics, rigid plastics, and flexible plastics; within these a further separation is done based on type and color.

While sorting plastic we need to separate different kinds from each other because while reprocessing the plastic waste we need to know which type of plastics we are dealing with. This is because plastic types have different properties and reprocessing them together will produce a product with poor properties, unexpected durability, and poor appearance. Some plastics may even look similar and need testing to find out their type. To identify the plastic type a chemical test such as infrared analysis could be used. However, experience can help in this field but in case of doubt we have to revert to testing (Lardinos and van de Klundert, 1995).

Some basic tests include:

- To distinguish between thermoplastics and thermosets, press a piece of wire just below red heat into the material. If it penetrates, it is a thermoplastic, if not, it is a thermoset (Vogler, 1984).
- To distinguish between the types of plastics, scratch the surface with the fingernail to see the material's flexibility. However, this test needs an experienced person and is not always reliable since the material could have been exposed to many weathering conditions and could become brittle and rigid and cannot be scratched. A thin material may seem flexible and thick (Vogler, 1984).
- To separate plastics from non-plastics, or to separate between plastics types, we can use the flotation test. This test is useful to distinguish between PP and HDPE, and between HDPE and LDPE. The test basically consists of differentiating between the different densities of the two plastics by mixing them with water and alcohol in certain proportions. It was found that in a mixture with an exact density of 0.925, HDPE will sink and PP will float and in a mixture with a density of 0.93, HDPE will sink and LDPE will float. However, for overlapping

densities of plastics like PP and LDPE, the fingernail test and visual inspection can be more reliable (Vogler, 1984).

In order to distinguish between PS and PVC, another flotation test is carried out using pure water and salt. The existence of salt in water forces the PVC and dirt to sink and the PS to float. However, the amount of salts need not be measured; it is achieved from experience based on trials (Vogler, 1984).

- A burning test is used to differentiate between the different types of plastics based on the color and smell of the flame produced. The test is carried out as follows: cut a 5 cm length of a strip of plastic material, tapered to a point at one end and 1 cm wide at the other. Hold the sample over a stove and light the tapered end. However, the person carrying out this test should be careful not to be too close to the sample under test and not to inhale the smoke as it might contain hazardous substances (Vogler, 1984).

The techniques used in testing plastics differ between the developed and developing countries. In the developing countries, the technology and expertise are not available in the informal sector. Whereas in the industrialized countries mechanical separation techniques are used. Instrumental analytical methods are becoming more available like infrared spectroscopy and thermal analysis (Lardinos and van de Klundert, 1995).

Some of the techniques developed by industrialized countries concerning the separation of plastics include a method of separating the plastic packaging materials using velocity – this plays an important role when identifying about six packaging plastics. In another technique velocity plays a minor role and is used to distinguish between about 30 different plastics, mainly those made from engineering plastics (Wogrolly *et al.*, 1995). Other methods include separating the plastics based on density, surface structure, ferromagnetism, conductivity, color, etc. Since the purity of the finished product depends on the sorting accuracy, we need to make a distinction between sorting due to density, flotation, electrostatic sorting, thermal separation, and sorting of plastics by the hand picking method discussed earlier (Wogrolly *et al.*, 1995).

Other separation methods include the bottle sorting mechanisms, which can be either manually as described earlier, semi-automatic or automatic (Wogrolly *et al.*, 1995). However, the problem of separation sometimes becomes more difficult when mixed plastics are involved. Therefore, the Society of Plastics Industry in the United States has developed a coding system shown in Figure 5.12 using symbols and numbers for the types of plastic used. This system has also been introduced in Europe and most countries worldwide. This coding system requires updating since a number of plastic types are included under the number "7".

The next step in the initial stage is the cleaning, which usually consists of washing and drying the plastics. It is important to wash the plastic before shredding it as this improves its quality. Usually washing takes place after cutting (cutting refers to splitting the plastics into two parts to ease the washing process) and sometimes even after shredding to obtain better results. Foreign materials like paper and covers are removed before cleaning. Washing can be carried out either manually or mechanically. Manually, the plastics are placed in a drum cut in half and the water is stirred with a paddle. The water is heated if the waste is greasy; soap and caustic soda are added to help remove the more difficult grease (Lardinos and van de Klundert, 1995).

As with washing, the plastics can be dried either manually or mechanically. Manually, the plastics are left to dry in the sun. This method works best for plastic films, which are hung on lines to dry. Shredded plastics are centrifuged and left to further dry in the sun. If left without being centrifuged, the plastics will need a longer time to dry, will require more stirring and a space of around 15–20 m^2 for 300 kg of shredded plastics (Lardinos and van de Klundert, 1995).

In developing countries, plastics are cut in two halves and washed in a hot water basin using burners. For 1 ton of plastics, 25 kg of caustic soda should be added to 2 m^3 of warm water. The plastics are then rinsed in cold water in two different basins to remove the caustic soda; they are then centrifuged and left to dry for 2–4 hours in the sun.

The second stage – plastics are prepared for further reprocessing

This stage involves size reduction in which waste plastics undergo cutting, shredding, agglomeration, and pelletizing. This process increases the cost of plastics waste since it eases their use in the manufacturing process and decreases transportation cost (Lardinos and van de Klundert, 1995). The smaller the size of the shredded plastics, the more regular their shape (as in the case of pelletizing) leading to wider market demand and higher price.

The first material transformation step is cutting the plastics into smaller pieces, as is the case with large bottles, cans, and buckets. These waste plastics are usually cut by a circular saw or with a bandsaw (Lardinos and van de Klundert, 1995). In most developing countries, the sorted and washed plastics are cut into two pieces by large scissors fixed on a wooden base. It is estimated that an average of three workers can cut up to 1 ton of sorted plastics per day.

Waste plastics cut in two or more pieces are fed into a shredder for further cutting. The plastic is cleaned before shredding and sometimes gets further cleaning after shredding according to manufacturer demand. The shredder machine used for thin film plastics is different from that used for rigid plastics. For rigid plastics a horizontal cutting machine is used where the blades are rotating on a horizontal axis and the shredded plastics pass through a grid into a collecting tray (Lardinos and van de Klundert, 1995).

The sizes of shredded plastics are from 5 to 10 mm. The motor that drives the shredder is approximately 30 hp and the shredded plastics are collected in bags to be further reprocessed or sold. Figure 5.13 shows a horizontal axis shredder used for rigid plastics waste. The end products of shredding are irregularly shaped pieces of plastic depending on the required final product and the type of industry that will use them. The shredded plastics could undergo further washing to ensure cleanliness, especially as the shredded pieces are more easily cleaned when they are in small sizes than in large ones (Lardinos and van de Klundert, 1995).

FIGURE 5.12 Plastic coding system

FIGURE 5.13 Horizontal axis shredder

In order to avoid feeding plastic films, bags, and sheets directly into a shredder, it is recommended to use an agglomerator that cuts, preheats, and dries the plastics into granules. The agglomerator will increase the material's density and quality which will ensure a continuous flow in the extruder and hence better efficiency. It is better to clean the plastics before agglomeration since foreign substances will be processed together with the plastics. In the process of agglomeration, heat is added indirectly through friction between plastic film and the rotating blade located at the bottom of the agglomerator. It is therefore important to rapidly cool the plastic film to obtain the crumb shape desired. This is achieved through adding a small cup of water. An example of an agglomerator is given in Figure 5.14.

In the second stage, pelletizing is considered to be the final process as shown in Figure 5.15. It is the process in which the shredded or agglomerated plastics are uniformly sized to produce a better quality product and increase the efficiency of the product manufacturing process, due to the higher bulk density of the pellets compared to the shredded and agglomerated ones. To reach the final product of plastics pellets, the shredded rigid plastics or the agglomerated films pass through extrusion and pelletizing processes. In the extrusion phase, the plastics undergo mixing, homogenization, compression, plasticization and melt filtration. Coloring pigments, for example, can be added to the process (Lardinos and van de Klundert, 1995).

FIGURE 5.14 Agglomerator machine

FIGURE 5.15 Pelletizing machine

The material is introduced in the hopper to the rotating horizontal screw in which there are heating elements that plasticize the waste and the screw compresses it. The extruder screw is surrounded with a water jacket to cool it. The heated mix then passes to a filter screen to remove solid particles and then to the extrusion head. This filter screen needs to be replaced every 2 hours to insure better quality and avoid blocking. The mix comes out of the die head as a hot spaghetti-like shape that gets cooled by passing through a water basin. The strings are then drawn into the pelletizer by passing first through the supporting rollers. The pelletizer chops the strings into short, uniform cylindrical pellets that are packed in bags to be sold to the manufacturer (Lardinos and van de Klundert, 1995).

The following steps can improve the quality of pellets:

- Adding virgin plastic pellets depending on the quality needed.
- Preheating the shredded plastics in a drying installation.
- Reducing moisture content of reprocessed pellets by further extrusion through a finer filter screen.
- Increasing the capacity of the process by introducing a rotating spiral gear wheel in the hopper that presses the material down to the screw and thus speeds up the process and increases production.

Mixing, extrusion and product manufacturing stage

In the third and last stage of plastic reprocessing, manufacturing processes take place, which include extrusion, injection molding, blow molding, and film blowing. The extrusion process is similar to the one used in the pelletizing unit except that the die takes the form of the output product. In the case of tube production the die is made of a steel plate pierced with a hole. The extruded material has to be cooled to solidify using a water bath. Similar to

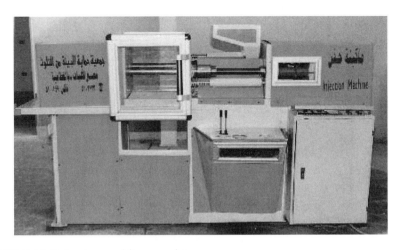

FIGURE 5.16 Injection molding machine

the extrusion process, the injection molding process as shown in Figure 5.16 has a spiral screw which is fed through the hopper and presses the mix to a strong, split steel mold (Lardinos and van de Klundert, 1995).

The rotating screw pushes the plastic while the heating element plasticizes it. The melt is pushed into a closed steel mold which is kept cool by a water jacket that accelerates the solidification process. Usually the process is manually stopped, this is achieved by allowing the mold to have one small hole allowing the exit of extra material, which indicates a full mold. The worker then stops the process, opens the mold, removes the piece and leaves it to cool. He then chops off all unwanted parts from the piece (Lardinos and van de Klundert, 1995).

The process of blow molding is used to produce bottles. The operation mainly takes place in two stages. First, the plastic is extruded in the form of a tube, then it passes through a mold which closes around the tube. Compressed air is then blown into a hollow mandrel which exists in the closed mold to force the tube to take the shape of the mold. The product is left to cool, solidify, and later removed from the mold: the process is then repeated (Lardinos and van de Klundert, 1995). Plastic bags can be made using the blow-molding machine as shown in Figure 5.17. The capacities of the blow-molding machines (film) vary between 100 and 200 kg per day depending on the power of the motor which ranges from 10 to 15 hp respectively.

In film blowing, the plastic is first extruded from a tubular die and moves upward to a film tower with a collapsing frame, guide rolls, and pull rolls driven by a motor. Air is compressed through the center of the die and inflates the tube. Air rings are mounted above the die to cool the outside surface. The tube is sealed and cut once it passes through the pull rolls.

FIGURE 5.17 Blow-molding machine (Film)

5.5 Recycling of Bones

Bone recycling is a simple process where useful products can be extracted. Minerals such as calcium powder for animal feed are extracted from the bone itself. The base material for cosmetics and some detergent manufacturing needs are extracted from the bone marrow.

The bone recycling process passes through seven stages starting from crushing and ending with packing. Figure 5.18 is a schematic diagram showing the bone recycling process which goes through the following steps:

- Crushing: Bones are conveyed through an auger from the receiving area to the crusher where bones are broken into pieces of about 10 cm in length.
- Cooking: In the cooker, crushed bones are subject to saturated steam supplied from a fire tube boiler via the steam line to cook bones with fat and protein and kill any bacteria or pathogens.
- Centrifugal separator: In the centrifugal separator unit, bone marrow and fats are expelled out of a perforated tank leaving the crushed bones in the bottom.
- Cooling: Crushed bones are cooled by circulating water in a cooler hopper. The circulating water is cooled in a forced draft-cooling tower.

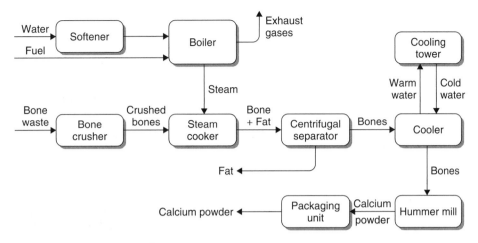

FIGURE 5.18 Bone recycling process

- Fine grinding: The cooled bones are transferred from the cooler to the hummer mill using an auger to obtain finer grains of calcium powder.
- Cyclone separator: A two-stage cyclone separator is used to separate the calcium powder before backing.
- Packing: At this stage the fine grains are weighed, packed, and are ready for marketing.

5.6 Recycling of Glass

Glass is a combination of sand and other minerals that are melted together at very high temperatures to form a material that is ideal for a wide range of uses from packaging and construction to fiber optics. As early as 4000 BC glass was used in the Middle East as a glaze to decorate beads. By 1550 BC, colored glass vessels were widespread and used for cooking and drinking. Until the late 18th century and early 19th century glass was very expensive and was used for limited applications, such as stained glass windows for churches. Large-scale glass manufacture began with the industrial revolution and the mass production of glass containers began at the onset of the 20th century.

Today the glass-making process can be used to make many different types of glass in infinitely varied colors formed into a wide range of products. Glass, chemically, is actually more like a liquid, but at room temperature it is so viscous or "sticky" it looks and feels like a solid. At higher temperatures glass gradually becomes softer and more like a liquid. It is this latter property that allows glass to be poured, blown, pressed, and molded into a variety of shapes.

Nowadays glass is much less expensive and is taken for granted as a packaging material in addition to its use in windows, doors, and other applications.

New glass is made from a mixture of four main ingredients: sand, soda ash, limestone, and other additives. These additives include iron for color (brown or green), chromium and cobalt for color (green and blue respectively), lead to alter the refractive index, alumina for durability, and boron to improve the thermal options.

Glass can be divided into different types according to composition and use:

- Container glass such as bottles and jars.
- Flat glass such as glass flat plates, automotive windscreens, and mirrors.
- Fiberglass: Two types of fiberglass can be produced, continuous and/or short fibers. Blanket fiberglass can be produced from continues fibers. The main applications are glass reinforced plastic (GRP), glass reinforced cements (GRC), and blanket fiber material used for insulation.
- Domestic glass such as domestic houseware, e.g. glasses and ornaments.
- Special glass such as cathode ray tubes as used in television, medical, and computer screens.

Glass manufacturing/recycled processes

The glass market has eight main categories, and within each category individual classes might exist. The eight categories are:

- Bottle applications.
- Building materials such as doors, windows, etc.
- Concrete applications in terms of concrete additives or aggregates.
- Industrial mineral uses such as sandblasting.
- Insulation applications such as fiberglass insulation.
- Paving applications as aggregate.
- Remelt applications such as art glass.
- Miscellaneous applications such as abrasive material, ceramic glazing, designer clothes, etc.

In all categories, a glass melt is prepared from silica sand and other raw materials such as lime, dolomite, soda, and cullet (broken glass). The use of recycled glass is increasing everywhere. It reduces the consumption of both raw materials and energy but necessitates extensive sorting and cleaning prior to batch treatment to remove impurities.

Glass used for new bottles and containers must be sorted by color and must not contain contaminants such as dirt, rocks, ceramics, etc. These materials, known as refractory materials, have higher melting temperatures than container glass and form a solid inclusion in the finished product.

A consideration in glass recycling is color separation. Permanent dyes are used to make different colored glass containers. The most common colors are green, brown, and clear (or colorless). In the industry, green glass is called emerald, brown glass is amber, and clear glass is flint. For bottles and jars to meet strict manufacturing specifications, only emerald or amber cullet (crushed glass) can be used for green and brown bottles, respectively. The glass is color sorted and sent to a glass crusher or ball mill.

Glass can be recycled indefinitely as part of a simple but hugely beneficial process, as its structure does not deteriorate when reprocessed. In the case of bottles and jars, up to 80% of the total mixture can be made from reclaimed scrap glass, called "cullet". Cullet from a factory has a known composition and is recognized as domestic cullet. From bottle banks it is known as foreign and its actual properties will not be known.

The cullet is then mixed with the raw material used in the production of glass. After the batch is mixed, it is melted in a furnace at temperatures ranging from 1,200°C to 1,400°C. The mix can burn at low temperature if more cullets are used. The melted glass is dropped into a forming machine where it is blown or pressed, drawn, rolled, or floated depending on the final products. The newly formed glass containers are slowly cooled in an annealing furnace.

The manufacture of glass uses energy in the extraction and transportation of the raw materials, and during processing as materials have to be heated together to a very high temperature. Large amounts of fuel are used and the combustion of these fossil fuels produces a lot of emissions. An efficient furnace will require 4 GJ of energy for each ton of glass melted.

The most important fuels for glass-melting furnaces are natural gas, light or heavy fuel oil, or liquefied petroleum gas. Electricity (frequently installed as supplementary heating) is also used in some special cases with low production capacity because energy requirements range from 3.7 to 6.0 GJ/t glass produced.

If recycled glass is used to make new bottles and jars, the energy needed in the furnace is greatly reduced. In addition recycling reduces the demand for raw materials. There is no shortage of the materials used, but they do have to be quarried from the landscape, so from this point of view, there are environmental advantages to recovering and recycling glass.

5.7 Foam Glass

Another technology for glass recycling is foam glass. There are numerous patents on foam glass production dating back to the 1930s. Even though there are numerous patents there are only a few that have been adopted on a commercial basis. Foam glass, also referred to as cellular glass, was originally manufactured from a specially formulated glass composition using virgin glass only. Currently, there are a number of foam glass production plants that are using up to 98% post-consumer waste glass in their product. The basic principle of foam glass manufacture is to generate a gas in glass at a temperature between 700°C and 900°C. The gas expands thus producing a structure of cells to form a porous body. The foam glass can be either made from molten glass or sintered glass particles. The latter process requires ground glass to be mixed with a foaming agent, then on heating the foaming agent releases a gas and expands the molten glass mass.

The main foam glass producers in Europe, Japan, and North America now use a high percentage of processed post-consumer glass in their products. Currently, there are three main product types of foam glass:

- Loose foam glass aggregate: Continuous production of sheets of foam glass that are then broken into loose foam glass aggregate and sized.
- Blocks and shapes: Generally continuous production of blocks and shapes in molds that are then cut and shaped. Can also be manufactured by a batch process.
- Pelletization: Continuous production of spherical pellets of foam glass that are then used in the manufacture of blocks, panels, and slabs.

Foam glass is best suited as a rigid insulation material. Due to its excellent structural properties, it is suitable for use as insulation in roofs, walls, and traffic areas such as flat roofs or floors, where other insulation products may be compressed resulting in an uneven surface and the loss of insulating properties. Foam glass has excellent fire resistant properties and its very low water absorption and water vapor transmission means that, unlike many other types of insulation, it tends to retain its insulating properties even when wet. It is also used as industrial insulation for a number of minor uses such as sandwich panels or is used as a product in extreme environmental conditions.

Foam glass has been manufactured for a number of years mainly in the USA and Europe as a lightweight high strength insulating material. The main driver for foam glass use has been the requirement of high energy efficiency standards for building construction which deals with energy conservation. The basic building block of all these regulations is the value of the overall heat transfer coefficient. This is the rate of heat loss, expressed in watts per square meter per degree temperature difference (w/m^2K). The use of foam glass in the construction of housing and buildings could greatly reduce energy consumption.

In addition to the potential energy saving from the use of foam glass, there are other less obvious advantages due to the lightweight nature of the material. These include design flexibility, construction productivity, reduced manual handling, lower transport costs, and lower foundation costs. Also it is rodent resistant, fire resistant, an effective sound absorber, non-toxic, and non-water absorbent.

Foam glass as a building construction material is in competition with insulating polymeric and fibre materials as it is a good insulator.

However, foam glass also has inherent strength properties. Foam glass characteristics of low flammability, thermal stability, and high chemical durability are a distinct advantage over polymeric materials, which have poor fire resistance. Foam glass also has the advantage that it contains no fibrous material. Fibrous insulation materials such as fiberglass require special handling procedures to protect the user from inhalation of fibers and skin irritation.

The desirable properties of foam glass are high strength, low density, and low thermal conductivity. Generally these properties are achieved by having a large number of small, evenly sized bubbles, with thin walls in between. As the product is made of glass it is naturally inert in most environments with respect to biological, thermal, chemical, and environmental degradation.

Foam glass mechanism

The principle of the foam glass process is that between 700°C and 900°C the glass powder forms into a viscous liquid and then the foaming agent decomposes to form a gas that in turn forms bubbles. The glass needs to have sufficient viscosity not to allow the gas bubbles to rise through the mass of the body but remain in position during the foaming heat cycle. If the temperature is too high the bubbles will rise and the body will collapse and not form a foam body. The control of the heating rate is one of the most important factors in optimizing the foam glass product. Rapid heating can cause the foam glass feedstock to crack, whilst slow heating will lead to early release of the gas from the foaming agent before the viscosity of the glass is low enough to allow the glass to expand.

A further complication is that the foam glass feedstock is relatively insulating due to a pack density of 80% and as the feedstock expands from the top surface this further insulates the materials below. Therefore, there is the potential to overheat the top surface in order to heat the bottom of the feedstock. This overheating can cause the top cells to collapse resulting in an inferior product.

The finely ground glass powder is mixed with the foaming agent which is the feedstock for the foaming furnace. Suitable foaming agents can be calcium sulphate ($CaSO_4$), coal, glass water, aluminum slag, or calcium carbonate ($CaCO_3$). The thermal conductivity will be lower if $CaSO_4$ is used: however, $CaCO_3$ is easier to work with. This is due to the production of sulfur gases from $CaSO_4$ during the foaming process. SO_2 has a lower thermal conductivity than CO_2; however, the formation of SO_2 requires more control as it is a noxious gas. Silicon carbide (SiC) is also used as a foaming agent, which gives controlled and precise cell sizes. It is thought that SiC is the most commercially used foaming agent for the reasons of control and reproducibility. The SiC reacts with the SO_3 within the glass structure to form CO_2 and S.

Gypsum is a readily available source of $CaSO_4$ and limestone is a readily available source of $CaCO_3$. If the air in the furnace at the foaming zone is replaced with either SO_2 or CO_2 then this will lower the thermal conductivity

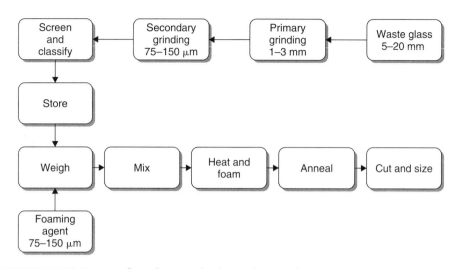

FIGURE 5.19 Process flow diagram for foam glass production

of the foam glass. Fly ash, which in itself is a waste mainly from coal-fired power stations, has been used as a foaming agent. The main constituents of fly ash are SiO_2, Al_2O_3, Fe_2O_3, CaO, and SO_3. SO_3 is the active foaming agent evolving SO_2 at temperature to form the foam glass. However, fly ashes are waste materials from incinerators and therefore can contain toxic compounds and heavy metals.

The foaming agent particle size affects the cell size, which will influence the performance and characteristics of the foam glass. This in turn determines the density of the product. The higher the cell size, the lower the product density, and the lower the thermal conductivity producing better thermal insulation properties. The smaller the cell size, the higher the compressive strength of the foam glass body. Therefore, compressive strength is inversely proportional to thermal insulation.

Foam glass process

Glass cullet is fed into a ball mill for grinding. Very fine glass powder ranging between 100 and 500 microns is mixed with a foaming agent and heated to the foaming temperature between 700 and 900°C. Figure 5.19 shows the process flow diagram from the waste glass "cullet" to the foam glass manufacture of a typical foam glass product. The following factors affect the properties of foam glass:

- Particle size: There is a direct proportional relationship between the density of foam glass products and the initial particle size of glass powder.

- Quantity of foaming agent in the initial mixture: There is an inverse proportional relationship between the density of the foam glass and the quantity of coal in the initial mixture.
- Time and temperature of heat: There is a direct dependence of foam glass formation and density on the time and temperature of heat.

5.8 Recycling of Aluminum and Tin Cans

Aluminum wastes are one of the most common items that can be recovered through municipal solid waste because they provide higher revenues than other recyclable materials. The recycling of aluminum cans uses 70–90% less energy than producing them from virgin materials.

Steel food cans make up 80–90% of all food containers and are often called tin cans because of the thin tin coating used to protect the containers from corrosion. Some steel cans, such as tuna fish cans, are made with tin-free steel, while others have an aluminum lid and a steel body and are commonly called bimetal cans. All these empty cans are completely recyclable by the steel industry and should be included in any recycling program. The collection vehicle discharges the solid wastes into a hopper bin, which discharges to a conveyor belt shown in Figure 5.5 in the transfer station. The conveyor transports the cans past an overhead magnetic separator where the tin cans are removed. The belt continues past a pulley magnetic separator, where any tin cans not removed with the overhead magnet are taken out. The aluminum and tin cans, collected separately, are baled for shipment through a hydraulic press located in the transfer station as shown in Figure 5.7. At the aluminum foundry, aluminum scrap is melted in a smelting process. Molten metal is formed into ingots that are transferred to manufacturing plants and cut into disks, from which cans and other products are formed. For more details, see Chapter 10 case studies.

5.9 Recycling of Textiles

Textile recycling has a long history, not for making new textile or returning textile to its original fibers or other textile products but for making paper (Dadd, 2004). Textile fiber can be classified into natural fibers such as cotton and wool and synthetic fibers. By recycling cotton wastes, we not only conserve landfill space but also reduce the amount of land, water, energy, pesticides, and human labor that goes into cotton production.

The textile recycling industry represents one of the most important recycling activities from solid wastes in developing countries because it is a labor intensive activity and can provide a lot of job opportunities as well as the availability of textile waste everywhere. Most of the textile recycling firms in developing countries are small, family-owned businesses with 5–20 semiskilled and marginally employable workers at the primary processing level including used clothing dealers, rag graders, and fiber recyclers.

Textile waste can be classified into two categories: pre-consumer and post-consumer (Dadd, 2004). The pre-consumer textile waste category consists of byproduct materials from the textile, fiber, and cotton industries. Pre-consumer textile waste can be recycled into new raw materials for the automotive, furniture, mattress, coarse yarn, female accessories, home furnishings, paper, and other industries. The post-consumer textile waste category consists of any type of garments or household article, made of some manufactured textile, that the owner no longer needs and decides to discard. These items are discarded either because they are worn out, damaged, or are no longer fashionable. Many items are made from fabric items recut to make new items, such as t-shirts cut to make cleaning cloths.

5.10 Recycling of Composite Packaging Materials

Food packaging has been essential since the birth of the industrial era when consumers moved into cities, away from the food production areas. But food preservation and storage has existed since ancient times. Natural materials like woven bamboo baskets, animal skins, and clay have been used to store, distribute, and protect food since the beginning of civilization.

Before the 19th century, food packaging was used mostly for transporting goods from the place of production to the consumer. But as people moved into ever-expanding towns and cities, during the early industrial age, the role of packaging widened from that of only containing products to one of protecting them too. So, with the beginning of industrialization, the search began for better ways of preserving foodstuffs and packaging became the main "component" for the development of long-life preserved foods. Today, "packaging is not of itself a product in the usual sense, it is a delivering tool to get a product to a consumer, safely, in excellent condition and fit-for-use; it is a fundamental part of any packaged goods" (Olah, 2004).

Glass making existed 6,000 years ago, but it was not until 200 BC that a Syrian discovered that he could blow molten glass with a pipe, and turn it to make a bottle (Planet Ark, 2005). So, before using the paper carton, the glass bottle was used for milk packaging. But glass bottles had some disadvantages as they are breakable and need to be handled with care; in addition, glass bottles are heavy and they need much energy for their transportation. Although they were reusable or refillable, they first had to be cleaned to a sterile state before reuse, and the cleaning process required large amounts of energy. Moreover, milk has always been difficult to preserve, due to its susceptibility to disease-causing agents which was one of the reasons for high infant mortality rates. Milk was also easily spoiled and quickly absorbed odors (Planet Ark, 2005). These conditions required localized dairy operations and intensive energy for distributing the milk packaged in glass bottles. Accordingly, other materials needed consideration for milk packaging for increasing efficiencies, transporting, and storing the milk products (Suriyage, 2005).

Milk packaged in paper cartons was first introduced in San Francisco in 1906, but using paper to package fluids was a difficult challenge and was not considered practical because of the perforated nature of paper and its absorption of moisture. To overcome this problem, milk cartons were coated with microcrystalline and paraffin waxes. But another problem encountered was to find a suitable bonding material for sealing the top, bottom, and sides of the package to form a box. Animal glues were used with varying degrees of success; they were not acceptable at first as they caused leaks and contamination of the contained milk. The paper cartons entirely disappeared from the market soon after their introduction (Planet Ark, 2005; Suriyage, 2005).

Containers made out of polyethylene terephthalate, a widely used type of plastic, became available to the market in 1977 as a packaging material for beverages (Berger, 2005). However, the energy-efficient production, reduction of carton weight, and improved sealing and opening techniques have kept the paper carton at the peak of packaging technology (Planet Ark, 2005).

Paper packaging products such as milk and beverage cartons are currently made of more than one material, where each layer is present for a specific reason. These multilayered cartons are generally classified into aseptic cartons and non-aseptic cartons. Non-aseptic cartons are composed of 89% paperboard and 11% polyethylene by weight (Charlier and Sjoberg, 1995), while aseptic cartons have different compositions due to the presence of an aluminum layer. Aseptic cartons are composed of several layers of three kinds of materials: two layers of paperboard representing about 74% of the material, four layers of polyethylene representing about 21% and a layer of aluminum foil of about 6.5 microns in thickness representing about 5% by weight (Charlier and Sjoberg, 1995; Olah, 2004).

Starting from the outside the first layer in non-aseptic cartons is made of polyethylene to act as a barrier for moisture and bacteria followed by a layer of paperboard to provide the carton with stiffness and strength. The last two layers are both made of polyethylene; the third layer is present as an adhesion layer while the fourth and last inner layer seals in the liquid contents. The adhesion layer is used to ensure that the layer of polyethylene in contact with the packed product remains intact. The first two layers in aseptic cartons are the same as those present in non-aseptic ones and have the same functions. The third layer is also made of polyethylene to act as an adhesion layer but it is present to bind the paperboard to the fourth layer of aluminum. This aluminum layer is essential to prevent oxygen and light from entering the carton. The last two layers are both made of polyethylene; the fifth layer is present as an adhesion layer while the sixth and last inner layer seals in the liquid contents. According to these structures, non-aseptic cartons are generally used for pasteurized products while aseptic cartons are used for long-life products. These aseptic packages allow liquid products to be distributed and stored without the need for preservatives for periods up to one year. By uniting the best attributes of paper, plastic, and aluminum, the multilayer aseptic package efficiently prevents light and air as well as water and microorganisms from entering the package during or after packaging. It also seals in nutrients and flavor, and removes the need for refrigeration for months contributing to energy savings (Aseptic Packaging Council, 2005; Olah, 2004).

Rodushkin and Magnusson (2005) studied the effect of aluminum (Al) foil contained in the laminated carton packages on the content of the package. They studied Al migration to orange juice contained in laminated paperboard packages during a storage time of one year and a storage temperature of 23°C by monitoring the Al content in the juice. Results demonstrated that Al concentrations in the filled juice did not change during its storage. They explained that the Al foil in the laminated package is covered by polymer coatings so there is no direct contact between the fill product and the foil. They further explained that Al foil is a solid metal that has negligible vapor pressure.

The Aseptic Packaging Council (2005) stated that the low density polyethylene used in the aseptic package is approved by the US Food and Drug Administration as a "food-contact surface material". The aseptic packaging industry tests have shown that no plastics leach into the product contained in aseptic cartons and no endocrine-mimicking chemicals are present in the products. The need for paper recycling worldwide as discussed before also includes the need for recycling used packaging carton wastes. This requirement also applies to composite packaging material which includes the recyclability of aseptic beverage cartons used for preserving liquids such as milk (Charlier and Sjoberg, 1995). Examples of beverages and liquid food products contained in these packages are milks, juices, cheese, tomatoes, etc. Tetra Pak is a multinational company that provides these carton packages for food production customers. Recycling these packages will reduce the volume of waste dumped in open dumpsites, and save natural resources. In addition, these cartons are made of very high quality virgin pulp, and accordingly they are valuable to recycle. However, the combined materials present for achieving high performance make recycling of these packages a difficult task.

Energy and material recovery

Anonymous (2001) described a technique for the recovery of the high energy content in both polyethylene and aluminum that is associated with the production of bauxite (aluminum oxide) material. The process of cement production requires the addition of aluminum oxide to guarantee adequate solidification of the cement. The residual Polyethylene and aluminum (PE/Al) materials can therefore be used as an alternative to burning coal at cement factories which requires large amounts of energy and at the same time provide the facility with the aluminum oxide material that is required as an additive for cement production. Emission measurements have demonstrated that there is no increase in air pollution when these residual materials are burnt in comparison with coal. This alternative is considered to be both economically and environmentally feasible.

The European Aluminum Foil Association (2005) stated that an innovative process is used by the Corenso recycling plant in Finland that recovers fiber, converts the plastic layers of used multilayer packaging into gas for energy, and recovers aluminum. The Corenso recycling plant processes 125,000 tons of mixed packaging waste into 250 GWh of heat and electricity per year. Around 3,000 tons of aluminum is recovered. Anonymous (2001) described the recovery of aluminum content together with the generation of steam and electricity from the polyethylene material. The PE/Al mixture is heated up to a temperature of 400°C in a gas reactor with air blown into it to make sure that there is about 40% air. The polyethylene content changes to gas that is used to heat a steam boiler up to 1,300°C, as a result of which process steam is produced and electricity is generated with the help of a back pressure turbine. Pure aluminum is recovered and separated with the help of a cyclone, then used to manufacture a variety of aluminum products. This process contributes to huge energy savings in addition to the production of significantly fewer air emissions in comparison to the use of other fuels such as coal or oil.

According to Alcao Aluminio Inc. (2005), a carton packaging recycling facility in Brazil uses groundbreaking plasma technology which enables the total separation of aluminum and plastic components contained in the aseptic packages. The application of plasma technology for the recycling of carton packaging utilizes electrical energy to produce a jet of plasma at 15,000°C to heat the plastic and aluminum mixture. Plastic is transformed into paraffin and the aluminum is recovered in the form of high purity ingot. The level of pollutants emitted during the recovery of the materials is minimal. The process is handled in the absence of oxygen, and it does not involve combustion; the energy efficiency rate is close to 90%. The plasma process thus provides another option for recycling, allowing for the return of all three components of the package to the productive chain as raw material closing the loop of materials. The paper, extracted during the first phase of the recycling process is transformed into cardboard. Alcoa Inc., which supplies thin-gauge aluminum foil for the manufacture of aseptic packages, uses the recycled aluminum to manufacture new foils, and paraffin is sold to the Brazilian petrochemical industry.

Thermal compression

Anonymous (2001) described a recycling process that uses the beverage carton as a whole to produce a new material called TECTAN. The used beverage cartons are chopped then pressed and heated to a temperature of 170°C with the help of Teflon-coated metal plates. The polyethylene content in the chopped cartons melts and bonds the chips together forming a plate.

The plate is then cooled and pressed in a cooling press to the desired thickness. The produced plate has moisture resistance and sound-deadening properties that are suitable for office furniture, flooring, interior decoration as well as other products.

Repulping of beverage cartons

Anonymous (2001) presented a pulping technique for processing used beverage cartons for pulp production. Beverage cartons are first shredded into palm-sized pieces then conveyed into a drum which is a metal cylinder about 22 m long with a diameter of about 3 m. This drum is divided up into two parts, a separation drum that is 10 m long followed by a sorting drum of about 12 m long. The shredded cartons are softened by water; rotating paddles inside the drum transport the material upwards, after which it drops back down. The impact causes the softened paper fraction in the shredded cartons to disintegrate into fibers and become detached from the other material; no chemicals are required. As a result of this process, a brown fiber pulp slurry is produced that is washed away through small holes, cleaned, concentrated, and then pumped to a paper machine to be converted into new paper or board products. The residual PE/Al mixture is discharged at the end of the drum and dried.

Charlier and Sjoberg (1995) presented a different mechanical pulping technique for recovering the fiber portion in the used beverage cartons in repulping facilities available in paper mills. Through the process of hydra-pulping, batches of used beverage cartons are mechanically heavily mixed with warm water leading to the separation of fibers from the other polyethylene and aluminum constituents of the laminates. The resulting fiber slurry is then separated and transferred to join the pulp and paper production route in the paper mill. A wet residue mainly of polyethylene and aluminum foil together with some fibers and foreign objects remains.

Process of paper pulp separation

The process of recycling paper constituted in used cartons requires the separation of the paper fraction from the aluminum and polyethylene layers. The process steps involved are cutting, soaking, repulping, and screening as illustrated in Figure 5.20. The different recycling stages are presented below in more detail.

Cutting: Used beverage cartons are cut by scissors into four pieces to decrease their size and increase their specific area. Cutting is necessary to facilitate the repulping process for paper recycling and reduce the time required to refine the fiber and form the pulp.

Soaking: The cut cartons are put in a tank filled with water and left overnight before pulping. This reduces the time required to refine the fibers for pulping and accordingly increases the pulper's capacity.

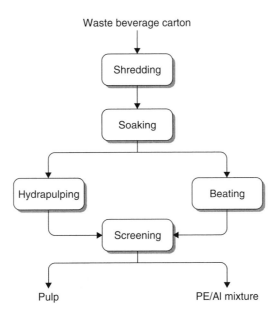

FIGURE 5.20 Composite packaging recycling process

Repulping: Two alternative technologies are used for repulping to separate pulp from Al and PE. The first alternative uses a hydrapulper which acts like a blender. The cut carton material and water are added in the hydrapulper where they are vigorously mixed resulting in friction forces between the materials that accordingly leads to the separation of the fibers from the polyethylene and aluminum material producing pulp. A schematic drawing of the hydrapulper is shown in Figure 5.21.

The second alternative uses a beater. The cut carton material and water are added in the beater and circulate in its oval tank discussed before as shown in Figure 5.9. As they circulate, they pass in between a rotating drum and a bed plate. This leads to the separation of the paper fibers due to impact and high friction. Continuous circulation of the material in the tank results in refining the fibers.

Screening: Screens are used to remove the pulp from the polyethylene and aluminum materials. Water and the repulped material are added to the screening device from the top. An air injection system is used to blow air through the water containing the suspended materials for a few minutes to avoid blocking the screens and to allow efficient separation of the materials from each other. The air flow is stopped and a water valve at the bottom of the device is opened for draining water. The pulp produced is obtained by being removed from the screen at the bottom. A schematic drawing of the screening device is shown in Figure 5.22.

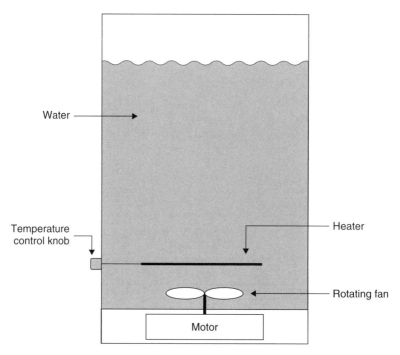

FIGURE 5.21 Schematic drawing of a hydrapulper

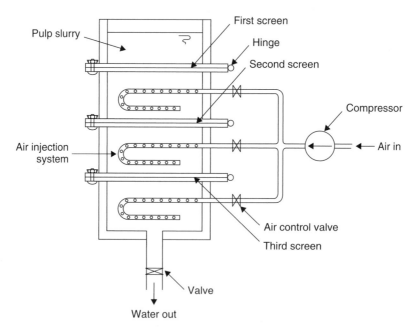

FIGURE 5.22 Schematic drawing of the air injection screening device

Changing the pulping technique has effects on the efficiency of pulp separation as well as effects on the properties of the paper produced. During beating, it was observed that as the carton pieces continuously pass between the rotating drum and the bed plate, small fine pieces of polyethylene and Al/PE layers are cut. These fine pieces were able to pass through the screen slots during the screening process and get mixed with the separated pulp fibers. Much fewer Al and PE impurities were present when the hydrapulper was used. At the same time, the rate of separating the pulp fibers using a beater was higher than that using a hydrapulper. As for the effect of changing the pulping technique on the paper properties the tensile, bursting, and tear strengths and stiffness of paper produced using the beater are compared to those produced using the hydrapulper. It can be concluded that the tensile and bursting strengths of paper are higher, while the tear strength and stiffness of paper are lower when the beater is used compared to the hydrapulper. So, depending on the paper product and its required paper properties the most suitable technique among the two studied techniques should be used. Tensile strength is necessary during printing or converting paper into useful products where the resistance to web breaking is important. Bursting strength is important for paper bags to withstand any applied pressures and avoid rupture. Tearing resistance is important for different products such as newsprint, books, and magazines as well as paper bags and wrapping paper. Stiffness is important for box boards and corrugating medium and·a certain degree of stiffness is required for printing papers.

5.11 Recycling of Laminated Plastics

Laminated plastics are polyethylene coated with a very thin aluminum layer and are used in chip bags, for example, and what is left from milk/juice cartons after fiber removal as discussed above. Charlier and Sjoberg (1995) considered the separation of the aluminum foil and the polyethylene film a critical task. They presented two different techniques for handling residues from the repulping facilities based on a technical feasibility study done by a researcher and a leading beverage carton producer. Both techniques aim at material recycling of aluminum and energy recovery of polyethylene starting with the removal of the undesired fibers and objects in the remaining residue. This can be achieved by first shredding the residue using monoshear type equipment followed by magnetic and eddy current separation. An aluminum-rich fraction representing about 50% in weight of the incoming material is obtained. The polyethylene film in the fraction is then thermally removed using one of two alternative techniques. One technique is thermal decoating based on a rotary kiln for decoating, where the incoming material is efficiently stirred inside the rotating kiln drum and directly heated by hot gases or by contact with the heated kiln wall. The other technique is thermal decoating based on a moving bed pyrolysis oven, where the incoming material is mainly indirectly heated by a combination of radiation and conduction.

For both techniques a low density, very fluffy, crumbled aluminum foil material is obtained. The decoated aluminum material is then remelted in an electrically heated crucible furnace. High quality remelted aluminum is produced which is suitable for the production of thin gauge foil.

Recycling of aluminum and polyethylene

A chemical process can be used to recover the aluminum content in the aluminum/polyethylene residual obtained from the repulping process discussed in the previous section. The process involves the addition of toluene or xylene solvents to the residual mixture contained in a crucible. The crucible is then heated to a temperature of about 80°C. As the mixture is heated, the toluene or xylene contained in the crucible causes the PE layer to separate from the aluminum foil. The resulting aluminum and polyethylene materials can be used further for recycling and the remaining solvent can be recovered.

Recycling of polyethylene

Alternative chemical processes are used to recover the polyethylene content in the aluminum/polyethylene residual obtained from the repulping process from milk packaging materials and produce aluminum compounds. The obtained polyethylene material is washed and repelletized to be suitable for producing recycled polyethylene products while the aluminum compounds can be used in different industries. Three different chemicals will be used independently to recycle polyethylene: hydrochloric acid, sulfuric acid or sodium hydroxide.

Using hydrochloric acid

This process involves the addition of hydrochloric acid (HCl) to the residual mixture. Diluted HCl is added to the PE/Al mixture. Hydrogen gas is produced as the aluminum content reacts with hydrochloric acid forming aluminum chloride. Polyethylene remains suspended in the resulting aqueous solution of aluminum chloride. The polyethylene is obtained by filtering the solution using filter paper.

Aluminum + Hydrochloric acid → Aluminum chloride + Hydrogen

$$2Al_{(s)} + 6HCl_{(aq)} \rightarrow 2AlCl_{3(aq)} + 3H_{2(g)}$$

Using sulfuric acid

This process involves the addition of sulfuric acid (H_2SO_4) to the residual mixture. Diluted H_2SO_4 is added to the PE/Al mixture. Hydrogen gas is produced as the aluminum content reacts with sulfuric acid forming aluminum sulfate. Polyethylene remains suspended in the resulting aqueous solution of aluminum sulfate. The polyethylene is obtained by filtering the solution using filter paper.

$$\text{Aluminum} + \text{Sulfuric acid} \rightarrow \text{Aluminum sulfate} + \text{Hydrogen}$$

$$2Al_{(s)} + 3H_2SO_{4(aq)} \rightarrow Al_2(SO_4)_{3(aq)} + 3H_{2(g)}$$

Using sodium hydroxide

This process involves the addition of sodium hydroxide (NaOH) dissolved in water to the residual mixture. Sodium hydroxide solution is added to the PE/Al mixture. Hydrogen gas is produced as the aluminum content reacts with sodium hydroxide forming sodium aluminate. Polyethylene remains suspended in the resulting aqueous solution and the polyethylene is obtained by filtering the solution using filter paper.

$$\text{Aluminum} + \text{Sodium hydroxide} \rightarrow \text{Sodium aluminate} + \text{Hydrogen}$$

$$2Al_{(s)} + 2NaOH_{(aq)} \rightarrow 2NaAlO_{2(aq)} + H_{2(g)}$$

The aluminum chloride resulting from the addition of hydrochloric acid is considered one of the most widely used aluminum compounds. It is employed in the manufacture of paints, antiperspirants, and synthetic rubber. It is also necessary for converting crude petroleum into gasoline, diesel and heating oil, and kerosene (BookRags Inc., 2005). The aluminum sulfate resulting from the addition of sulfuric acid is also known as papermaker's alum. It is used in the papermaking industry as it can react with small amounts of soap on paper pulp fibers producing gelatinous aluminum carboxylates, which coagulate the pulp fibers into a hard paper surface. It is also used for making aluminum hydroxide, which is used at water treatment plants to filter out impurities, as well as to improve the taste of the water (Wikipedia, 2005). The sodium aluminate resulting from the addition of sodium hydroxide is used in water softening systems. It is also used as a coagulant aid, to remove dissolved silica, and is used in the construction industry to accelerate the solidification of concrete and in the paper industry for producing fire brick and alumina (Wikipedia, 2005).

5.12 Recycling of Food Waste

Food waste recycling can take place through aerobic fermentation (composting) or anaerobic fermentation (biogas). Composting is the recommended method for recycling food wastes. Composting is a process that involves biological decomposition of organic matter, under controlled conditions, into soil conditioner (El-Haggar *et al.*, 1998). Aerobic fermentation is the decomposition of organic material in the presence of air. During the composting process, microorganisms consume oxygen, while CO_2, water, and heat are released as result of microbial activity as shown in Figure 5.23.

Factors affect the composting process

Four main factors control the composting process: moisture content, nutrition (carbon:nitrogen ratio of the material), temperature and oxygen (aeration).

Moisture content: The ideal percentage of the moisture content is 60% (El-Haggar *et al.*, 1998). The initial moisture content should range from 40 to 60% depending on the components of the mixture. If the moisture content decreases less than 40%, microbial activity slows down and becomes dormant. If the moisture content increases above 60%, decomposition slows down and odor from anaerobic decomposition is emitted.

Carbon to nitrogen ratio: Microorganisms responsible for the decomposition of organic matter require carbon and nitrogen as a nutrient to grow and reproduce. Microbes work actively if the carbon:nitrogen ratio is 30:1. If the carbon ratio exceeded 30, the rate of composting decreases. Decomposition of the organic waste material will slow down if C:N ratios are as low as 10:1 or as high as 50:1.

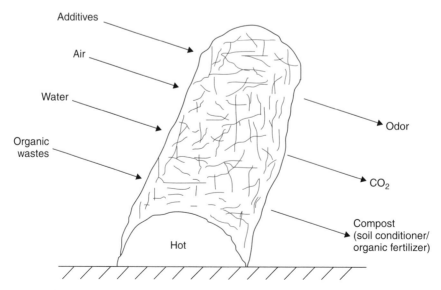

FIGURE 5.23 Composting process

Temperature: The activity of bacteria and other microorganisms produce heat while decomposing (oxidizing) organic material. The ideal temperature range within the compost for it to be efficient varies from 32°C to 60°C. If the temperature is outside this range, the activity of the microorganisms slows down, or might be destroyed.

The increase of temperature while composting above 55°C, kills weeds, ailing microbes, and diseases including Shengella and Salmonella; this helps to reduce the risk of disease transmission from infected and contaminated materials. The outside temperature has an effect on the composting process. In winter, the composting process is slower than in spring and summer.

Oxygen (aeration): A continuous supply of oxygen through aeration is a must to guarantee aerobic fermentation (decomposition). Proper aeration is needed to control the environment required for biological reactions and achieve the optimum efficiency. Different techniques can be used to perform the required aeration according to the composting techniques. The most common types of composting techniques are natural composting, forced composting, passive composting, and vermi-composting.

Natural composting

Piles of compost are formed along parallel rows as shown in Figure 5.24 and continuously moisturized and turned. The distance between rows can be determined according to the type and dimension of the turning machine (Tchobanoglous *et al.*, 1993). Piles should be turned about three times a week in summer and once a week in winter to aerate the pile and achieve homogeneous temperature and aeration throughout the pile. This method needs large areas of land, many workers, and has high running costs.

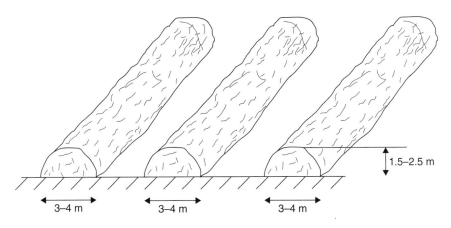

FIGURE 5.24 Natural composting process

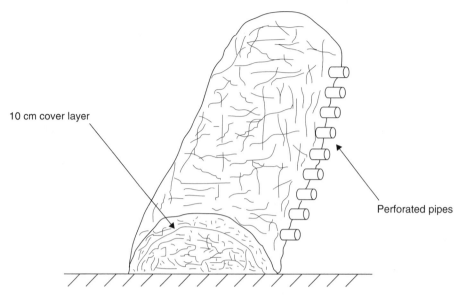

10 cm cover layer

Perforated pipes

FIGURE 5.25 Passive aeration process

Advantages

- Natural aeration
- Low cost

Disadvantages

- Odor emission
- Needs long time for maturation
- Needs labor for turning

Passive composting

Parallel rows of perforated high pressure PVC piping are placed at the bottom, of the compost. The pipes are perforated with 10 cm holes to allow air to enter the composting piles as shown in Figure 5.25. The pipe manifold helps in distributing the air uniformly. This system is better than the natural system because of the limited flow rates induced by the natural ventilation. This method needs limited areas of land, has lower running costs, and does not need skilled workers. This method is recommended for its cost effectiveness. Therefore, it is the most suitable method for the developing communities that want to achieve maximum benefit from food recycling with the minimum capital investment. This process, known as passive composting, produces good quality soil conditioner and organic fertilizer if NPK (nitrogen:phosphorous:potassium) is adjusted with natural rocks (El-Haggar *et al.*, 2004a).

Advantages

- Natural convection
- Low running cost
- Less maturation time
- Odor can be controlled by adding a top layer of finished compost

Disadvantages

- Relatively high cost compared with natural aeration

Forced aeration

Forced aeration works like the previous system except that the ends of plastic pipes are connected to blowers that force (or suck) the air through the compost at a specific rate and velocity. Otherwise if the air rate exceeded a certain limit, the temperature inside the compost pile decreases affecting the microbial activity. Also, the air velocity during the day should always be higher compared with the air velocity at night. This system needs higher technology with air velocity control and more energy consumption, which is why it is less economic compared to the other two systems and it is not recommended for rural or developing countries that want to make a profit out of all recycling processes. This method needs capital investment, skilled workers, and has high running costs.

Advantages

- Odor can be controlled by adding a top layer of finished compost
- Less maturation time compared with natural aeration

Disadvantages

- Needs electrical source near compost
- High capital cost and velocity control
- Requires skilled workers

Vermi-composting

Vermi-composting is an ecologically safe and economic method that depends on worms' characteristic of transforming the organic wastes to fertilizers that are extremely beneficial to the earth. There are two types of earthworms that are used due to their insensitivity to environmental changes:

the red wiggler (*Eisensia foetida*) and
the Red Worm (*Lumbricus rebellus*).

Under suitable aeration, humidity, and temperature, worms feed on organic wastes and expel their manure (worm castings) that break up soil providing it with aeration and drainage. The process also creates an organic soil conditioner as well as a natural fertilizer. Worm castings have more nutrients than soil conditioner in terms of nitrogen, phosphorus, etc.

A mature worm will produce a cocoon every 7 to 10 days which contains an average of seven baby worms that mature in approximately 60 to 90 days, and in one year each 1,000 worms produce 1,000,000 worms (El-Haggar, 2003a).

Vermi-composting can be used internally easily by using a special container (worm bin) that can be placed in any place that is not subjected to light such as a kitchen, garage, or basement. The organic waste is put in this container with the worms. The worms are odorless and free from disease.

5.13 Rejects

The rejects problem starts at the municipal solid waste sorting stage where recyclable wastes can be collected leaving unrecyclable wastes (rejects) without collection to be disposed off. The plastic bags that are contaminated with organic material are not collected because cleaning them is very costly and recycling of such waste is not cost effective. Also items such as small pieces of glass can be harmful to workers during manual sorting, so they are not collected. In a typical municipal solid waste composting plant, the incoming raw materials are food waste mixed with contaminated plastic bags and other impurities. The compost piles, after reaching maturation stage, are driven by a conveyor belt to a drum-like machine that separates the organic material from the rejects as shown in Figure 5.26. This drum rotates compost inside and the cutting tools located at the inner surface decompose the organic material further. As the drum rotates the organic material penetrates through the openings along the body of the drum while the rejects, shown in Figure 5.27, that rotate and move along the axis of the drum pass by an exit at the end of the drum (Sawiries *et al.*, 2001). Figure 5.26 shows a drum in a typical composting plant in Egypt, with the piles of compost at the front and the rejects at the other end. Figure 5.27 shows the rejects accumulated at the side of the composting plant. These rejects represent 20–25% of the total solid waste generated (El-Haggar and Toivola, 2001). Rejects can be disposed of using incineration, landfills, or recycling if possible.

Incineration

Rejects can be incinerated to be converted into ash. Incineration is the process of thermally combusting solid waste using a thermal treatment process. There are various types of incinerators as discussed in Chapter 1 and the type used depends on the type of waste to be burnt. Conceptually, incinerators are not recommended for sustainable development because

FIGURE 5.26 Composting plant

FIGURE 5.27 Rejects

they deplete the natural resources and might pollute the environment if they are not managed properly.

Landfill

Rejects and/or ash produced from the incineration process should be landfilled. Although landfill is still the most widely used method, it depletes the natural resources and uses the unsustainable solid waste management approach as discussed in Chapter 1.

In conclusion, recycling has proved to be the most suitable solution to the problem of rejects for conservation of natural resources compared with incineration and/or landfill according to the last 2Rs of the 7Rs Golden Rule discussed in Chapter 1. Recycling of rejects will be discussed in detail in Chapter 6.

Questions

1. Discuss the cost/benefit analysis of the traditional method of recycling glass and converting glass cullet into foam glass.
2. Discuss the sustainability of municipal solid waste management in your community. In other words what is the percentage of MSW landfilled?
3. If the percentage of MSW landfilled in question 2 is not zero, what constitutes the remaining MSW that cannot be recycled?
4. What kind of technology do you recommend using to recycle the remaining MSW discussed in question 3? Recommend some recycled products you can use in your community.
5. Compare between natural composting and passive composting to recycle organic waste.
6. Some countries might recommend biodegradable plastics over polymeric plastics for environmental protection; comment.
7. What are the advantages and disadvantages of vermi-composting? Do you recommend vermi-composting in your community? Why?

Chapter 6

Recycling of Municipal Solid Waste Rejects

6.1 Introduction

The potential use of Municipal Solid Waste (MSW) plastic rejects resulting from composting rotating screens as discussed in Chapter 5 or from dumpsites in the production of a composite material for different applications in construction, sanitary, mechanical, costal zones, etc. is very important for approaching a cradle-to-cradle concept for MSW in order to reach sustainable development. This is a new approach worldwide and has been implemented with full success at The American University in Cairo (AUC) as well as The Association for the Protection of the Environment (APE). Both AUC and APE are non-profit organizations and the cooperation among universities, research institutes, and NGOs (non-governmental organizations) represents a unique model for sustainability. The evaluation of using MSW rejects as a new material will be based on both mechanical and service properties of the proposed material together with an environmental assessment associated with the production and use of this composite material.

Rejects recycling

Recycling of solid waste rejects was developed at AUC using three steps. First, to innovate new technology for unrecyclable materials like rejects. Second, to develop different valuable and economical products from recycled rejects according to demand and needs. Third, to conduct a market study, which is a very important step in this approach to guarantee the sustainability of the project. Without these three steps, the recycling of rejects will never be sustainable and continually developed.

Advantages of reject recycling:

- Using a resource that would otherwise be wasted.
- Reducing or preventing the amount of waste going to landfill.

- Reducing the costs involved in the disposal of waste, which ultimately leads to savings for the community.
- Providing employment.
- Protecting natural resources.
- Reducing pollution.

6.2 Reject Technologies

The rejects recycling system (El-Haggar, 2004c) as shown in Figure 6.1 consists of a screen separator to separate the rejects from the organic waste. The screened rejects will be agglomerated to cut the plastic into small pieces. After the rejects have been agglomerated, they are mixed with sand and plastic additives to adjust the properties and appearance and heated indirectly to 140–240°C depending on the mixture and required applications. The hot paste is

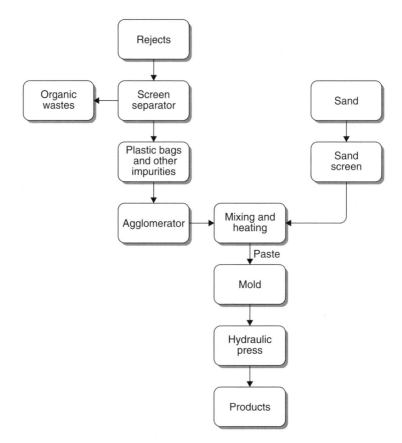

FIGURE 6.1 Schematic diagram of silica-plast "rejects" products

then transferred to the mold according to the required shapes and applications. The mold will be placed in a hydraulic press to be pressed into bricks, interlocks, table toppings, wheels, manholes, road ramps, and other products as will be discussed later. The manufacturing process of rejects (silica-plast) products consists of the following steps and as shown in Figure 6.1:

- Sieving the rejects to remove organic wastes and returning the organic waste to the composting process.
- Agglomerating the rejects (contaminated plastic waste).
- Sieving of sand to remove oversize grain.
- Mixing the agglomerated rejects with sieved sand and heating the mix indirectly to the required temperature.
- Pouring the hot mix into molds and pressing it in a hydraulic press to reach the required density and shape.
- Cooling the product.

The hot mix can be easily molded into any form and any decorative shape according to the shape of the mold. The key issue behind this technology is the continuous mixing of sand and plastics to guarantee homogeneous distribution of materials and good quality products. Reject technology consists mainly of the following machines:

- Sand screen
- Reject screen
- Agglomerator
- Indirect flame furnace
- Hydraulic press to form the products according to the mold shape
- Overhead crane
- Conveyor system

The two key machines used in reject technology are the agglomeration machine and the indirect flame furnace. Details of both machines will be given in the following sections.

Agglomeration machine

The agglomeration machine shown in Figure 5.13 consists of a cylinder with four stationary and four rotating knives as shown schematically in Figure 6.2. Two pulleys are used to transmit the motor power to the disk with the rotating knives. Plastic is fed to the machine through the upper lid. The disk containing the four knives rotates via the power transmitted from the motor. The four stationary knives on the cylinder are used in association with the rotating knives to shred/agglomerate the plastics into small pieces. A door is used near the bottom of the machine for the release of the product "agglomerates".

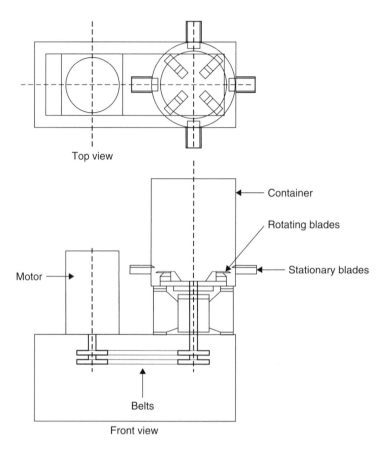

Top view

Container

Rotating blades

Motor

Stationary blades

Belts

Front view

FIGURE 6.2 Schematic diagram of the agglomeration machine

Indirect flame furnace

The indirect flame furnace consists of a cylindrical chamber equipped with an electric motor with a gearbox that transmits its power output to a three gear system located at the top of the cylindrical chamber shown in Figure 6.3. The gears rotate mixing blades in two different directions to stir a homogeneous mixture of heated sand and other additives. The fuel burner is located at the bottom of the chamber to heat the mix indirectly. A 1/3 hp blower is attached to the mixing chamber to collect gases emitted from the paste during the heating process into the combustion chamber to minimize the air pollution generated during the process. The temperature inside the chamber is controlled by a thermostat and a control panel.

The plastic from the agglomeration machine is mixed with sand and fed through the upper lid into the furnace. The machine operates for about 30–40 minutes in order to reach the required temperature. The stirring action starts

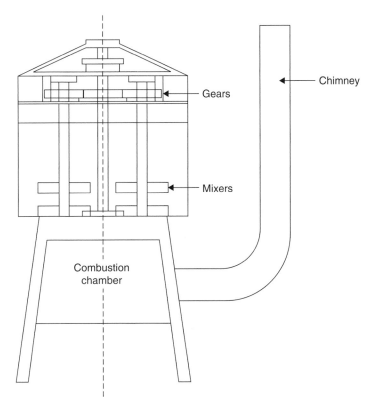

FIGURE 6.3 Mixing and heating furnace

when the temperature reaches 90°C. When the mixture is hot enough, the motor is then started to operate the gears rotating the two blades inside the cylinder to stir the mixture. The mixture is stirred for about 10–15 minutes, and then the sliding door at the bottom is moved to drop the paste into the mold, which is then pressed using the hydraulic press. The blower collects the gases (continuously during charging of raw material, during operation, and during discharging the paste to the mold) from the mixing chamber and uses them as a secondary fuel in the combustion chamber.

6.3 Product Development from Rejects

Product development from waste recycling, in general, is important to maximize the benefits. This leads to a number of economic opportunities to remanufacture products with the recovered material. Just as market forces cannot be ignored when introducing a new product, they must also be taken into account when introducing remanufactured products. The market analysis for these products may be more difficult than new ones because customers are more concerned with product durability and cost of maintenance. The product development from recycling MSW rejects is more important and challenging than the product development of recycling of waste because the properties of products from rejects will change from one time to another due to the heterogeneous nature of rejects. Therefore, continuous checking is vital with appropriate quality control and quality assurance.

6.4 Construction Materials and Their Properties

Bricks

Bricks were the first application of rejects because they were the easiest to produce. The major problem with bricks is the adhesion problem since they are made from plastics and sand. Adhesive materials are very expensive, which add to the cost of the brick. There is a cheap way to bind them together using pins as shown in Figure 6.4 for easy assembly and disassembly, but the density of the brick is very high so the weight of the brick from rejects was heavy compared with ordinary bricks. Producing bricks was found to be uneconomic, compared to the ordinary bricks including adhesives. One useful property of these bricks was their resistance to bullets and they were therefore used for military purposes.

Interlocks

The idea of bricks developed and interlocks were produced from rejects by changing the shape of the mold. There were several applications for interlocks. They were used in pavements, gardens, factory floors, backyards, etc.

FIGURE 6.4 Development of bricks

The production of interlocks was much more profitable than brick, as they did not need adhesives. The cost of silica-plast interlocks is 30–50% of the traditional types of interlock depending on top layer coating. The appearance of the interlocks can be improved by coating the top layer of the interlock with better quality plastic waste. The top layer of plastic waste shown in Figure 6.5 is a cheap coating layer since it is made out of plastic waste. The plastic waste that was used were clothes hangers that were crushed and melted in the heater, put in the mold, and pressed with the reject paste to produce this top layer.

Bricks proved to be unstable due to the difficulty to maintain adhesion because of water absorption. Interlocks proved to be very competitive, as the plastic coat was optimum because it is made out of clean waste plastic. This plastic coat completely covered the black color of the rejects and it was proved to be durable.

The brick/interlock material was composed of a mixture of plastic rejects and natural sand. The utilized plastic rejects included unsorted thermoplastic wastes from MSW, the plastics recycling industry, thermoplastic wastes from the packaging industry, etc. The major waste constituents were LDPE (low density polyethylene) and HDPE (high density polyethylene), which together comprised about 80% of the total waste. Small fractions of PS (polystyrene), PET (polyethylene terephthalate) and PVC were also present at a weight percentage of 10–15% while the remaining balance contained other minor waste constituents such as paper, wood, cardboard, small pieces of glass,

FIGURE 6.5 Development of interlocks

food scrap, cloth fragments, and small metal chips. This rejects are produced because of the unsorting behavior of MSW at the source which is typical in most developing countries and some developed countries. Inorganic siliceous sand with a fineness modulus (ASTM, Siere Analysis, 1998) of 3.29 was used as filler in the brick/interlock production. Two different sand particle sizes were employed, fine particles passing a 1.18 mm (No. 16) ASTM sieve and coarser particles passing a 2.36 mm (No. 8) ASTM sieve. The brick/interlock material was heated up while mixing to two temperatures, 185°C and 240°C. The mix was then dispatched to a steel mold and compacted using a hydraulic press.

A total of 12 different mixes were produced incorporating the different variables, sand content, sand sieve and mixing temperature as shown in Table 6.1.

There are no standards or codes to describe the testing procedures for recycled MSW composite material. Therefore, the standards for plastics testing, pedestrian and light traffic paving bricks, and cement tiles were used as guidelines for evaluating the properties of the produced composite.

Morphology

Morphological analysis (Abou Khatwa *et al.*, 2005) was conducted on various samples of the recycled MSW rejects composite using stereomicroscopy and scanning electron microscopy. The material showed an intact morphology for the different sand concentrations with the existence of high volume fractions of voids of 6.3% for the 40% sand concentration, and 4.6% voids for the 60% sand concentration but with larger sizes in the form of cavities for 60% sand. For the 20% sand concentration, the volume fraction of voids was 2.3% which

TABLE 6.1
Mix-Design Matrix for Sample Production (Abou Khatwa *et al.*, 2005)

Mix no.	Sand content, % (by weight)	Sand sieve size, mm (ASTM)	Mixing temperature, °C
Mix 1	20	1.18	185
Mix 2	40	1.18	185
Mix 3	60	1.18	185
Mix 4	20	2.36	185
Mix 5	40	2.36	185
Mix 6	60	2.36	185
Mix 7	20	1.18	240
Mix 8	40	1.18	240
Mix 9	60	1.18	240
Mix 10	20	2.36	240
Mix 11	40	2.36	240
Mix 12	60	2.36	240

is less than that observed in the mixes prepared at a temperature of 185°C. However, a 60% sand concentration prepared at 240°C revealed a volume fraction of voids of 7% which is higher than that observed in the mixes prepared at 185°C. For more details see Abou Khatwa *et al.* (2005). A comparison between samples prepared at a temperature of 185°C and 240°C revealed a better mix homogeneity for the high mixing temperature. The dispersion of the sand particles in the polymer matrix was enhanced and there were fewer areas of polymer segregation in the composite material. In addition, polymer segregation was much more pronounced in mixes containing high sand concentrations as observed by the scanning electron micrographs of these segregated polymeric phases. It can also be observed that weak interfacial bonds existed between the segregated areas and the polymeric matrix as revealed from the inclusions present around the segregates. Moreover, the incomplete melting of the recycled rejects can also be observed, which resulted in lack of diffusion between the fragmented plastic wastes.

Mechanical properties

Compressive strength

The compressive properties for all the mixes as shown in Table 6.2 were evaluated using the standard test method ASTM D 695M-91 (1998). The compressive strength values in this case can only be used as an indication of the stability of the material investigated.

The average compressive strength values ranged between 10.2 and 22.2 MPa at a strain of 0.1. Mix 3 (60% sand, sieve 1, and temperature 185°C), mix 6 (60% sand, sieve 2, and temperature 185°C), mix 9 (60% sand, sieve 1, and temperature 240°C), and mix 12 (60% sand, sieve 2, and temperature

TABLE 6.2
Mechanical Properties of Prepared Mixes (Abou Khatwa *et al.*, 2005)

Mix no.	Compressive strength (MPa)	Flexural strength (MPa)	Shore D hardness no.
Mix 1	10.154	12.241	58.28
Mix 2	11.897	11.184	63.37
Mix 3	15.622	10.447	66.20
Mix 4	12.821	15.191	59.43
Mix 5	15.502	11.307	64.17
Mix 6	16.478	10.725	68.53
Mix 7	10.992	16.463	60.40
Mix 8	13.125	21.239	67.73
Mix 9	20.250	11.260	68.80
Mix 10	13.122	13.550	61.88
Mix 11	17.981	16.345	69.12
Mix 12	22.247	13.936	69.87

240°C) exhibited the highest compressive strength values ranging from 15.6 to 22.2 MPa, while mix 1 (20% sand, sieve 1, and temperature 185°C), mix 4 (20% sand, sieve 2, and temperature 185°C), mix 7 (20% sand, sieve 1, and temperature 240°C), and mix 10 (20% sand, sieve 2, and temperature 240°C) revealed the lowest compressive strength values ranging from 10.2 to 13.1 MPa. On the other hand, mix 2 (40% sand, sieve 1, and temperature 185°C), mix 5 (40% sand, sieve 2, and temperature 185°C), mix 8 (40% sand, sieve 1, and temperature 240°C), and mix 11 (40% sand, sieve 2, and temperature 240°C) had compressive strength values ranging from 11.9 to 18 MPa.

It can be noticed that there is a direct proportionality between the compressive strength value and the sand content as shown in Figure 6.6. This agrees with the behavior experienced with filled elastomeric systems, where the filler particles reinforce the matrix by diverting the path of rupture and hence increasing the energy required to propagate a crack (Holliday, 1966).

Another observation is the increase in the compressive strength values associated with the increase in the sand particle size as shown in Figure 6.6. This suggests better dispersion and wetting characteristics associated with the large filler particles leading to stronger interfacial bonds. Moreover, it is known that for particle-filled composites, as the size of the filler decreases, their surface area increases causing an increased particle–particle interaction leading to the formation of filler clusters (Nielson and Landel, 1994). Clusters represent weak points in the material as they detach easily upon load application creating voids and cavities. Moreover, particle clusters usually contain entrapped air and it is established that strength and modulus decrease with the increase in the amount of entrapped air (Nielson and Landel, 1994).

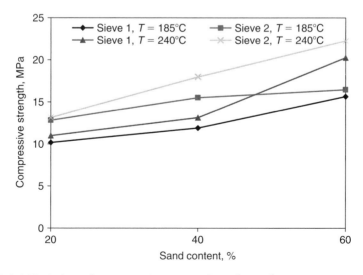

FIGURE 6.6 Variation of compressive strength with sand content, sieve size, and mixing temperature (Abou Khatwa *et al.*, 2005)

Figure 6.6 also indicated an increase in compressive strength values with the increase in mixing temperature. This resulted from the better mix homogeneity associated with high mixing temperature. Fewer polymer segregates were observed for mixes prepared at 240°C in comparison to the 185°C mixing temperature as revealed by the optical micrographs. Polymer segregates will split apart from the matrix upon load application forming microcracks at the interface, hence causing premature failure.

Following the ASTM specification C 936-96 (1998k), the average compressive strength for solid concrete interlocking paving units should not be less than 55 MPa for all test samples with no individual unit less than 50 MPa. However, it was shown during trafficking tests in South Africa that the behavior of block pavements was unaffected by changes in compressive strength ranging from 25 to 55 MPa (Shackel, 1990). Hence, mix 12 prepared at a temperature of 240°C with the large sand particles (sieve 2) at 60% sand content has the potential for use as interlocking paving blocks. Moreover, the same mix also satisfies the ASTM C 902-95 (1998j) specification for pedestrian and light traffic paving bricks entitling a minimum compressive strength value of 20.7 MPa.

Flexural properties

The flexural strength for all the mixes was determined by test method I of the ASTM D 790M-93 (1998). The results of all the flexure tests are summarized in Table 3.2. The average modulus of rupture of the prepared mixes ranged from 10.4 to 21.2 MPa. The ASTM standard C 410-60 (1998i) for industrial floor bricks specifies a minimum modulus of rupture of 13.8 MPa for type M and L industrial floor bricks, and 6.9 MPa for type T and H. About half the investigated mixes satisfied the requirement for type M and L, and all the mixes satisfied the type T and H minimum value. In addition, the modulus of rupture of the produced composite material was higher than the modulus of rupture of clay bricks ranging from 3.5 to 10.5 MPa (Cordon, 1979). These results show the potential use of the composite material in industrial flooring applications, as well as structural applications.

The effects of sand content, sieve size, and mixing temperature on flexural strength was investigated graphically in Figure 6.7. It is evident from the figure that for low mixing temperatures of 185°C, the flexural strength decreases with the increase in the sand content. On the other hand, for a high mixing temperature of 240°C, the flexural strength values increased with the increase in sand content from 20% to 40% with a maximum value for the 40% sand content of 21.2 MPa and 16.3 MPa for sieve size 1 and 2, respectively. However, increasing the sand content beyond this level resulted in a decrease in the flexural strength. This decrease in the flexural strength with the increase in sand content is attributed to the stress concentrations induced by the filler particles. This stress concentration will promote failure upon load application.

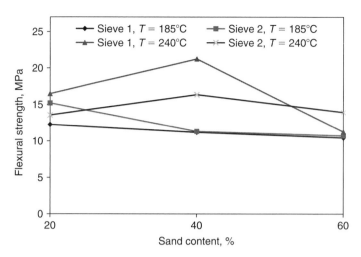

FIGURE 6.7　Variation of flexural strength with sand content, sieve size, and mixing temperature (Abou Khatwa *et al.*, 2005)

Figure 6.7 also showed an increase in flexural strength with the increase in mixing temperature. The reason behind this follows the same explanation discussed earlier for compressive strength; higher mixing temperature will result in a more homogeneous mix with fewer polymer segregates. Thus, the possibility of flaws and cracks decreases leading to an increase in flexural strength.

Durometer hardness

The indentation hardness of the mixes was evaluated using the ASTM standard test method D 2240-97 (1998). The use of this test method was intended only for comparison purposes and was selected due to its wide range of applications.

The results of the shore D hardness tests are presented in Table 6.2. The hardness values for all the investigated mixes ranged between 58 and 70. These numbers place the investigated composite material in the moderately hard plastics category having a shore D hardness ranging from 65 to 83 (Cordon, 1979). These hardness values also correspond to a value of 1 on Moh's scale of hardness implying a weak scratch resistance capability in comparison to most plastics that have a Moh's value ranging from 2 to 3.

The effect of sand content, sieve size, and mixing temperature on hardness was investigated graphically in Figure 6.8. It is clear that the hardness of the material increased with the increase in sand content. This is attributed to the increased hardness of the filler over that of the polymer matrix. An increase in hardness was also encountered with the increase of the sand particle size from sieve 1 to sieve 2. However, such increase was very small reaching a maximum of 3.5% between mix 3 (60% sand, sieve 1, and temperature 185°C)

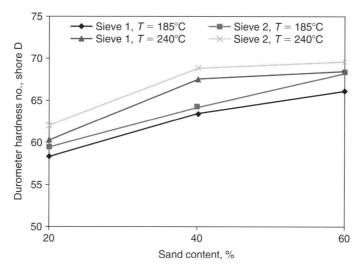

FIGURE 6.8 Variation of shore D hardness no. with sand content, sieve size and mixing temperature (Abou Khatwa *et al.*, 2005)

and mix 6 (60% sand, sieve 2, and temperature 185°C). The average increase in hardness values associated with increasing the sand particle size for mixes prepared at 185°C and 240°C were 2.25% and 2.02%, respectively. On the other hand, increasing the mixing temperature showed a more significant effect in increasing the hardness values. The average increase for mixes prepared with sieve 1 was 4.82%, while mixes prepared with sieve 2 revealed an average increase of 4.6%. The increase in hardness values associated with the increase in mixing temperature can be related to a better adhesion between the polymeric matrix and the sand particles.

Based on the above evaluated mechanical properties, mix 12 prepared at 60% sand content using sieve 2 at a mixing temperature of 240°C was selected as the optimum mix design regarding the proposed paving and tiling applications. This mix revealed the highest compressive strength and hardness values, together with a moderate flexural strength that remains higher than the requirements for clay and industrial floor bricks. Therefore, this mix will be the main focus for the further service properties evaluation.

Service properties

Abrasion resistance
The abrasion resistance of the bricks to Egyptian Standard for Cement Tiles, was determined in accordance with the Egyptian standard test method ES 269-1974 (1974). The testing apparatus consisted of a rotating disk, a specimen-mounting plate holder, a hopper, and distributor for feeding fresh abrasive. The loose abrasive utilized was quartz sand graded to pass a 0.9 mm (No. 25) sieve and retained on a 0.6 mm (No. 28) sieve.

The results for abrasion tests conducted on the group of samples prepared at 240°C using sieve 2 with different sand contents are shown in Table 6.3. The average loss in thickness ranged from 0.05 to 0.24 mm. Mix 10 (20% sand, sieve 2, and temperature 240°C) revealed the highest abrasion resistance value and this value decreased with increasing the sand content as shown in Figure 6.9.

TABLE 6.3
Abrasion Resistance Results of Prepared Mixes (Abou Khatwa *et al.*, 2005)

Mix no.	Sample no.	Weight before abrasion (g)	Weight after abrasion (g)	Bulk density (g/cm³)	Thickness loss (mm)	Average thickness loss (mm)
Mix 10	1	128	127	1.08	0.19	0.05
	2	164	164	1.08	0	
	3	156	156	1.08	0	
	4	122	122	1.08	0	
Mix 11	1	243	243	1.06	0	0.1
	2	183	182	1.06	0.2	
	3	177	176	1.06	0.19	
	4	179	179	1.06	0	
Mix 12	1	205	203	1.08	0.37	0.24
	2	196	195	1.08	0.19	
	3	209	208	1.08	0.19	
	4	217	216	1.08	0.19	

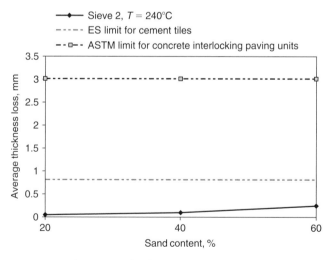

FIGURE 6.9 Variation of average thickness loss with sand content (Abou Khatwa *et al.*, 2005)

Abrasive wear is the natural consequence of the shearing of junctions between surfaces caused by the introduction of abrasive particles from outside; by using soft plastics, the abrasive particles will sink below the surface resulting in no further harm (Cordon, 1979). Thus, increasing the polymeric content on behalf of the filler reduces abrasive wear and seizure can be completely avoided. Moreover, when there is a weak adhesive bond between the filler and the polymeric matrix, fillers will increase the rate of wear of the material due to the ease with which the filler gets separated from the matrix. Hence, increasing the sand content will decrease the resistance to abrasion.

Regardless of the differences encountered in the average thickness loss between the mixes, the highest value of 0.24 mm for mix 12 (60% sand, sieve 2, and temperature 240°C) was 70% lower than the Egyptian standard limit for cement tiles, and 92% lower than the ASTM standard limit for solid concrete interlocking paving units. Hence, all mixes reveal an excellent abrasion resistance.

Density and water absorption

The density of all the prepared mixes was determined using procedure A of the ASTM standard test method D 792-91 (1998). The relative rate of absorption of water for all mixes was evaluated by the 24 hour immersion procedure of the ASTM standard test method D 570-95 (1998) for water absorption of plastics.

The average density and water absorption results of the different mixes investigated are shown in Table 6.4. All mixes resulted in low densities ranging between 1.12 and 1.68 g/cm³. The increase in the sand content is due to the high density of sand as opposed to that of a polymeric material.

Table 6.5 lists the maximum water absorption percentages entitled by the ASTM and the Egyptian standards for various types of bricks, tiles, and paving

TABLE 6.4
Average Densities and Water Absorption Results of Prepared Mixes
(Abou Khatwa *et al.*, 2005)

Mix no.	Density (g/cm³)	Water absorption (%)
Mix 1	1.12	1.6
Mix 2	1.27	0.71
Mix 3	1.63	1.51
Mix 4	1.13	0.7
Mix 5	1.29	0.94
Mix 6	1.59	1.1
Mix 7	1.12	1.04
Mix 8	1.39	0.64
Mix 9	1.59	1.29
Mix 10	1.16	0.98
Mix 11	1.41	0.83
Mix 12	1.68	0.85

TABLE 6.5
Maximum Water Absorption Percentages for Construction Bricks and Tiles (Abou Khatwa *et al.*, 2005)

Application	Standard specification	Water absorption (%)
Industrial floor brick	ASTM C410-60	1.5–12
Solid concrete interlocking paving units	ASTM C936-96	5
Pedestrian and light traffic paving brick	ASTM C902-95	11–17
Cement tiles	ES 269-1974	12
Recycled plastic waste composite material		0.64–1.6

units, compared to those produced by the composite material. It is evident from the table that even mix 1 (20% sand, sieve 1, and temperature 185°C), holding the highest average water absorption value over all the investigated mixes of 1.6%, easily satisfies all the standard requirements for industrial floor bricks, interlocks, paving bricks, and cement tiles. It is evident from Table 6.5 that the investigated plastic waste composite reveals superior water absorption resistance when compared to cementitious materials. However, the values are slightly higher than those expected for most plastics (less than 1% absorption). This could be related to the presence of imperfections in the prepared specimens, either in the form of cavities, minor waste constituents such as paper or pores resulting from the use of sand as filler material.

Resistance to chemical reagents
To assess the serviceability of the recycled plastic material when exposed to outdoor environment conditions, the ASTM standard practice D 543-95 (1998) for evaluating the resistance of plastics to chemical reagents was employed. Three reagents were selected for the tests, benzene, sulfuric acid (3%), and sodium hydroxide solution (10%). Testing was conducted for 7, 14, and 28 day's immersion.

Tables 6.6–6.8 list the average increase in weight over an immersion period of four weeks for the three reagents. It is apparent from the results that the material possesses excellent chemical resistance to acids and alkalis. The average increase in weight after 28 days was 1.37% and 3.38% for the sulfuric acid and the sodium hydroxide solution, respectively. However, the increase in weight after seven days was nearly half the values for the total immersion time. On the other hand, the increase in weight associated with the benzene reagent was much higher starting at a 5.59% increase after the first seven days and reaching 9.85% after 28 days. It is well known that most polymers exhibit very high resistance to chemical attacks by acids and alkalis. However, they are less resistive to organic solvents which react with the carbon atom chains

TABLE 6.6
Chemical Resistance of Samples Subjected to 10% Sodium Hydroxide Solution
(Abou Khatwa *et al.*, 2005)

Mix no.	Sample no.	Weight (g)				Increase in weight (%)		
		Initial	7 days	14 days	28 days	7 days	14 days	28 days
12	1	39.08	39.66	39.93	40.14	1.48	2.18	2.71
	2	33.99	34.75	35.05	35.26	2.24	3.12	3.74
	3	37.32	38.11	38.38	38.7	2.15	2.84	3.7
	Average increase in weight (%)					1.95	2.71	3.38

TABLE 6.7
Chemical Resistance of Samples Subjected to 3% Sulfuric Acid (Abou Khatwa *et al.*, 2005)

Mix no.	Sample no.	Weight (g)				Increase in weight (%)		
		Initial	7 days	14 days	28 days	7 days	14 days	28 days
12	1	30.34	30.58	30.73	30.74	0.79	1.29	1.32
	2	40.05	40.37	40.52	40.62	0.8	1.17	1.42
	3	37.9	38.22	38.4	38.42	0.84	1.32	1.37
	Average increase in weight (%)					0.81	1.26	1.37

TABLE 6.8
Chemical Resistance of Samples Subjected to Benzene (Abou Khatwa *et al.*, 2005)

Mix no.	Sample no.	Weight (g)				Increase in weight (%)		
		Initial	7 days	14 days	28 days	7 days	14 days	28 days
12	1	30.46	32.37	33.48	34.06	6.27	9.91	11.81
	2	33.5	34.66	35.28	35.81	3.46	5.31	6.9
	3	32.83	35.14	36	36.39	7.04	9.66	10.84
	Average increase in weight (%)					5.59	8.29	9.85

and their pending side groups. This explains the increase in absorption rates
for the benzene reagent. A visual analysis was also performed on the samples
and revealed no evidence of decomposition, discoloration, swelling, or cracking for the total immersion time.

Vicat softening temperature
The Vicat softening temperature was determined using the ASTM standard
test method D 1525-96 (1998). The Vicat softening temperature test was performed on three samples from the optimum mix 12. The results indicated a

softening temperature between 93 and 110°C, which is much higher than the expected service temperature of 60°C.

Environmental related test

The health hazards of the investigated waste material will be determined according to the water leaching test DIN 38414-S$_4$ (Kourany and El-Haggar, 2001). Fresh deionized water media were used to determine the heavy metals content and leaching characteristics of the produced bricks. In this test, three specimens measuring $50 \times 50 \times 50$ mm from the selected mixes were immersed in 300 ml of deionized water media for 28 days. At the end of the immersion period, the media were analyzed for chromium, cadmium, and lead concentrations using a Perkin Elmer SIMAA 6000 spectrometer. A blank sample of the deionized water was also tested to determine the existence of any background levels of heavy metals.

Table 6.9 lists the results of the leaching tests by water for 28 days together with the World Health Organization (WHO) guideline values (Kourany and El-Haggar, 2001) for heavy metal content in drinking water. The results indicate very low concentrations of heavy metals for the various mixes which are expected since most of the wastes are plastic rejects with only minor potential sources of heavy metal contamination from the garbage dumpsites. It is obvious that the samples are nearly free from any cadmium (Cd) and contain very small amounts of lead (Pb) and chromium (Cr). The only exception was mix 12 which revealed high concentrations of 2.2 μg/L and 3.2 μg/L, for lead and chromium, respectively. Nevertheless, all mixes showed concentrations far below the WHO guideline values, which proves the safety of using recycled plastic waste composite material.

In conclusion, the morphological analysis of the composite material reveals a general intact structure with the presence of polymer segregation phases and particle filler (sand) clustering phases. The tendency of polymer

TABLE 6.9
Heavy Metal Content by Water Leaching for 28 Days (Abou Khatwa *et al.*, 2005)

Mix no.	Element concentration (μg/L)		
	Lead (Pb)	*Chromium (Cr)*	*Cadmium (Cd)*
Mix 4	0.1	0.2	0
Mix 5	1.4	1.5	0
Mix 6	0	0	0
Mix 10	0.4	0	0
Mix 11	0.1	0	0
Mix 12	2.2	3.2	0.1
Blank sample	0.2	0.8	0
WHO guidelines	10	50	3

segregation decreases with increasing temperature and decreasing sand concentrations. Clustering of sand particles depends on the concentration of the filler employed as well as its size; the clustering tendency increases by increasing the filler concentration and decreasing the particle size.

The compressive strength was 7.25% higher than the ASTM limit for pedestrian and light traffic paving bricks; however, it failed to meet the ASTM limit for solid concrete interlocking paving units. The modulus of rupture was also higher than the standard limits for clay bricks and industrial floor bricks. The abrasion resistance of the composite material was 70% higher than the Egyptian standard limit for cement tiles allowing for its use in tiling applications. The composite material exhibited very low density and water absorption rate in comparison to cement tiles and paving units. However, such absorption rates may be detrimental in applications where the material is subjected to tensile loads since water is a plasticizer for many polymers and can result in a decrease in the elastic modulus. Water can also swell the polymeric waste creating stress concentrations. The softening temperature of the composite material was much higher than the maximum expected summer service temperatures. In addition, the material showed high chemical resistance to acidic and alkaline attack. Based on the compressive strength, flexural strength, abrasion, chemical resistance, and leaching tests of the composite material, it can be safely used to manufacture paving bricks, large tiles, or sheets for use in industrial facilities where physical requirements are not of great importance. The major problem with the composite material remains its susceptibility to environmental attack by ultraviolet radiation and microorganisms and further investigation of this matter together with tests for the leaching of organic compounds should be conducted.

6.5 Manhole

A manhole or maintenance hole, sometimes called an inspection cover is the top opening to an underground vault used to house an access point for making connections or performing maintenance on underground and buried public utility and other services including sewers, telephone, electricity, storm drains and gas. It is protected by a manhole cover designed to prevent accidental or unauthorized access to the manhole. Manhole base and cover are usually made out of case iron or Reinforced Fiber Plastic (RFP) or can be produced out of MSW rejects. The main advantages of manholes made out of MSW rejects are less cost, less energy consumption, durability and their resistance to acid. The weight can be adjusted by adding sand to the mix. Sand will increase the required specific weight up to a certain percentage (28%) to avoid any strength impact. The carrying load for a 40 cm manhole is 1.8 tons, which is slightly higher than FRP manholes and can be increased by adding steel bars.

A manhole cover is a removable plate forming the lid over the opening of a manhole, to prevent someone from falling in and to keep unauthorized persons out. Manhole covers usually weigh more than 100 lb (50 kg), partly because the weight keeps them in place when traffic passes over them, and partly because they are often made out of cast iron, sometimes with infills of concrete. This makes them expensive and strong, but heavy. They usually feature pick holes in which a hook handle is inserted to lift them up.

Manhole covers and manholes are round, but could also be designed to any shape according to the required dimensions. Round tubes are the strongest and most material-efficient shape against the compression of the earth around them. A circle is the simplest shape whereby the lid cannot fall into the hole. Circular covers can be moved around by rolling and they need not be aligned to put them back.

A manhole and its cover is a very important application to be produced from rejects, from social, economical, and technical points of view. Rejects can be used instead of ordinary cast iron with all the different geometries and dimensions as shown in Figure 6.10.

FIGURE 6.10 Sand-plast manholes

6.6 Breakwater

Water waves are one of the main problems facing coastal zones and shore protection. Reinforced concrete blocks and stones are always used as a breakwater to protect the shores and coastal zones. They cost a lot of money and should be replaced frequently depending on wave velocity and wave height. HDPE is being used to manufacture breakwater structures especially floating breakwaters (Whisperwave, 2005). Such technological advancement has been utilized to afford protection to marinas, beaches, and private property subject to destructive or annoying wave/wake forces. The design of the module enables it to be filled with or evacuated of water (with the help of a standard air compressor) to precisely adjust its buoyancy. The module can be "puncture proofed" by filling it with marine grade buoyant foam.

Accordingly, the most favorable scenario would be producing a shoreline erosion protection structure using plastic recycled material. As mentioned earlier, the main aim is utilizing solid waste rejects in producing shoreline erosion structures, preferably floating. The alternative material is made out of rejects mixed with sand to produce breakwater or floating breakwater by adjusting the percentage of sand without sacrificing the properties. The high percentage of plastics is always recommended to increase the lifetime of the sand-plast breakwater. The use of rejects in breakwater products might require further analysis for leachate to make sure there is no water contamination.

Breakwaters are generally shore-parallel structures that reduce the amount of wave energy reaching a protected area. Breakwaters are built to reduce wave action through a combination of reflection and dissipation of incoming wave energy. When used for harbors, breakwaters are constructed to create sufficiently calm waters for safe mooring and loading operations, handling of ships, and protection of harbor facilities. Breakwaters are also built to improve maneuvering conditions at harbor entrances and to help regulate sedimentation by directing currents and by creating areas with differing levels of wave disturbance. Protection of water intakes for power stations and protection of coastlines and beaches against tsunami waves are other applications of breakwaters (Burcharth and Hughes, 2003).

When used for shore protection, breakwaters are built in near-shore waters and usually oriented parallel to the shore like detached breakwaters. The layout of breakwaters used to protect harbors is determined by the size and shape of the area to be protected as well as by the prevailing directions of storm waves, net direction of currents and littoral drift, and the maneuverability of the vessels using the harbor. The cost of breakwaters increases dramatically with water depth and wave climate severity. Also poor foundation conditions significantly increase costs. These three environmental factors heavily influence the design and positioning of the breakwaters and the harbor layout (USNA, 2006).

Until World War II breakwater armoring was typically either made of rock or of parallel-epipedic concrete units (cubes) (Bakker *et al.*, 2003). Armor units were typically either randomly or uniformly placed in single or double layers. The governing stability factors are the units' own weight and their interlocking. Breakwaters were mostly designed with gentle slopes and relatively large armor units that were mainly stabilized by their own weight. A large variety of concrete armor units has been developed in the period 1950–1970.

Floating breakwater

Floating breakwaters are usually made of HDPE, recycled plastic, or rejects without sand content resulting in the following (Ibrahim, 2006):

- The zero percent sand content rejects float over water and have a very low absorption rate compared to other mixing ratios, 67% less than the following absorption rate of the 30% sand mix.
- The density of the zero percent sand content rejects is higher than the HDPE by 2%.
- The zero percent MSW reject mix is less stiff by 59% thus reacting better to sudden loads.
- The ultimate strength of the zero percent sand content rejects is higher than the lower limit of HDPE by 23%.

Accordingly, usage of rejects with zero percent sand mix in the production of a floating breakwater is recommended. However, foam injection could be investigated and might be utilized due to its superior properties for floating breakwater.

Breakwater

Breakwaters or armor units used in breakwaters as mentioned before are made of concrete and new improvements include the usage of HDPE in particular designs, so the following concerns were noticed while comparing MSW rejects with the concrete properties (Ibrahim, 2006):

- This usage requires higher specific gravity as illustrated in the 50% sand mix which is 33% higher than the 40% sand mix and 65% higher than the 30% sand mix.
- The modulus of elasticity of the 50% sand mix is 6% higher than the 40% sand mix and 8% higher than the 30% sand mix; temperature negatively impacts this property as the 40% sand mix modulus decreases by 9.5 and 11.5% respectively at 160 and 180°C.
- In regards to strength, the 40% mix is found to be 19% higher than the 50% sand mix and only 5% higher than the 30% sand mix; but temperature did not add much to this property.

- Specific gravity/density of the mixes is less than that of concrete by 25% (compared to the 50% sand mix, the highest density).
- The absorption is less than concrete by more than 95% (compared to the 50% sand mix, the highest density).
- The 40% sand mix is higher than the lower limit required for strength of concrete by 17%; the 50% sand mix is lower than the lower limit by 2.2%; however, temperature increase would only lead to further deterioration of the property as the 40% sand mix decreased by 18% at an increase in temperature of 20°C.

While the 50% sand mix is superior in terms of specific gravity and stiffness modulus, the compressive strength is more important. So the 40% sand mix is suitable, while increasing the temperature to increase specific gravity. To overcome the weight issue, usage of perforated armor units should be further investigated. It should be noted that the specific gravity of irregular shapes is 5% less than compacted MSW reject paste.

Beach/canal revetment

Revetment tiles are used to stabilize beaches and river/canal banks. The revetments are sometimes made of rock, concrete bricks, or molded bricks, so conclusions on usage were based on comparison of the MSW rejects to the concrete properties. The strength of the 50% sand mix is compared to data for loading perpendicular to the plane, and the result is higher than the required strength of pavement tiles by a minimum of 30%. In addition to the superior performance a wet/dry test revealed a very slight weight gain of 0.001% overcoming the specified loss of 0.5%. Thus, the 50% sand mix is suitable for such usage (Ibrahim, 2006).

6.7 Other Products

Table toppings

After improvements were made to interlocks, manholes, and breakwaters the idea of table topping emerged. A mold was manufactured with the required dimensions to produce the topping of the table as shown in Figure 6.11. Polyurethane coating is an ideal coating to use to improve the appearance and protect the environment from any emissions as shown in Figure 6.12.

A grani* top layer, a mixture of rocks and granite granules from industrial wastes of rocks and marble industry, can be used to improve the quality and appearance of the top layer as shown in Figures 6.12 and 6.13. Rocks are

---

*Grani is a mixture of granite powder and an organic binder

FIGURE 6.11 Table topping from rejects

 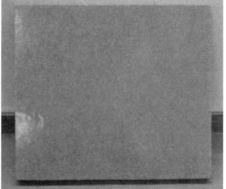

FIGURE 6.12 Different top layers grani industrial waste (mixture of granite powder with organic binder)

crushed using ball crushers (all mill) followed by a process of sieving mixed with organic polymer and poured in to another mold with a 1 cm space all around.

Table toppings proved to be very competitive. The table topping had many coatings, each coating was used in a different application and the grani layer has proved to be the best due to its durability and appearance. This led to the production of plates with thicknesses ranging from 10 mm to 30 mm used for fences, garden furniture, etc.

FIGURE 6.13 Final shape of the table

FIGURE 6.14 Wheels of different sizes

Wheels

Wheels are another application of the rejects as shown in Figure 6.14. The wheels are used instead of standard wheels in slow moving cars. The cost of these wheels is 10% of the standard wheels. However, one disadvantage is

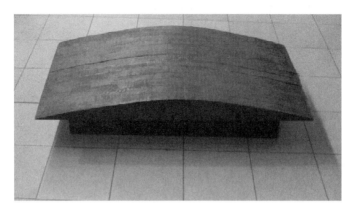

FIGURE 6.15 Road ramp

that these wheels tend to embed themselves in unstable ground when carrying heavy loads, so they are best used on paved roads.

Road ramp

Road ramps are used to control vehicle speeds in residential areas, schools and hospitals. Traditionally, road ramps are made from cast iron, rubber, or asphalt mix. The new material made out of sand-plast mix can be used to replace these traditional materials as shown in Figure 6.15. It provides an excellent alternative from technical, economical, and environmental points of view.

Questions

1. Estimate the quantity of MSW generated in your country and state the recycling percentage of the total MSW.
2. Explain the method of disposing of the MSW in your country and how much it costs per 1 ton of waste.
3. Discuss the types of MSW disposed in your country as well as the quantity of such waste.
4. What are rejects?
5. Compare between a manhole cover and base produced in your country and that produced from rejects, from technical, social, economical, and environmental points of view.
6. What type of products do you recommend to be produced from rejects according to the demand and need concept in your community?
7. Explain the rejects' recycling technology. Develop a feasibility study, technical study, and environmental study to develop such rejects' recycling plant.

Chapter 7

Sustainability of Agricultural and Rural Waste Management

7.1 Introduction

Historically, energy supply was based on biomass and forest products. Despite the shift to the conventional fossil fuel in many parts of the world, the traditional fuel wood and charcoal still persist in many parts of the world, especially in Asia and Africa where the majority still live in rural areas. Today, fuel wood is of marginal importance to the industrialized world, where fuel wood consumption is an insignificant percentage of the world total energy. It is estimated that in the developing countries as a whole, wood accounts for 21% of the total energy consumption; and in Africa as much as 58% (ABC Hansen, 2001). While using wood for fuel is becoming more common for various economic reasons the fact that replanting (afforestation) is not occurring at a sufficient pace is invariably making fuelwood an increasingly limited energy resource.

In many parts of the world, tons of agricultural wastes in the form of straws, shells, stalks, husks, wood and forest residue, etc. are disposed of by burning in fields. This is usually so mainly because other economical and efficient uses have not been identified for the vast volume of residue produced. However, burning of agricultural waste is increasingly being discouraged for environmental, ecological, and health reasons. Agricultural wastes are recognized as having "hidden" economic value. About 1.2 billion tons of oil equivalent (approx. 15% of the world energy consumption) is assumed to be hidden in unutilized biomass reserves including agricultural wastes (ABC Hansen, 2001).

Some of the uses found for plant residues include: mulch for soil cover; compost for soil conditioner; fodder for animals; construction/building materials, e.g. roofing in the rural areas, strengthening fiber/particle board for panel doors and furniture, and insulating materials in walls and ceilings.

Other uses of agricultural residue include direct burning as fuel for domestic and industrial cooking/heating and production of biogas and biomass for power generation. In addition to its use in raw forms as fuel, agricultural wastes are a form of renewable biomass and can be processed into other solid forms (e.g. briquettes, charcoal, pellets) or into liquid fuel through pyrolysis or gaseous fuel through gasification or biogas. It is hoped that by putting agricultural residues into different uses in the developing countries, jobs can be created and earnings guaranteed. This may go a long way in alleviating environmental problems that come with indiscriminate disposal.

The amount of agricultural waste generated in developed and developing countries is huge and causes a lot of environmental pollution if it is not properly utilized. It also represents a very important natural resource which might provide job opportunities and valuable products. Some of the agricultural waste is used as animal fodder, others are used as a fuel in very primitive ovens causing a lot of health problems and damage to the environment, and others still are burnt in fields causing air pollution problems. The type and quantity of agricultural waste in developing countries changes from one country to another, from one village to another and from one year to another because farmers always look for the most profitable crops suitable for the land and the market. The main agricultural waste with the highest amount generated are the rice straws, corn stalks, wheat/barley straws, cotton stalks, bagasse, etc.

Agriculture biomass resources in developing countries are huge. Fifty percent of the biomass is used as a fuel in rural areas by direct combustion in low efficiency traditional furnaces. The traditional furnaces are primitive mud stoves and ovens that are extremely air polluting and highly energy inefficient.

One of the main agricultural wastes is a cotton stalk. The available amount of cotton stalk in Egypt is estimated at 1.6 million ton/yr (corresponding to 740,000 ton oil equivalent "TOE"/y). Egyptian Ministry of Agriculture regulations and resolutions commit farmers to dispose of the cotton plant residues through environmentally safe disposal methods immediately after harvesting (within 15 days). The easiest and cheapest method is to burn the cotton stalks as soon as possible in the field. The reason behind this regulation is to kill insects and organisms carrying plant diseases. The cotton stalks will be stored for a long time giving the chance for the cotton worms to complete the worm life cycle and attack the cotton crop in the next season. Such a process leads to a total energy loss estimated at 0.74 mTOE/y which accounts for its high money value in addition to the negative environmental impacts due to releasing vast amounts of greenhouse gases.

Moreover, the traditional storage systems for plant residues in the farms and roofs of the buildings in rural communities gives the opportunity for insects and diseases to grow and reproduce as well as being a fire hazard.

7.2 Main Technologies for Rural Communities

Different agricultural residues have varying physical and chemical composition or properties which make them suitable for different applications.

Application may also depend on location and type of economic activities in the region. The organic matter content tends to give priority to using agricultural wastes as compost and as animal fodder respectively over their use as fuel. Other compositions such as cellulose, hemicellulose, and lignin contents qualify as agricultural residues for use in the production of chemicals, resins, and enzymes (Khedari *et al.*, 2003). While some agricultural residues have found priority in some applications, it is believed that surplusses usually exist and can be processed into solid-fuel briquettes.

The four cornerstone technologies for agricultural waste suitable for rural communities are animal fodder, briquetting, biogas, and composting (ABBC technologies). These technologies can be developed based on demand and needs. In principle three agricultural waste recycling techniques can be selected to be the most suitable for the developing communities. These are animal fodder and energy in a solid form (briquetting) or gaseous form (biogas) and composting for land reclamation. There are some other techniques, which might be suitable for different countries according to their needs such as gasification, fiber boards, chemicals, silicon carbide, pyrolysis, etc. These techniques might be integrated into a complex as will be discussed later in this chapter which combine them together to allow 100% recycling for the agricultural waste according to a cradle-to-cradle approach. Such a complex can be part of the infrastructure of every village or community. Not only does it get rid of the current practice of disposing of harmful agricultural waste, but it is also of great economical benefit.

The amount of agricultural waste varies from one country to another according to type of crops and farming land. This waste occupies the agricultural lands for days and weeks until the farmers get rid of it by either burning it in the fields, or storing it in the roofs of their houses, both practices affecting the environment by allowing fires to start and diseases to spread. The main crops responsible for most of these agricultural wastes are rice, wheat, cotton, corn, etc. These crops were studied and three agricultural waste recycling techniques were found to be the most suitable for these crops. The first technology is animal fodder that allows the transformation of agricultural waste into animal food by increasing the digestibility and nutritional value. The second technology is energy, which converts agricultural wastes into energy in a solid form (briquetting) or gaseous form (biogas). The briquetting technology that allows the transformation of agricultural waste into briquettes can be used as useful fuel for domestic or industrial furnaces. The biogas technology can combine both agricultural waste and municipal wastewater (sewage) in producing biogas that can be used for generating electricity, as well as producing organic fertilizer. The third technology is composting, which uses aerobic fermentation methods to change agricultural waste or any organic waste into soil conditioner as explained in Chapter 5. The soil conditioner can be converted into organic fertilizer by adding natural rocks to control the nitrogen:potassium:phosphorus (NPK) ratio as will be explained in the case studies. Agricultural waste varies in type, characteristics,

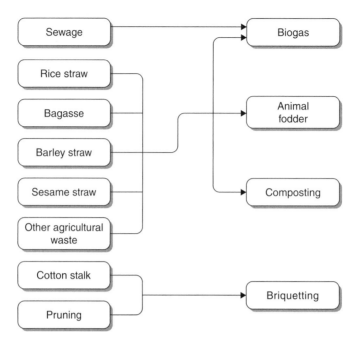

FIGURE 7.1 Matching diagram between output technology and agricultural/rural waste

and shape, thus for each type of agricultural waste there is an appropriate technique as shown in Figure 7.1.

A complex combining these four techniques is very important to guarantee each waste has been most efficiently utilized in producing beneficial outputs like compost, animal food, briquettes, and electricity. Having this complex will not only help the utilization of agricultural waste, it will help solve the rural community problems that face most of the developing countries and some parts of developed countries, as a certain percentage of the sewage will be used in biogas production and composting techniques to adjust the carbon:nitrogen ratio and water content. An efficient collection system should be well designed to collect the agricultural waste from the land and transport it to the complex in the least time possible to avoid having these wastes occupying valuable agricultural land. These wastes are to be shredded and stored in the complex to maintain a continuous supply of agricultural waste to the system and in turn continuous output in terms of energy, animal fodder, and organic fertilizer.

7.3 Animal Fodder

Using agricultural waste as animal feed, fish feed, or as constituent in feed preparation is a waste to wealth initiative. However, many agricultural wastes are unsuitable for direct consumption by animals as they need to be treated mechanically and chemically to make them edible. Roughage and fiber residue are often low in nutritional value and need supplements to enrich them.

Ojewole and Lange (2000) substituted (up to 30%) maize for cowpea hull and maize offal residue in chicken feed. This reduced the feed cost (due to escalating cost of maize) and the yield (egg weight) also improved. Experimental study of feeding Tilapia fish (Otubusin, 2001) with agro-industrial waste – corn bran, rice bran, and brewers' waste – in single and in combination showed that corn bran:rice bran (1:1) gave best fish weight, best feed conversion ratio, and specific growth rate.

The deficiency of animal foodstuff in developing countries causes raw material to be imported with inherent high cost and reduction in animal production. Transforming some of these wastes into animal foodstuffs will help a great deal in overcoming this deficiency. These wastes have high fiber content that makes them difficult to digest. The size of the waste in its natural form might be too big or tough for the animals to eat. To overcome these two problems several methods were used to transform the agricultural waste into a more edible form with a higher nutritional value and greater digestibility.

Mechanical and chemical treatment methods were used to transform the shape of the roughage (waste) into an edible form. The further addition of supplements can enrich the foodstuffs with the missing nutritional contents. The mechanical treatment method consists of chopping, shredding, grinding, moistening, soaking in water, and steaming under pressure. The mechanical method has been proved to give good results with high digestion by animals but they were never widespread because of high cost and therefore were unfeasible for small farms.

The chemical treatment method with urea or ammonia is more feasible than the mechanical treatment method. The best results were obtained by adding 2% ammonia (or urea) to the total mass of the waste. It is recommended to cover the treated waste with a wrapping material usually made from polyethylene at a 2 mm thickness. After 2 weeks (summer) and 3 weeks (winter), the treated waste is uncovered and left for 2–3 days to release all the remaining ammonia before use as animal feed.

7.4 Briquetting

Agricultural wastes burn so rapidly that it is difficult to maintain a steady fire due to difficulty in controlling the combustion process. Also, wastes do not fit in form and structure for traditional coal pots and stoves. While recycled wood wastes had found some use as fuel by burning them directly in retrofitted industrial boilers, direct burning of loose bulky agricultural wastes is inefficient. They have low energy value per volume and hence are uneconomical; they also cause problems for collection, transportation, storage, and handling.

One of the approaches that is being pursued in some parts of the world, for improved and efficient utilization of agricultural residues, is their densification into solid fuel pellets or briquettes. This involves reducing the size by pressing the bulky mass together. The ease of storing and transporting such an improved solid fuel briquette (usually in log form) of high specific weight makes them attractive for use at home and in industry. Unlike the loose and bulky form, combustion of briquettes can be more uniform. This could make it possible for briquetted materials to be burned directly as fuel in somewhat similar fashion as the fuel wood and coal in domestic (perhaps retrofitted) stoves and ovens. Some developing countries, e.g. India, Thailand, and a few places in Africa, have had experience of substituting fuel briquettes for fuel wood and coal to reduce the problems of firewood shortage and farmwaste disposal (Bhattacharya *et al.*, 1989).

Briquetting improves the handling characteristics of the combustible material, increases the volumetric value, and makes it available for a variety of applications – domestic and industrial. Materials that can be briquetted and used as fuel in industry are not only limited to agricultural wastes. There is a combination of varied forms of material including waste wood, sawdust, agro-industrial residue, plastic, rubber, and various other forms of combustible material which can be compressed by powerful industrial press machines.

The briquetting process is the conversion of agricultural waste into uniformly shaped briquettes that are easy to use, transport, and store. The idea of briquetting is using materials that are unusable, due to a lack of density, and compressing them into a solid fuel of a convenient shape that can be burned like wood or charcoal. The briquettes have better physical and combustion characteristics than the initial waste. Briquettes will improve the combustion efficiency using the existing traditional furnaces, in addition to killing all insects and diseases as well as reducing the destructive fire risk in the countryside. Therefore, the main advantages of briquetting are that they:

- Get rid of insects
- Decrease the volume of waste
- Produce efficient solid fuel of high thermal value
- Have low energy consumption for production
- Protect the environment
- Provide job opportunities
- Are less risk hazardous.

The raw materials suitable for briquetting are rice straws, wheat straws, cotton stalks, corn stalks, sugar cane waste (bagasse), fruit branches, etc. However, in the suggested complex explained later in this chapter cotton stalks and fruit branches are best utilized by briquetting. The briquetting process starts with collection of wastes followed by size reduction, drying, and compaction by extruder or press.

Briquetting quality parameters

Different agricultural residues have different structural and chemical properties. Briquetting agricultural wastes for fuel is meant to improve the residue value as well as environmental criterion; burning them in the field is being discouraged. Properties of the residue and briquetting process determine briquette qualities – combustion, durability, stability, etc. Among the parameters with which briquette quality is measured are bonding or compressive strength, porosity, density, calorific value, and ash content.

Among the variable parameters that have been investigated by various authors (El-Haggar *et al.*, 2005) on various residues that thrive in different localities are briquetting applied pressure, the material's moisture content, particle size, and temperature.

Applied pressure influences briquette density; the higher the density, the higher the calorific value in kJ/kg. High pressure is assumed accompanied by some inherent rise in temperature. Ndiema *et al.* (2002) stated that when the temperature of the material to be briquetted is elevated (preheat) beyond the natural state, a low pressure would be required for densification.

Increase in density, however, reduces ease of ignition (i.e. pre-combustion) of the solid fuel; increasing density reduces porosity. The particle size of the material could have an effect on the resulting briquette density and compressive strength. The nature of plant residue suitable for briquettes is categorized into fine, coarse, and stalk types (Tripathi *et al.*, 1998).

The level of moisture in the material at compression is an important processing parameter. The significance of moisture content on biomass compaction was reported by numerous researchers (Faborode and O'Callahan, 1987; Hill and Pulkinen, 1988). Excess moisture or inadequate drying of residue decreases the energy content of the briquette. Studies revealed that briquetting agricultural residues within a range of moisture content could improve a briquette's stability, durability, and strength. On the other hand, excess moisture could hinder briquette processing, lead to poor briquettes and increase energy requirement for grinding or drying the material.

Another important quality determinant is the presence or absence of binding material. Briquetting is done either with binder or is binderless. A binding agent is necessary to prevent the compressed material from "springing" back and eventually returning to its original form. In binderless briquetting, applied pressure and temperature ooze out the natural ligneous material (binder) present in the material which helps in bonding.

When a residue lacks the natural lignin that helps in bonding (or the percentage of lignin is low) the introduction of a binder will be necessary to improve briquette quality. However, appropriate selection and amount of binder need to be made in order to prevent smoke, or emission of volatile material that negatively impacts humans and the environment. Also, material that lacks the natural binder can be mixed with those that have. Materials with the natural binder include cotton stalk, saw dust, corn stalk, among others. Some artificial binders include tar, starch, molasses, or cheap organic materials.

In conclusion, briquette quality can be determined by the following:

- Stability and durability in handling, transportation and storage; these are measurable by changes to the weight, dimension, and ultimately the relaxed density and strength of the briquettes.
- Combustion (energy value) or ease of combustion, and ash content.
- Environmental concern, i.e. the toxic emissions during burning.

Parameters that determine briquette quality are:

- Pressure and/or temperature applied during densification.
- Nature of the material:
  - Structure (e.g. size, fibrous, non-fibrous, etc.)
  - Chemical (e.g. lignin-cellulose content)
  - Physical (e.g. material particle size, density, and moisture content)
  - Purity (e.g. trace of element (sulfur), etc.).

Parameters that determine stability and durability are:

- Compressive strength, impact strength.
- Compressive time.
- Relaxation: Moisture, length, density (post-briquetting parameter).

Briquetting process

Apart from the inherent properties of the raw material (agricultural waste), the briquetting process could also have an effect on briquette quality (Ndiema *et al.*, 2002). Briquettes from different materials or processes differ in handling and combustion behavior; briquettes from same material under different conditions can have different qualities or characteristics. Moreover, the feed material, the storage conditions, the briquette geometry, its mass and the mode of compression all have a bearing on the stability and durability of briquettes (Ndiema *et al.*, 2002).

Briquettes with low compressive strength may be unable to withstand stress in handling, e.g. loading and unloading during transfer or transportation. Stability and durability of briquettes also depend on storage conditions. Storing briquettes in high humid conditions may lead to briquettes absorbing moisture, disintegrating and subsequently crumbling. This disintegration is sometimes referred to as relaxation characteristic. The briquetting process may be responsible for briquette relaxation. Drying may be accompanied by shrinkage; expansion (increase in a briquette's length or width) is also possible.

The briquetting process primarily involves drying, grinding, sieving, compacting, and cooling. The components of a typical briquetting unit are (1) preprocessing equipment; (2) material handling equipment; and (3) briquetting press. Preprocessing equipment includes a cutter/clipper, and drying equipment (dryer, hot air generator, fans, cyclone separator, and drying unit). Among material handling equipment are screw conveyors, pneumatic conveyors and holding bins.

In briquetting agricultural residue (or a blend of residues) for fuel purposes, optimum combinations of parameters that meet desired briquette qualities for a particular application (domestic or industrial fuel) should be the target. Effort needs to be made to determine a set or a range of parameters (moisture content, particle size, and applied pressure or/and temperature) which can bring about optimum or desired briquette quality (combustion, durability and stability, smoke/emissions level).

Briquetting technology

Studies on briquette production cover availability of agricultural wastes (husks, stalks, grass, pods, fibers, etc.) and agro-industrial wastes, and the feasibility of the technology and processes for converting them into briquettes in commercial quantity. The technology used to compress the biomass or agricultural waste are piston, screw extruder, pellet presses, and hydraulic presses.

Much research has investigated the optimum properties and processing conditions of converting agricultural residues (either alone or in combinations with other materials), with or without binders, into quality fuel briquettes. The desired qualities for briquettes as fuel include good combustion, stability and durability in storage and in handling (including transportation), and safety to the environment when combusted. Measures of these properties include the energy value, moisture content, ash content, density or relaxed density, strength, ease of ignition, smoke and emissions.

In piston presses, pressure is applied by the action of a piston on material packed into a cylinder against the die. They may have a mechanical coupling and flywheel or use hydraulic action on the piston. The hydraulic press usually compresses to lower pressures.

In the screw extruder, pressure is applied continuously by passing the material through a cylindrical screw with or without external heating of the die and conical screws. The heat helps in reducing friction and the outer surface of the briquette is somehow carbonized with a hole in the center. In both piston and screw technology, the application of high pressure increases the temperature of the biomass, and the lignin present in the biomass is fluidized and acts as binder (Tripathi *et al.*, 1998).

In the pellet presses, rollers run over a perforated surface and the material is pushed into the hole each time a roller passes over. The dies are made out of either rings or discs. Other configurations are also possible. Generally, presses are classified as low pressure (up to 5 MPa), intermediate (5–100 MPa), and high pressure (above 100 MPa).

Al Widyan *et al.* (2002) examined parameters for converting olive cake (12% moisture) into stable and durable briquettes; olive cake being an abundant residue byproduct of olive oil extraction in Jordan. Durability and stability were believed to be influenced by briquetting pressures and moisture content of the material.

Cake of varying moisture was compacted into a 25 mm diameter cylindrical shape by hydraulic press under varied pressures (15–45 MPa) and dwell times (5–20 seconds). Through Design of Experiment (DOE), and Analysis of Variance (ANOVA), significance of applied pressure, moisture content, and dwell time were tested. A briquette's stability was expressed in terms of relaxed density (mass to volume ratio) of the briquette after sufficient time (about 5 weeks) had passed for their dimensions (diameter and length) to stabilize. For the relative durability test each briquette was dropped four times from a height of 1.85 meters onto a steel plate. Durability was taken as the ratio of final mass retained after successive droppings. The method was noted as unconventional; relaxed density was taken as a better quantitative index for stability.

Ndiema *et al.* (2002) carried out an experimental investigation of briquetting pressure on relaxation characteristics of rice straw using a densification plunger at differing pressures between 20 and 120 MPa. Relaxation characteristics were taken as percentage elongation and fraction void volume of sample at time *t* after briquette ejection from the die. Laboratory condition was between 50 and 60% relative humidity. Time *t* was fixed at 10 seconds and 24 hours after ejection from the die. Both expansion and fraction void volume were noted to decrease with increase in die pressure until a die pressure of about 80 MPa was reached. Beyond 80 MPa compression, no significant change in briquette relaxation was noticed. The study concluded that for a given die size and storage condition, there often exists a maximum die pressure beyond which no significant gain in cohesion of the briquette can be achieved.

7.5 Biogas

Another use of plant residue is biogas[1] production. Biogas is a good source of energy. Certain types of agricultural waste including rice straws, wheat straws, malt straw, ground cotton stalk, and corn stalk can be used for biogas. Biogas has been found suitable as fuel in internal combustion (IC) engines, and can provide opportunities for small power generation to meet the needs of areas which are yet to be connected to the grid. It is estimated that the gas from biomass is capable of substituting conventional fuels required in India to the

---

1. Biomass gasification is conversion of solid biomass (wood waste, agricultural residues, etc.) into a combustible gas mixture normally called producer gas (methane). Biogas is a mixture of gases mainly methane and carbon dioxide that result from anaerobic fermentation of organic matter by the action of bacteria. The process involves subjecting duly dried agro-residue to thermal decomposition in the presence of limited air.

extent of 60–70%. The methane can have a high calorific value of approx. 9,000 kcal/m^3. The net heat value of biogas is approx. 4,500 to 6,300 kcal/m^3. Biogas is not yet popular in many developing countries.

Thakur and Singh (2000) carried out bench scale studies on some agricultural wastes and weeds to assess their possible use for biogas production. It was found that banana stem was more suitable giving 95 L/kg total solid (TS) compared to cow dung, 70 L/kg TS. Biogas from maize stone, paddy straw, wheat straw, bagasse, water hyacinth, *cannabis sativa*, *Croton sparsiflorus* and *Parthenium hysterophorus* gave 72–80 L/kg TS. A higher retention time of 45 days was required for maximum methane production for the wastes and weeds.

Biogas is the anaerobic fermentation of organic materials by microorganisms under controlled conditions. Biogas is a mixture of gases mainly methane and carbon dioxide that results from anaerobic fermentation of organic matter by the action of bacteria. Biogas is ranked low in priority in some developing countries because of lack of energy policy. Most developing countries have no energy policies to utilize biogas and realize its potential of being a significant part of the country's total energy production.

Huge amounts of organic wastes are generated in rural communities such as agricultural waste, municipal solid waste, sludge from municipal treatment plants, and organic waste from garbage, food processing plants as well as animal manure and dead animals. All these can be considered as a biomass that is an organic carbon-based material, which could be an excellent source for biogas and fertilizer.

7.6 Composting

Composting is the aerobic decomposition of organic materials by microorganisms under controlled conditions. The process improves organic waste and kills pathogen organisms in the organic waste product in order to produce a rich soil. In 1876 Justus von Liebig (Epstein, 1997), a German chemist, had figured out that northern African lands that were supplying two thirds of the grains consumed in Rome were becoming less fertile, losing their quality and productivity. After brief research, he knew the reason behind such a phenomenon. It was due to the fact that when crops were exported from North Africa, their waste never returned; on the contrary it was flushed into the Mediterranean. The agricultural waste that comes from rice, cotton, corns, etc. are rich in organic matter. This matter is given by the soil and now the soil wants it back in order to continue producing healthy crops. However, this was never the case. In von Liebig's opinion this was breaking the natural loop that should give the land back its nutrients. He named this phenomenon "direct flow". As a consequence, von Liebig began producing artificial fertilizers. Although artificial fertilizers were meant to compensate the soil for its loss of organic matter, they were never the same as natural fertilizers.

Organo-mineral fertilizers, a product from composted animal and crop residue (poultry manure, cow dung, sawdust, shear nut cake, and palm kernel cake) and sorted city refuse enriched with local mineral improved maize proved to be an effective waste management strategy (Ogazi *et al.*, 2000). The degradable city refuse reduced the composting cycle from 84 days to 55 days, and the application of the fertilizers on maize/cassava intercrop increased maize yield by 60% over zero fertilizer plot, and 20% over mineral fertilizer plot. Cassava yield improved by 200% over zero fertilized plot, and 40% over mineral fertilizer plot.

Composting is one of the more popular recycling processes for organic waste to close the natural loop. The major factors affecting the decomposition of organic matter by microorganisms are oxygen and moisture. Temperature, which is a result of microbial activity, is an important factor too. The other variables affecting the process of composting are nutrients (carbon and nitrogen), pH, time, and physical characteristics of raw material (porosity, structure, texture, and particle size). The quality and decomposition rate depends on the selection and mixing of raw material.

Aeration is required to recharge the oxygen supply for the microorganisms. The passive composting method (El-Haggar, 2003a) is the recommended technique for developing communities from the technical and economical factors as explained before in Chapter 5. The main advantages of composting are the improvement of soil structure by adding the organic matter and pathogens structure as well as utilizing the agricultural wastes that can cause high pollution if burned.

Because compost materials usually contain some biologically resistant compounds, a complete stabilization (maturation) during composting may not be achieved. The time required for maturation depends on the environmental factors within and around the composting pile. Some traditional indicators can be used to measure the degree of stabilization such as decline in temperature, absence of odor, and lack of attraction of insects in the final products.

Compost can be adjusted by adding natural rocks such as phosphate (source of phosphorus), feldspar (source of potassium), dolomite (source of magnesium), etc. to produce organic fertilizer for organic farming (El-Haggar *et al.*, 2004b and c). Organic farming results in better taste, no effect on people's health, and it is less harmful to the environment. Organic farming seeks to reduce external cost, produce good yields, save energy, maintain biodiversity, and keep soil healthy. For more details see case studies.

7.7 Other Applications/Technologies

Construction industry

Rice byproducts have been utilized in pulp and paper production as well as other applications such as erosion control, enhancing the geo-technical properties of soil, as a supplementary material for cement mortars, and for housing applications. India, Pakistan, China, and other countries that have high production of rice utilize rice straw and rice husks as an alternative raw material not only as a fuel alternative but also as an additive to building materials and other uses.

Many other uses have been found for agricultural residue especially in the building and construction industry. Rice husk ash (RHA) has been found to be silica rich with pozzolanic behavior that reacts actively with lime and water yielding hydraulic cements (Boateng, 1990; Chandrasekhar *et al.*, 2002). Proportioning RHA and lime or RHA with ordinary Portland cement can yield mortar of acceptable compressive strength but takes longer to cure.

Tests on concrete made from coarse artificial aggregate of palm kernel shell (Okpala, 1990) obtained from palm nut showed a density range of 1,450–1,750 kg/m^3 which classifies it as adequate lightweight structural concrete based on American Standards for Testing Materials (ASTM); the compressive strength increased with longer curing periods of up to 90 days. The concrete had good sound absorption capacity and low thermal (heat) conductivity.

An experimental investigation on peanut shell ash mortars (concrete) classified it as class C pozzolana (Nimityongskul and Daladar, 1995) according to ASTM standards, with some resistance against acidic attack.

Okpala (1993) experimented with partial replacement of cement with rice husk ash in production of sandcrete blocks – a major cost component of most common buildings in Nigeria. The study found that rice husk ash has a specific gravity of 1.54; and that the chemical and constituent met BS 3892 and ASTM for pozzolanas. Sandcrete mix (cement:sand) of 1:6 ratio with up to 40% cement replacement and 1:8 ratio with cement replacement up to 30% were found adequate for sandcrete block productions suitable for building in Nigeria.

In a similar study (Mannan and Ganapathy, 2002), it was observed that the concrete formed when oil palm shell was included in the aggregate had sufficient strength to qualify it as structural lightweight concrete.

The lignocellulosic[2] characteristic in woody and non-woody agricultural wastes (sugarcane bagasse, cereal straw, cotton stalks, rice husks, etc.) was tested for the manufacture of binderless panel boards (Shen, 1991). These panels can replace expensive, synthetic, petrochemical-based resin adhesives commonly used. The elimination of formaldehyde emission also makes the panels particularly appropriate for indoor use.

---

2.   The hemicellulose portion of the lignocellulosic material is first converted to low molecular weight water-soluble carbohydrate and other decomposition products as the resin binder through thermal hydrolysis. By subjecting the lignocellulosic material to heat and pressure to polymerize and thermoset the resin binder in situ the binderless panel board produced had strong mechanical strength; good dimensional stability; and was boil-proof such that it surpassed the Canadian national standard of exterior grade for construction use.

The result of pyrolyzed and leached rice husk showed that the resultant product – 40% carbon and 56% silica – can be used as filler or reinforcement in rubbers (Jain *et al.*, 1994). The test revealed a tenfold increased tensile strength and modulus of elasticity with 100 bar. An electronic grade potassium silicate chemical can also be produced when the leached char is digested with 10–15% KOH solution at 303–373 K for about 1–10 h.

A study by Raghupathy *et al.* (2002) on paperboard production using arecanut leaf sheath indicated that a 2:1 and 3:1 combination of arecanut and waste paper increased the board weight, tear strength, tensile strength, and bursting strength based on Bureau of Indian Standards. The more arecanut sheath materials in the composition, the more the resistance to water absorption.

In a similar study, Sampanthrajan *et al.* (1992) tested some farm residues for the manufacture of low density particle boards in a hot press using urea formaldehyde as the binding material. Maize cob board was reported to be superior to other boards in mechanical and screw/nail holding strength while paddy-straw and coconut-pith boards were found suitable for insulation purposes due to their low thermal properties.

Akaranta (2000) test-produced 1.2 cm thickness particleboard from rubber seed pod, cashew nut shell, and their blends using adhesive resin from cashew nut shell liquid – a lignocellulosic material. The boards satisfied the ASTM specifications for building board grade; and the bending strength, water resistance, and swelling ratio of the boards were reported to be better than those obtained for commercial boards.

Russell (1990) evaluated the use of cereal (grain) straws – wheat, barley, and flax – as feedstock for industrial particleboards using isocyanate bonding chemistry. The panels were reported to have exceptionally high strength properties suitable for interior and exterior applications, and met water-soak swelling requirements for exterior grade waferboard based on Canadian Standard Association criteria for grade R: high quality furniture core.

Particleboards from agricultural waste of tropical fruit peels with low thermal conductivity was reported (Khedari, 2003) to have low mechanical properties, but was suitable for specific applications such as insulating ceilings and walls.

Another use for agricultural processing waste, found in Japan, is production of ceramics from rice bran (Iizuka *et al.*, 2000). Phenol resin was used as adhesive and the formed ceramic was carbonized in nitrogen or vacuum at above 900°C. Mechanically stable ceramic with compressive strength of about 60 MPa, bending strength 18 MPa, and fracture strength of about 0.6 MPa was achieved.

Power generation

Of all the technologies that can employ biomass for energy generation, direct combustion of residues (agricultural, agro-industrial and forest wastes), which are wasted or suboptimally utilized, has been recognized as an important route for generation of grid quality power. The biomass power industry in the United States is reported to have grown from less than 200 MW in 1979 to more than 6,000 MW in 1990; and was projected to reach 22,000 MW by 2010 (Bain, 1993). In 1999, 28.5 MW capacity of biomass power projects were commissioned in India (Mohan, 2000). Biomass power generation is described as CO_2 neutral and thus environmentally safe, limiting the greenhouse effect.

Silicon carbide

Silicon carbide is a crystalline compound that is extremely hard and heat resistant. The hardness deems it suitable for use as an abrasive or as reinforcement for plastics or light metals thus increasing their strength and stiffness. The high resistance of silicon carbide to heat makes it suitable as a heat refractory material and thus utilized in the production of rods, tubes, and firebrick. Also, extremely pure crystals of silicon carbide are utilized in the production of semiconductors.

The production of silicon carbide using rice husks was first performed by Cutler in 1973 (Singh, 2002). Utilizing this extremely low cost agricultural waste in producing this valuable powder has since gained significant attention. The fact that silicon carbide has superior chemical and physical properties as well as being an industrially important ceramic material has caused significant research to be done on this topic that not only offers a solution to a pending environmental problem but also presents a lucrative economic proposal. Much more research has been conducted on the synthesis of silicon carbide from rice husks rather than rice straw. This is possibly due to the larger percentage of silicon dioxide found in rice husks. Rice husks entail 15–20% of silicon dioxide as opposed to 10–13% found in rice straw. Concerning storage and primary treatments, rice husks also prove advantageous due to their less abrasive and tough nature as opposed to straw thus leading to easier cutting and briquette making.

The structure of rice husks and straw and their constituent components deem them excellent raw material for the production of silicon carbide. The constituents of silicon carbide (silicon and carbon) are both found in abundance in rice husks. The silicon is found in the husks in hydrated amorphous form. This silica is localized in the epidermis of the rice husk/ash. It has been shown not to dissolve in alkali and it can withstand relatively high temperatures (Sun and Gong 2001). The carbon, on the other hand, is found within the large amounts of cellulose in rice husks. The cellulose, when thermally decomposed, will yield carbon. Thus, it can be concluded from the chemical composition that the necessary constituents for the yield of silicon carbide are present. Also, the close proximity of these constituent elements aids the reaction's occurrence. A third contributing factor would be that rice husks have a very large surface area. This large surface area provides a good ground for the occurrence of a reaction and aids in speeding up the reaction rates to a large degree. Thus, rice husks are not only suitable for the synthesis from a chemical point of view, but they are also appropriate precursors due to their physical properties.

Silicon carbide technology

Two main technologies were found to produce silicon carbide from rice husk or rice straw using pyrolysis or plasma reactor. The following sections will explain the pyrolysis technology as well as plasma reactor technology to produce silicon carbide from rice husk or rice straw.

Pyrolysis technology
The original method through which silicon carbide was obtained from rice husks was the pyrolysis method. After the rice husks are washed to remove clay and rock impurities they are cut and pressed into briquettes. The briquettes are then coked at temperatures ranging from 500 to 900°C thus producing reduced husks. This step of coking process is not necessary for the formation of silicon carbide; however, incentives to carry out such a step are that problems and the high expense of transportation of the bulky husks are thus eliminated. Also, in India, Pakistan, and other eastern countries the usage of husks as fuel is widespread. Thus, the disposal of this rice husk ash (coked rice husks) is also an advantage of performing the coking step. Despite these incentives, however, there have been studies showing that silicon carbide forms more readily using raw rice husks rather than pretreatment (Sujirote, 2003). After coking, the reduced husks are pyrolyzed at temperatures ranging from 1,500 to 1,600°C for optimum results. During pyrolysis an inert gas (usually argon) is injected into the chamber to physically displace the carbon monoxide being produced. The presence of the carbon monoxide slows down the reaction rate and therefore its steady displacement throughout the process is important for the productivity rates of the synthesis. This is the basic procedure that will yield silicon carbide from the precursor rice husks.

Plasma reactor technology
The second technology to convert rice husk into silicon carbide is plasma reactor. The main incentive for using plasma reactors is the extremely high temperatures that can be reached in relatively short time. Plasma reactors have temperature climbing rates of approximately 10^6°C/min. In addition to high temperature, another aspect that makes plasma reactors advantageous is that they facilitate continuous production versus batch production using the pyrolysis method. Plasma reactors have been utilized to produce high quality silicon carbide that is suitable for the production of semiconductors. Alongside the usage of plasma reactors pretreatments and modifications similar to those used for the pyrolysis method have been found also to improve the quality of the produced silicon carbide. The usage of plasma reactors in the synthesis has been researched and the results have been especially promising (Nayak et al., 1996).

However, many modifications concerning chemical additives have been studied in an attempt to find catalysts and to optimize the silicon to carbon ratio for the process. There are many pretreatments that have been researched. These are applied to the rice husks prior to their pyrolysis. These methods include:

- Acid leaching: This has been found to decrease the content of impurities in the produced silicon carbide.
- Enzymatic treatments: The application of enzymes removes excess carbon content found in the rice husks. The obtained rice husks have been shown to have optimum silicon to carbon ratio to facilitate the reaction thus greatly increasing the rate of reaction.

In addition to the above pretreatments there have been many compounds that have proved to induce catalytic effects of the reaction and thus speed up the production of silicon carbide. Among these compounds are metallic compound additives which have been especially successful. These compounds are iron sulfate ($FeSO_4$), oxides of iron and nickel, $CoCl_2$, nitrates of iron and nickel, and sodium silicate. Post-treatments that have been investigated have also proved successful in increasing the level of purity of silicon carbide.

7.8 Integrated Complex

Rural villages in developing countries are connected to the drinking water supply without a sewer system. Other places in urban and semi-urban communities have no sewage treatment networks. Instead under each dwelling there is a constructed septic tank where sewage is collected or connected directly to the nearest canal through a PVC pipe. Some dwellings pump their sewage from the septic tank to a sewer car once or twice a week and dump it elsewhere, usually at a remote location.

In general, a huge amount of sewage and solid waste, both municipal and agricultural are generated in these villages. Because of the lack of a sewer system and municipal solid waste collection system, sewage as well as garbage are discharged in the water canals. This and the burning of agricultural waste in the field cause soil, water, and air pollution as well as health problems. Some canals are used for irrigation, other canals are used as a source of water for drinking.

Rural communities have had agricultural traditions for thousands of years and future plans for expansion. In order to combine the old traditions with modern technologies to achieve sustainable development, waste should be treated as a byproduct. The main problems facing rural areas nowadays are agricultural wastes, sewage, and municipal solid waste. These represent a crisis for sustainable development in rural villages and to the national economy. However, few studies have been conducted on the utilization of agricultural waste for composting and/or animal fodder but none of them has been implemented in a sustainable form. This chapter combines all major sources of pollution/wastes generated in rural areas in one complex called an eco-rural park (ERP) or environmentally balanced rural waste complex (EBRWC) to produce fertilizer, energy, animal fodder, and other products according to market and need.

The idea of an integrated complex is to combine the above-mentioned technologies under one roof, a facility that will help utilize each agricultural waste with the most suitable technique that suits the characteristics and shape of the waste. The main point of this complex is the distribution of the wastes among the basic four techniques – animal fodder, briquetting, biogas, and composting (ABBC) – as this can vary from one village to another according to the need and market for the outputs. The complex is flexible and the amount of the outputs from soil conditioner, briquettes, and animal food can be controlled each year according to the resources and the need.

The distribution of these wastes on the four techniques (ABBC) should be based on:

- The need to utilize all the sewage (0.5–1.0% solid content) using the biogas technique.
- Adding agricultural waste to the sewage to adjust the solid contents to 10% in the biogas system.
- Generating biogas to operate the briquetting machine and other electrical equipment.
- Mixing fertilizer from the biogas unit (degraded organic content) with the compost to enrich the nutritional value.
- Using cotton stalks in the briquetting technique because they are hard and bulky for all other techniques and have a high heating value.

Based on the above criteria, an environmentally balanced rural waste complex (EBRWC) will combine all wastes generated in rural areas in one complex to produce valuable products such as briquettes, biogas, composting, animal fodder, and other recycling techniques for solid wastes, depending upon the availability of wastes and according to demand and need.

The flow diagram describing the flow of materials from waste to product is shown in Figure 7.2. First, the agricultural waste is collected, shredded, and stored to guarantee continuous supply of waste into the complex. Then according to the desired outputs the agricultural wastes are distributed among the basic four techniques. The biogas should be designed to produce enough electrical energy for the complex; the secondary output of biogas (slurry) is mixed with the composting pile to add some humidity and improve the quality of the compost. And finally briquettes, animal feed, and compost are main outputs of the complex.

The environmentally balanced rural waste complex (EBRWC) shown in Figure 7.3 can be defined as a selective collection of compatible activities

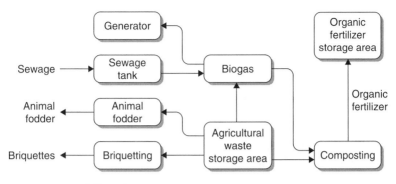

FIGURE 7.2 Material flow diagram

located together in one area (complex) to minimize (or prevent) environmental impacts and treatment cost for sewage, municipal solid waste, and agricultural waste. A typical example of such a rural waste complex consists of several compatible techniques such as animal fodder, briquetting, anaerobic digestion (biogas), composting, and other recycling techniques for solid wastes located together within the rural waste complex. Thus, EBRWC is a self-sustained unit that draws all its inputs from within the rural wastes achieving zero waste and pollution. However, some emission might be released to the atmosphere, but this emission level would be significantly much less than the emission from the raw waste coming to the rural waste complex.

The core of EBRWC is material recovery through recycling. A typical rural waste complex would utilize all agricultural waste, sewage, and municipal solid waste as sources of energy, fertilizer, animal fodder, and other products depending on the constituent of municipal solid waste. In other words, all the unusable wastes will be used as a raw material for a valuable product according to demand and need within the rural waste complex. Thus a rural waste complex will consist of a number of such compatible activities, the waste of one being used as raw materials for the others generating no external waste from the complex. This technique will produce different products as well as keep the rural environment free of pollution from the agricultural waste, sewage, and solid waste. The main advantage of the complex is to help the national economy for sustainable development in rural areas.

A collection and transportation system is the most important component in the integrated complex of agricultural waste and sewage utilization. This is due to the uneven distribution of agricultural waste that depends on the harvesting season. This waste needs to be collected, shredded, and stored in the shortest period of time to avoid occupying agricultural lands, and the spread of disease and fire.

Sewage does not cause transportation problems as it is transported through underground pipes from the main sewage pipe of the village to the system. Sewage can also be transported by sewage car which is most common in rural areas since pipelines may prove expensive.

Household municipal solid waste

Household municipal solid waste represents a crisis for rural areas where people dump their waste in the water canals causing water pollution or burn it on the street causing air pollution. The household municipal solid waste consists of organic materials, paper and cardboard, plastic waste, tin cans, aluminum cans, textile, glass, and dust. The quantity changes from one rural community to another. It is very difficult to establish recycling facilities in rural areas where the quantities are small and change from one place to another. It is recommended to have a transfer station(s) located in each community to separate the wastes, and compact and transfer them to the nearest recycling center as explained in Chapter 5. The transfer station consists of a sorting conveyer belt that sorts all valuable wastes from the organic waste, which is then compacted by a hydraulic press. The collected organic waste can be mixed with other rural waste for composting or biogas as explained above.

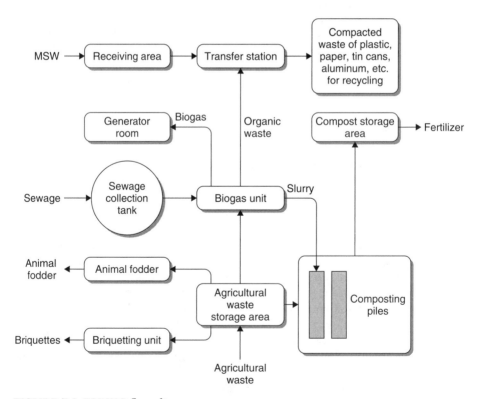

FIGURE 7.3 EBRWC flow diagram

The outputs of the EBRWC are valuable and needed goods. EBRWC is flexible and can be adjusted with proper calculations to suit every village; moreover inputs and outputs from the complex can be adjusted every year according to the main crops cultivated in the village, which usually varies from year to year. The key element to the success of this solution lies in the integration of these ABBC technologies to guarantee that each type of waste is most efficiently utilized.

7.9 Agricultural and Rural Waste Management Case Studies

Two case studies will be demonstrated in this chapter. One of the case studies is dealing with the production of organic fertilizer from agriculture waste through composting processes with some natural additives. This case study is very important for organic farming to be able to produce good quality agriculture products and reduce the use of chemical fertilizer. The other case study is the development of eco-rural parks to combine all types of wastes generated in a rural community in one complex such as agricultural wastes, municipal solid wastes, and sewage in order to achieve cradle-to-cradle in rural communities for conservation of natural resources as well as protecting the environment.

Organic fertilizer

The main objective of organic farming is to reduce external costs, improve produce yields and quality, save energy, maintain biodiversity, and improve soil fertility. Organic products mean better taste, no effect on people's health, and it is less harmful to the environment. The use of agricultural wastes during the composting process to produce organic fertilizer will increase the worms' and insects' cycles, which might decrease the use of pesticides, which kill beneficial species and pose a risk to farm workers and potentially to consumers.

Organic farming sometimes called "green farming", "organic agriculture" or "organic gardening" is based mainly on an agriculture system that fulfills the nourishment of soil fertility without using any toxics, pesticides, or chemical fertilizers to maintain food integrity. The National Organic Standards issued a definition of organic agriculture as "Organic Agriculture is an ecological production management system that promotes and enhances biodiversity cycles and soil biological activity. It is based on minimal off-farm inputs and on management practices that restore, maintain and enhance ecological harmony". In other words, organic farming is in beat with nature's laws and fulfills the environmental sustainability ecosystem for all beings. The philosophy of organic farming aims to:

- Reduce soil degradation and improve soil quality.
- Recycle agricultural and rural wastes.

- Protect air and water from pollution.
- Maintain biodiversity.
- Improve product quality as well as yield.
- Decrease the use of pesticides.
- Prevent the use of chemical fertilizers.
- Approach environment sustainability.

The key element in organic farming is the organic fertilizer. Organic fertilizers are defined as any material that was in its origin wholly or partially a living creature, or produced by a living creature such as its waste or its decomposed dead material. Organic fertilizers in a way demonstrate a recycling process within nature in which the decomposed organic matter is used as food for bacteria and microorganisms and ultimately contribute to bringing forth a new cycle of plant life. There are many benefits and advantages associated with the use of organic fertilizers, some of which are:

- They protect the environment and ecosystem from toxicity, contamination, or pollution.
- They represent a continuous supply of organic nutrients and nourishment elements.
- They represent a continuous supply of amino acids and fatty acids whenever needed.
- They restore the needed vitamins and mineral content to the soil.
- They are safe to use in many applications at a very low cost.
- They do not cause harm to the microorganisms necessary for the soil.
- They allow for the restoration of unproductive soil.

The world consumption of nutrient chemical fertilizers in 1920/21 was 1.7 million tons. In 1960/61, the world consumption was 30.04 million tons. In 1980/81, the world consumption was 117.21 million tons. While in 1998/99–2000/01 the world consumption reached 137.96 million tons (IFIA, 2002). This means that the need for fertilizers is increasing. In order to study the advantages of organic fertilizers, one has to examine the long-term effects of the chemical fertilizers and pesticides used in the conventional agricultural techniques. Effects to be examined are:

- Environmental effects on natural resources: soil degradation, ground waters, and air.
- Effects on living organisms.
- Effects on human beings.

Since the chemical fertilizers are very soluble, they are transported directly through the plants and cause no improvement to the soil's biological activity. On the contrary, organic natural fertilizers must be first transformed in composites, so plants can absorb them. Chemical fertilizers cause degradation of the soil's nutritious role and this in turn causes the production of denatured fruits and vegetables. Moreover, due to the continuous decrease of the humus rate, the toxic effect of heavy metals such as lead, cadmium, etc. used in some pesticides and fertilizers increases. When the percentage of heavy metals increases over a certain limit, plants absorb the metals and therefore can be observed in the human diet.

Objective

The main objective of this case study is to produce a good quality organic fertilizer because organic fertilizer is environmentally, economically, socially, and hygienically beneficial and applicable. From an environmental point of view, organic fertilizers will:

- Overcome soil degradation and improve soil fertility.
- Improve desertification phenomena.
- Maintain biodiversity and the ecosystem.
- Get rid of the huge organic wastes that may contaminate the environment.
- Reduce air, water, and soil pollution.
- Allow relatively high crop yields that may, in the long run, provide enough crops to cover population demands.
- Increase the quality of the crops.

From an economical point of view, organic fertilizers will:

- Save money spent on importing the huge amounts of chemical fertilizers.
- Increase the quality of the crops and reach international standards.
- Reduce expenses of waste handling.
- Increase income due to the recycling of waste materials.

From a social point of view, organic fertilizers will:

- Improve the social status of the individuals and the community.
- Create motivation for people to live in the countryside by providing job opportunities and business plans.

Finally, from a hygienic point of view, organic fertilizers will:

- Produce chemical-free crops which will improve people's health.
- Reduce diseases due to the effects of chemicals either directly or indirectly.
- Reduce the danger of lung diseases and other diseases resulting from burning the organic wastes in the field.
- Protect children and new generations from chronic diseases.

This case study will focus on using the co-composting technique or aerobic fermentation to convert agricultural residues, animal manures, and poultry manures (organic wastes) to organic fertilizers with the application of natural rocks to adjust N:P:K.

Organic farming worldwide (Yussefi and Willer, 2003)

Organic farming is currently implemented in more than 100 countries of the world and the area under organic management is growing every day. Almost 23 million hectares are managed organically worldwide. Australia has the major part of its land area managed organically, i.e. around 10.5 million hectares, followed by Argentina with 3.2 million hectares, and then Italy with more than 1.2 million hectares.

The European Union (EU) and other European countries all have more than 5 million hectares under organic management; i.e. about 2% of the total agricultural land. Organic land in many countries in Latin America is about 0.5%. In North America, more than 1.5 million hectares are organically managed. Organic farming in Asia and Africa is very low, only around 600,000 hectares represent the total organic area in Asia and more than 200,000 hectares in Africa are managed organically.

Methodology

Different compost mixtures were examined (El-Haggar *et al.*, 2004a,b), using animal manure (cow dung), poultry manure, cane sugar waste (pith), and natural rocks mixed together with different ratios. Table 7.1 illustrates mixtures of composted material. Each mixture was made separately. The materials of each mixture were mixed together thoroughly. Mixtures were turned weekly and moistened for about 60% of their water holding capacity. Physical and chemical properties were monitored periodically during the fermentation period. The composting period extended for 10 weeks after the fermentation period. Samples were taken every 2 weeks. The composite samples were picked, after turning each mixture thoroughly, from five different spots.

TABLE 7.1
Different Organic Mixtures

Material		*Mixture*				
		1	2	3	4	5
Poultry manure	(%)	0	25	50	75	100
	(kg)	0	50	100	150	200
Animal manure	(%)	100	75	50	25	0
(cow dung)	(kg)	200	150	100	50	0
Pith in (kg)		50	50	50	50	50

The five mixtures (1, 2, 3, 4, 5) composed of animal manure (cow dung), poultry manure, and pith were mixed with the natural rocks (rock phosphate, feldspar, sulfur, dolomite, and bentonite) shown in Table 7.2.

Analyses of input raw materials
A sample of the input raw materials used in this case study was taken from each material for analysis before mixing with each other. The input materials are: animal manure (cow dung), poultry manure, cane sugar waste (pith) and natural rocks (rock phosphate, feldspar, sulfur, dolomite, and bentonite).

Animal manure (cow dung)
The analysis of animal manure (cow dung) is shown in Table 7.3. It was noticed, through analysis, that the bulk density of the animal manure (cow

TABLE 7.2
Percentages and Weights of Natural Rocks

Natural rock	kg	%
Rock phosphate	6.25	2.5
Feldspar	6.25	2.5
Sulfur	2.5	1
Dolomite	6.25	2.5
Bentonite	25	10

TABLE 7.3
Analysis of Animal Manure (Cow Dung)

Test	Animal manure (cow dung)
Density (kg/m^3)	500
Moisture content (%)	58.4
pH (1:10)	9.35
EC (1:10) (dS/m)	3.57
Total – nitrogen (%)	1.56
Ammoniacal – N (ppm)	346
Nitrate – N (ppm)	216
Organic matter (%)	68.36
Organic carbon (%)	39.65
Ash (%)	31.64
C/N ratio	25.4:1
Phosphorus (%)	0.4
Potassium (%)	1.78
Parasites	*Gardia lamblia* + *Entamoba hystolytica*
Fecal CF	—
SS	—

TABLE 7.4
Analysis of Poultry Manure

Test	Poultry manure
Density (kg/m^3)	450
Moisture content (%)	13.4
pH (1:10)	8.43
EC (1:10) (dS/m)	3.63
Total – nitrogen (%)	3.45
Ammoniacal – N (ppm)	9.80
Nitrate – N (ppm)	87
Organic matter (%)	68.10
Organic carbon (%)	39.50
Ash (%)	31.9
C/N ratio	11.4:1
Phosphorus (%)	0.72
Potassium (%)	1.44
Parasites	—
Fecal CF	—
SS	—

dung) is $500\,kg/m^3$, organic matter percentage is 68.36%, and organic carbon is 39.65%. Animal manure (cow dung) is considered the main source of soil fertilizing material and plays an important role in soil sustainability.

Poultry manure
The analysis of poultry manure is shown in Table 7.4. It was noticed, through analysis, that the bulk density of the poultry manure is $450\,kg/m^3$, organic matter percentage is 68.10%, and organic carbon is 39.50%. It should be noticed that there are many factors affecting the nutrient elements of the used manure such as: type and amount of feed, methods of collecting storage, and the method of application.

Pith
Pith is a waste from sugarcane or a byproduct of the sugar industry. The analysis of pith is shown in Table 7.5. It is noticed, through analysis, that the organic matter percentage is 96.59%, organic carbon is 56.02%, and total nitrogen is 0.56%. This material will be used as a bulking agent to adjust the C/N ratio to about 30:1.

Natural rocks
Phosphate rocks: Phosphorus, an element of phosphate, strengthens roots and helps them to mature early. Table 7.6 shows the rock phosphate analysis which contains 26.7% P_2O_5. Rock phosphate is recommended in organic farming as a source of phosphorus for plants.

TABLE 7.5
Analysis of Pith

Tests	Pith
Weight of 1 m^3 (dry – kg)	145
Weight of 1 m^3 (humid – kg)	300
Moisture content (%)	51.9
Total nitrogen (%)	0.56
Organic matter (%)	96.59
Organic carbon (%)	56.02
Ash	3.41
C/N ratio	100:1
Total phosphorus	0.06
Total potassium	0.02
Water holding capacity (%)	255

TABLE 7.6
Chemical Analysis of Rock Phosphate

Oxides	Rock phosphate (%)
SiO_2	2.43
Al_2O_3	1.2
Fe_2O_3	4.6
MgO	9.5
CaO	41
K_2O	0.5
Loss of ignition	10.44
P_2O_5	26.7
Copper*	0.2

*Heavy metal.

Phosphorus is concerned with the vital growth process in plants. It is a constituent of nucleic acid and nuclei which are the essential parts of all living cells. Any deficiency of this element will restrict the growth of the plant, the metabolism of fats, and the function of the processes in root development and the ripening of seeds (Merck Index, 1952).

Importance of phosphorus (Mahmoud, 2003):

- Main constituent of NPK (nitrogen, phosphorus, potassium).
- Main constituent of RNA.
- Main constituent of plasmatic tissues in plant cells.
- Contributes to the inhalation process of plants.
- Helps the formation of chlorophyll.
- Increases the growth of flowers and nodes.

TABLE 7.7
Chemical Analysis of Feldspar

Oxides	Feldspar (%)
SiO_2	68
Al_2O_3	17
Fe_2O_3	2.5
MgO	0.1
CaO	0.4
K_2O	11
Loss of ignition	0.6
P_2O_5	0.01

- Plays an important role in the formulation of cell boundaries.
- Activates some enzymes efficiently.
- Helps in nitrogen absorption.
- Reduces the poisonous effects of excess doses of boron.
- Slows the element degradation which enables the plants to benefit from the elements for a longer time. However, it increases the size of the fruits, roots, and leaves.

Feldspar: Used for treating and protecting plants from lack of potassium. Feldspar raw material contains potassium. Table 7.7 shows feldspar analysis which contains 11% K_2O, 68% SiO_2, 17% Al_2O_3.
    Advantages of potassium (Mahmoud, 2003):

- Increases the osmosis pressure and in turn water absorption.
- Increases plants' mechanical tissues which increase the solidarity of the cell boundary.
- Activates protein representation.
- Activates enzymes.
- Increases photosynthesis; lack of potassium causes slower photosynthesis.
- Helps in decomposing starch into sugar.
- Regulates the water content and water loss through the cell.
- Increases plants' absorption of other mineral elements.

Dolomite: Contains magnesium and calcium elements as shown in Table 7.8. Dolomite contains 21% MgO and 31.8% CaO. Dolomite was used in this case study as a source of magnesium and calcium.
    Importance of magnesium (Mahmoud, 2003):

- Constituent of chlorophyll.
- Carrier of phosphoric acid inside the plants.

TABLE 7.8
Chemical Analysis of Dolomite

Oxides	Dolomite (%)
SiO_2	0.08
Al_2O_3	0.01
Fe_2O_3	0.01
MgO	21
CaO	31.8
K_2O	0.01
Loss of ignition	43.83

- Plays an important role in the growth and constitution of oily seeds.
- Plays a main role in enzymes' activities inside the plants.

Sulfur: Categorized as a natural pesticide in Europe and Egypt for many years. It contributes to the protection of plants from pest infections and is considered one of the main fertilizing elements for plants.
Importance of sulfur (Mahmoud, 2003):

- Improves the muddy soil.
- The acidic properties of sulfur compensate for the alkalinity of the soil, which is why it is used for alkali soils.
- Eliminates all pests in the soil.
- Provides appropriate media for the roots so they can absorb other elements such as phosphorus, iron, zinc, manganese, and copper from the soil.

Bentonite: One of clay minerals, its chemical composition contains potassium silicate, magnesium, calcium, and iron. Bentonite provides coherence between soil particles which results in improving the absorption of other elements and keeping a reserve of water. Bentonite also improves soil fertility (Mahmoud, 2003). Table 7.9 shows the analysis of bentonite which contains 2.4% K_2O, 0.6% MgO, 3.7% CaO, and 7% Fe_2O_3.
Importance of bentonite constituents:

- Potassium silicate is a cofactor for more than 40 enzymes. Potassium acts as an osmo-regulator and in maintenance of electro-neutrality in cells. Also, potassium is involved in carbohydrate metabolism. May be involved in the pumping process by which sucrose is trans-located (Brecht, 2003).
- Magnesium is the main constituent of chlorophyll. It is required for ribosome integrity and involved in phoshorylating reactions of carbohydrate metabolism and phosphate transfers from ATP. Thus,

TABLE 7.9
Chemical Analysis of Clay Mineral Bentonite

Oxides	Bentonite (%)
SiO_2	55
Al_2O_3	20
Fe_2O_3	7
MgO	0.6
CaO	3.7
K_2O	2.4
Loss of ignition	10
P_2O_5	0.8

photosynthesis, respiration, and nitrogen fixation all require manganese (Brecht, 2003).

- Calcium is a silver-colored, moderately hard metallic element; it constitutes approximately 3% of the earth's surface. Calcium is a basic component of animals and plants. Calcium exists in forms of fluorite, limestone, gypsum, and its compounds.
- Iron is a metallic silver-white element, malleable, ductile, and exists in the form of compounds with other metals. Iron occurs naturally in soil and it is necessary for the growth and health of plants. The absorption of iron depends on the soil alkalinity pH. If the soil is too alkaline, a plant may not be able to absorb iron. Iron improves the surface of the soil and the color of the leaves.

Quality of organic fertilizer

Different parameters were tested to investigate the performance of the composting process to produce a good quality fertilizer. Physical and chemical parameters were monitored such as: bulk density, moisture content, temperature, pH, carbon dioxide, electrical conductivity, organic matter, organic carbon, total nitrogen, C/N ratio, ammonia, nitrate, total phosphorus, soluble phosphorus, total potassium, soluble potassium, soluble calcium, soluble magnesium, and microbiological detections (acid producing bacteria, total fecal coliform bacteria, fecal bacteria, salmonella and shigella, and nematode). The following findings were observed:

- Bulk density increased with time during the composting process due to the decrease of the volume and the increase of composted material breakdown. The percentage increase of bulk density was 20–32%.
- Moisture content decreased with time during the composting process due to the evaporation of water from the compost. The percentage decrease of moisture content was 45–52%.

- Composting temperature increased with time during the composting process and reached the maximum values after 2 weeks due to the decomposition of the composted material as a result of microorganism activity. The maximum values of temperature were 54.5–70.5°C. The temperature kept high values until week 4, and then started to decrease until week 6. From 6 to 10 weeks temperature decreased gradually down to 40°C.
- pH of the compost decreased with time during the composting process and reached a final range of 6–8.
- Carbon dioxide emitted from composting piles increased with time during the composting process due to the accumulation of CO_2 as a result of microorganism activity. CO_2 reached the maximum percentage ranges after 2 weeks, fluctuated during the period 4–6 weeks, then decreased gradually. CO_2 maximum values were ~30–38%.
- Electrical conductivity of the compost increased with time during the composting process due to the discharge of ions and acids within the compost. The electrical conductivity values fluctuated till week 6 then decreased and reached final values (higher than the initial values). The percentage increase of electrical conductivity was 43–104%.
- Organic matter decreased with time during the composting process due to the degradation of organic material. The percentage decrease ranges of organic matter were 32–40%.
- Organic carbon decreased with time during the composting process due to the degradation of organic carbon. The percentage decrease ranges of organic carbon were 33–44%.
- Total nitrogen increased with time during the composting process due to the destruction of organic matter. The percentage increase ranges of total nitrogen were 31–64%.
- C/N ratio decreased with time during the composting process then stabilized gradually due to losses of organic matter and evolution of CO_2. C/N reached a final range of 10.5:1–14.2:1.
- Ammonia decreased with time during the composting process due to the conversion of ammonia to nitrate during the nitrification process. Ammonia values increased and reached the maximum values after 2 weeks, then tended to decrease. Ammonia final value ranges were ~142–492 ppm.
- Nitrate increased with time during the composting process due to the accumulation of nitrate as a result of the conversion of ammonia to nitrate during the nitrification process. Nitrate values fluctuated until week 6, then increased and reached final value ranges ~635–1152 ppm.
- Total phosphorus increased with time during the composting process due to the release of macro-elements. The application of natural rocks increased final total phosphorus percentage to 20.61%.
- Total potassium increased with time during the composting process due to the release of macro-elements. The application of natural rocks increased final total potassium percentage to ~17.29%.

- Soluble phosphorus increased with time during the composting process due to the transformation of macro-elements to soluble form. The application of natural rocks increased final soluble phosphorus value to 33.49%.
- Soluble potassium increased with time during the composting process due to transformation of macro-elements to soluble form. The application of natural rocks increased final soluble potassium value to 63.08%.
- Soluble calcium increased with time during the composting process due to transformation of macro-elements to soluble form. The application of natural rocks increased final soluble calcium value to 150.65%.
- Soluble magnesium increased with time during the composting process due to transformation of macro-elements to soluble form. The application of natural rocks increased final soluble magnesium value to 147.8%.
- Acid producing bacteria increased with time during the composting process due to the activity of microorganisms until the maximum values were reached in week 4, then they decreased to the minimum. The maximum values of acid producing bacteria range between 111×10^4 and 351×10^4 cfu.
- The fecal bacteria, total coliform bacteria, nematode, Salmonella and Shigella decreased with time during the composting process; by week 3, the pathogenic detection started to approach zero. When the temperature increases to the maximum, pathogenic bacteria are destroyed.

Proposed eco-rural park for rural development

This case study is very important for the sustainable development of rural communities in order to protect the environment and conserve the natural resources. Rural communities in developing countries are very poor where people live below the international social standard without basic needs. Therefore, this case study will provide not only a better environment but also job opportunities to increase the social standard.

The estimated amount of agricultural waste in Egypt is 33.5 million dry tons per year (El-Haggar, 2004c) as shown in Table 7.10. Some of the agricultural waste is used as animal fodder, others are used as a fuel in very primitive ovens causing a lot of health problems and damaging the environment. The rest are burnt in the field causing air pollution problems. The type and quantity of agricultural waste in Egypt changes from one village to another and from one year to another because farmers always look for the most profitable crops suitable for the land and the environment. The main crops with the highest amount of waste are the rice straws, corn stalks, wheat/barley straws, cotton stalks, banana waste, and bagasse.

Most rural villages in Egypt were connected to a septic tank where sewage is collected. The houses pump their sewage from the septic tank to a

TABLE 7.10
Estimated Quantities of Agricultural Waste
in Egypt (Million Ton/Yr)

Type	Quantity
Rice straw	3.6
Corn (maize) stalks	4.5
Cotton stalks	1.6
Sugarcane, field waste	1.86
Sugarcane, bagasse	5.030
Wheat straw	6.9
Barley straw	0.2
Sugar beet	0.32
Fruit tree wastes	1.68
Legumes, vegetables	0.71
Banana waste	1.685
Beans waste	0.427
Parks and gardens	1.141
Sorghum stalks	1.2
Sesame stalks	0.56
Palm trees	0.66
Potato waste	0.317
Tomato waste	1.11
TOTAL	33.5

sewer car once or twice a week and dump it elsewhere, usually in a remote location. As a result of this practice, the dumpsite of the sewage will be polluted (air, water, and soil).

Municipal solid waste (garbage) management in rural communities is very poor where people throw their garbage in the nearest water canal polluting the water or burn it in an open area polluting the air. The mismanagement or lack of a municipal solid waste management program in rural communities will lead to depleting the natural resources as well as polluting the environment.

In general, a huge amount of sewage, agricultural waste, and municipal solid waste is generated in rural communities. Because of the lack of a sewer system, garbage collection system, and waste management in these communities, the villages soon turn into disaster areas.

Eco-industrial parks/eco-rural parks

Eco-industrial parks (EIP), discussed in detail in Chapter 3 and used all over the world within industrial communities, are defined as "Industrial community of manufacturing and service companies to enhance their eco-efficiency through improving their economic and environmental performance by collaboration among each other in the management of the natural resources." EIP proved that it is the most valuable approach in the industrial zone from economical and environmental points of view. In other words, EIP is a collection of a compatible activities located together in one area (park or complex) to minimize (or prevent) environmental impacts and treatment cost and conserve the natural resources.

Similarly, an eco-rural park (ERP) is a collection of compatible activities located together in one area (park) which complies with environmental regulation and utilizes the wastes generated – sewage, municipal solid waste agricultural waste, etc. – to enhance the eco-efficiency through improving the economic and environmental performance by collaborating with each other in the management of the natural resources. The compatible technologies within the eco-rural park consist mainly of briquetting technology, biogas technology "anaerobic digestion", composting technology, animal fodder technology, and other recycling techniques for municipal solid wastes. The ERP is a part of the rural infrastructure and should be located within the rural community in order to minimize transportation cost to and from the ERP. Thus, ERP is a self-sustained unit that draws all its inputs from within the rural wastes approaching a cradle-to-cradle concept. ERP should be located down wind and far from the rural residential area by 1–3 km in order to avoid any emission that might be released to the atmosphere as a result of mismanagement of the ERP. The main objective of the ERP is to help the sustainable development of the rural community by providing the basic needs with high quality as well as job opportunities in a sustainable manner.

Agricultural wastes utilization techniques
Agriculture biomass resources in Egypt are estimated to be around 34 million tons (dry matter) per year. Fifty percent of the biomass is used as a direct supplement to fodder and as a fuel source in rural areas by direct combustion in low efficiency traditional furnaces similar to most developing countries. The traditional furnaces are primitive mud stoves and ovens that are extremely air polluting and highly energy inefficient. Moreover, the traditional storage systems for plant residues in the farms and roofs of the buildings provide opportunities for insects and diseases to grow and reproduce as well as being a fire hazard.

Briquetting
During the First and Second World Wars briquettes were discovered to be an important source of energy for heat and electricity production by using simple technologies. The briquetting process is the conversion of agricultural waste into uniformly shaped briquettes that are easy to use, transport, and store (El-Haggar, Adeleke and Gadallah, 2005). One of the recommended technologies is the lever operating press (mechanical or hydraulic press). Nevertheless, briquetting allows ease of transportation and safe storage of

wastes as the wastes will have a uniform shape and will be free from insects and disease carriers.

The idea of briquetting is using otherwise unusable materials due to a lack of density, compressing them into a solid fuel of a convenient shape, and burning them like wood or charcoal. The briquettes have better physical and combustion characteristics than the original waste. Briquettes will improve the combustion efficiency using the existing traditional furnaces, and in addition kill all insects and diseases as well as reduce the destructive risk of fire. The suitable raw materials for briquetting are rice straws, wheat straws, cotton stalks, corn stalks, sugarcane waste (bagasse), fruit branches, etc. The briquetting process starts with collection of wastes followed by size reduction, drying, and compaction by extruder or press.

Composting
Composting is the aerobic fermentation process of organic waste by the action of aerobic bacteria under controlled conditions such as pH, air, moisture content, particle size, C:N ratio, etc. Composting is one of the better known recycling processes for organic waste to be converted into soil conditioner and close the natural loop reaching a cradle-to-cradle approach. This organic matter was given by the soil to the plant and now the soil wants it back in order to continue producing healthy crops for sustainable development. Although the farmer provides the soil with the chemical "artificial" fertilizer to compensate the soil for its loss of organic matter, the crops will never have the same quality as the natural fertilizer and soil fertility will suffer.

Aeration is required to recharge the composting pile with the required oxygen for the microorganisms to grow using natural aeration with a windrow turning machine or passive composting with embedded perforated pipes within the pile or forced aeration assisted with a blower. A passive composting method is the recommended technique for the Egyptian environment from technical, environmental, and economical points of view. The main advantages of passive composting are less capital cost compared with forced aeration, less running cost compared with natural aeration and it is odor free because the composting pile can be covered with finished compost to protect the environment. The time required for maturation might be faster in the passive composting technique depending on the environmental factors within and around the composting pile. Maturation or degree of stabilization can be measured by indicators such as decline in temperature, absence of odor, and lack of attraction of insects in the compost pile.

Biogas
Biogas is an anaerobic fermentation process to convert organic waste into energy "biogas" and fertilizer. Different types of organic wastes are generated in Egypt such as agricultural waste, sludge from municipal treatment plants, and organic waste from garbage as well as animal manure and dead animals. Table 7.11 shows a sample of the main types and quantities of organic

TABLE 7.11
Sources and Quantities of Organic Wastes in Egypt

Organic waste source	Quantity
Agricultural waste	34 million tons of dry material per year
Municipal solid waste	8 million tons of dry organic waste per year
Sewage treatment plants	4.3 million tons of dry sludge per year

wastes generated in Egypt. All these can be considered an excellent source for biogas technology to produce energy and fertilizer.

Biogas activities in Egypt until now have focused mainly on small-scale plants with a digester volume of 5–50 m³, with a few exceptions such as Gabel Al-Asfar Plant in Cairo. The biogas potential in Egypt was evaluated in 1995 by the DANIDA team. The team found that Egypt has a substantial amount of biomass resources, which could be used for biogas plants. The total energy potential of centralized biogas plants with a 50 to 500 ton/day input was estimated to be about 1 million TOE. If the total technical potential was exploited, it was estimated that Egypt could produce 40% of its present electricity consumption from biogas and save a substantial amount of chemical fertilizer.

A realistic potential was that 4% of the present electricity consumption could be covered by biogas applications. The potential sites for large biogas plants were identified by the team as being large cattle and dairy farms, communities in old and new villages, food processing industries, sewage treatment plants, waste treatment companies regarding solid organic municipal waste, new industrial cities, and tourist villages. In the short term, the large farms were seen as having the greatest potential.

Animal fodder
The deficiency of animal foodstuff in Egypt reaches more than 3 million tons of energy a year. Transforming some of these wastes into animal feedstuffs will help a great deal in overcoming this deficiency. These wastes have high fiber content which makes them difficult to digest. The size of the waste in its natural form might be too big or tough for the animals to eat. To overcome these two problems several methods were used to transform the agricultural waste into a more edible form with a higher nutritional value and greater digestibility.

A combination of mechanical and chemical treatment methods was used to transform the shape of the roughage (waste) into an edible form. The further addition of supplements can enrich the foodstuffs with the missing nutritional contents. The mechanical treatment method consists of chopping, shredding or grinding and moistening with water or wastewater rich with nutrients such as industrial wastewater from the sugarcane industry, as will be explained in Chapter 10, or industrial wastewater from the dairy industry, as explained in Chapter 2. The chemical treatment method can be used to replace the addition of liquid waste to enrich the agricultural waste and increase digestibility. Chemical treatment can use urea or ammonia by adding 2% to the total mass of the waste. It is recommended to cover the treated waste with a wrapping material usually made out of polyethylene of 2 mm thickness to avoid any air within the pile causing an "anaerobic fermentation process". After 2 weeks (summer) and 3 weeks (winter), the treated waste is uncovered and left for 2–3 days to release all the remaining ammonia before being used as animal feed.

Municipal solid waste
Municipal solid waste represents a major crisis for rural communities because of the lack of awareness of the effects of people dumping their waste in the water canals causing water pollution as well as visual pollution. Others might burn the MSW in the streets causing air pollution as well as visual pollution. The municipal solid waste consists of organic waste, waste paper, plastic waste, tin cans, aluminum cans, textile, glass, etc. as discussed in Chapter 5. It is always recommended to establish a transfer station in rural communities because of low quantity. The transfer station consists of a conveyer belt for sorting, a hydraulic press for compacting paper, tin cans, textile, etc., and a ball mill to crush glass waste. It is always recommended to establish centralized recycling facilities among a number of rural communities.

Eco-rural park approach
The rural waste can be utilized using the four ABBC (animal fodder, biogas, briquetting, composting) techniques mentioned above as well as MSW techniques mentioned in Chapter 5 and as shown in Figure 7.4. The distribution of these wastes on the main four techniques should be based on the following:

- The need to utilize all the sewage generated in the community (0.5–1.0% solid content) using the biogas technique.
- Some agricultural waste such as rice straw will be added to the sewage to adjust the solid contents to 10%.
- Biogas generated will be used to operate the briquetting machine as well as other electrical equipment and the lighting system.
- Fertilizer from the biogas unit (degraded organic content) will be mixed with the compost to enrich the nutritional value. Natural rock might be used to adjust the quality of organic fertilizer.
- Cotton stalks will be utilized using the briquetting technique because they are too hard and bulky for the other techniques and have a high heating value.

Based on the above criteria, the eco-rural park or rural complex will combine all wastes generated in rural areas in one complex to produce valuable products such as briquettes, biogas, composting, animal fodder, and other recycling

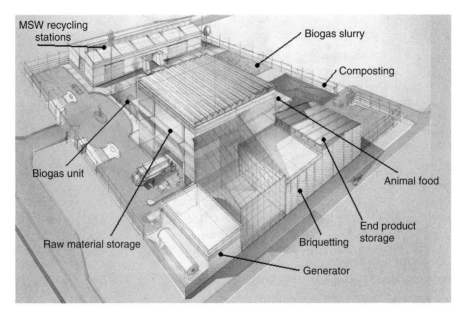

FIGURE 7.4   Eco-Rural park

techniques for solid wastes. The main outputs of the ERP are fertilizer, energy, animal fodder, and other recycling materials depending upon the availability of wastes and according to demand and need.

Questions

1. Estimate the types and quantity of agricultural wastes generated in your country.
2. What kinds of technologies are used to utilize the agricultural waste in your country? Try to develop a matching diagram between the technology used and the type of agricultural waste. What is the current situation of agricultural waste utilization/disposal in your country?
3. Can silicon carbide be produced from agricultural waste? What are the applications of silicon carbide in your country?
4. Discuss the briquetting technology and the parameters affecting the quality of briquettes.
5. Can agricultural waste be used to produce charcoal? Explain the concept of a charcoal kiln.
6. Do you recommend using agricultural waste as animal fodder or should it be treated before use? Why?
7. Develop a simple technology to utilize agricultural waste as a construction material suitable for the nearest community to your house.

Chapter 8

Sustainability of Construction and Demolition Waste Management

8.1 Introduction

Construction and demolition (C&D) wastes are considered the major wastes produced by the construction and demolition industries, and according to Clark County Code (CCC), these wastes comprise two categories. The first waste is the construction waste which is defined by the same code as the "materials that are generated as a direct result of building construction activity; such waste includes, but is not limited to, concrete, rubble, fiberglass, asphalt, bricks, plaster, wood, metal, caulking, paper and cardboard, roofing wastes, tar-paper, plastic, plaster and wallboard and other similar materials". Not only those materials but also various paints, sealants, adhesives, and fasteners used in construction. Demolition waste is considered to have almost the same definition but with different materials which result from the demolition activity. Construction and demolition waste may include packaging material and land-clearing debris. As a result of disposing of such wastes, there is a minute impact to the environment. First, it is unsightly and can lead to huge economic losses. The second impact is groundwater quality, which is classified as two types. The first type is contamination with a trace of hazardous chemicals such as organic compounds and heavy metals resulting from either applied construction materials or improper disposal of demolition waste. The second impact is contamination with non-hazardous materials such as chlorine, sodium, sulfates, and ammonia resulting from leaching of C&D primary waste.

According to the Wisconsin Department of Natural Resources, the C&D wastes consist of many materials:

- Aluminum siding
- Architectural antiques

- Asphalt
- Brick/Masonry
- Carpet
- Carpet pad
- Concrete
- Gas pipe/Metal pipe
- Lumber
- Porcelain plumbing fixtures
- PVC pipe
- Site clearance vegetative woody debris
- Steel: structural or rebar
- Vinyl siding
- Wallboard/Drywall (gypsum)

8.2 Construction Waste

Construction waste is defined as relatively clean, heterogeneous building materials generated from the various construction activities (Tchobanoglous *et al.*, 1993). Possible sources of generating construction waste can be classified under six main categories (Al-Ansary *et al.*, 2004a; Gavilan, 1994), namely: design source, procurement source, handling of materials source, operation source, residual source and other sources. However, quantity and quality of construction waste generated from any specific project would vary depending on the project's circumstances and types of materials used as shown in Figure 8.1 and Table 8.1. The annual production rate of construction and demolition waste

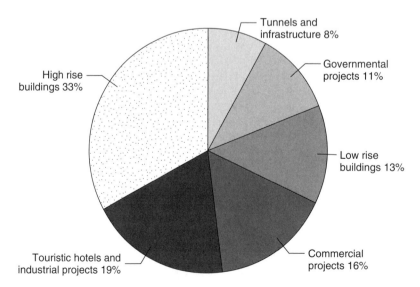

FIGURE 8.1 Cumulative percentages of projects generating construction waste in Egypt

from the whole planet is around 3 billion tons (Elliott, 2000). A possible method of resolving this problem is to develop and implement a comprehensive and practical sustainable waste management strategy that manages the amount and the types of the construction waste. Sustainable development for the construction industry can be developed through the entire life cycles of a building from cradle to cradle; including the early planning phase, the architectural and structural design phases, the construction phase, and the in-use phase.

8.3 Construction Waste Management Guidelines

Developing a good strategy for construction waste management will rectify the course of action of the construction industry towards achieving the sustainability targets. Various environmental benefits could be attained as a result of the construction waste management including:

- Decreasing the non-renewable waste and its impacts on the environment.
- Conservation of natural resources.
- Prolonging or preventing in the long run the landfill/disposal site life span (Ferguson *et al.*, 1995; CIRIA, 1993).

TABLE 8.1
Estimated Range of Wastes by Material Type from the Egyptian Construction Sites (Al-Ansary *et al.*, 2004a)

Material	*Min (%)*	*Average (%)*	*Max (%)*
Wood/Lumber	7	11.5	15
Excavated soils	25	36	48
Steel	6	8	10
Concrete	6	7	9
Mortar	7	10	12
Bricks	7	9	11
Concrete blocks	7	10	13
Plastics	3	4	5
Ceramics	6	9.5	12
Chemicals	2	2.5	3
Minerals	0	2.5	5
Prefabricated units	1	5	8
Mixed waste	N/D*	25	N/D
Marble/Granite	N/D	2	N/D
Cables, ducting and pipes	N/D	17.5	N/D
Corner bead	N/D	1	N/D
Glass	N/D	0.5	N/D
HVAC insulation	N/D	4	N/D

* N/D means no data is available.

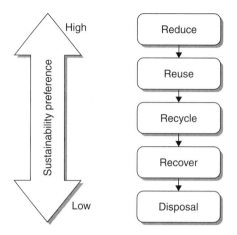

FIGURE 8.2 4Rs Golden Rule of waste management hierarchy

Proper construction waste management will provide economic benefits by decreasing the cost of the project through proper implementation of a waste management program. The true cost of waste is hidden and is at least equal to the costs of disposal/tipping fees, purchase cost of materials, the handling/processing cost, lost time and revenue, and any potential liabilities and risks put together (Environmental Agency for England and Wales, 2001). Finally, social benefits can be accomplished through the generation of potential work opportunities as well as preventing visual pollution as a result of mismanagement of construction wastes.

The quantity of construction waste might change from one country to another or from one project to another. Table 8.1 shows the estimated waste percentages from the Egyptian construction sites. Also, Figure 8.1 illustrates the percentage of construction waste generated from different construction projects (Al-Ansary *et al.*, 2004a).

The traditional construction waste management hierarchy consists of five main categories: Reduce, Reuse, Recycle, Recover, and Disposal. Each category exhibits strategies and roles of the project team to mitigate construction waste.

Reduction at the source

The best approach to manage construction waste is to minimize it at the source before it becomes a physical problem. Reduction at the source could be implemented during the following project phases.

The planning phase

The main participants during the planning phase are the owner and the engineers, i.e. the owner's team. The owner's team has to establish the project's goals during the early planning stages and consider waste minimization as a required target within the project. The designer(s) should incorporate these goals through the owner/designer agreement.

The design phase

The main participant during the design phase is the designer(s). The role of the designer is to optimize the material use (i.e. decreasing the leftovers). High quality and durable materials should always be specified to reduce rejects and wastes.

The tender and contract formulation phase

During this phase the main participants are (i) the owner's team and (ii) the contractor's team. The owner/engineer team should avoid evaluating the contractor's bid on the lowest price but on the lowest responsible bid in which quality of material and construction is taken into consideration. The contractor should include the associated costs of implementing the waste management plan in the price quotation. If the contractor fails to submit the waste management plan within the tender/bid documents, the contractor should be held to be irresponsible, and should hence be disqualified. The contractor should submit within the bid documents a draft of the waste management plan. The waste plan should include a summary of the following requirements:

- Cost/benefit analysis identifying estimated savings or costs of using different waste management techniques than the conventional method of waste disposal.
- A list of all the used materials in the project that could potentially produce wastes.
- Estimated quantities of each waste stream. A flow chart of construction wastes processing is suggested, as shown in Figure 8.3.
- Organizational/responsibility chart of the contractor's waste management team.
- List of on-site separation/segregation/sorting techniques for each waste stream.
- Procedures of material handling for waste materials.
- Techniques for on-site storage for each waste stream.
- Transportation method with destinations for each waste stream.
- Estimated tipping fees for the inevitable waste streams.
- List of responsible subcontractors for transporting wastes to the final destinations.
- Document templates that would be used to assure on-site waste management.
- A draft of the project's waste management meetings agenda addressing training and updates on waste management requirements.
- Value engineering proposals to modify the design that would minimize the waste.

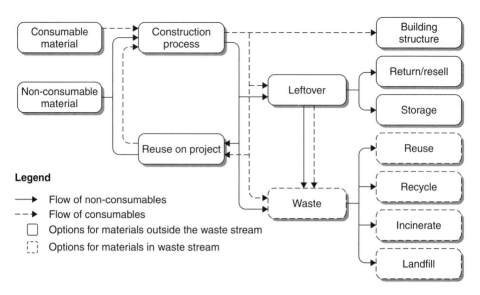

FIGURE 8.3 Construction waste process flow diagram (Peng *et al.*, 1997)

- Training program for the construction personnel, site workers, and
  subcontractors on waste management techniques. Training program
  could include:
  - Techniques of waste separation and sorting at various stages of
    the project.
  - Implementation of waste management practices throughout the
    project phases.
  - Document control and quality assurance of waste management
    techniques.

The contractor should appoint planners, at this stage, to provide assistance
in plotting the preliminary time schedules illustrating the time frame of
potential waste generation from the project's materials and activities. Planners
should allocate the resources needed to fulfill the construction waste management. The resources should consist of a waste management team, the equipment needed, and the allocated costs.

During the construction phase
During this phase, the main participants are (i) owner's team, (ii) designer's
team, and (iii) contractor's team. Their recommended roles and responsibilities are as follows:

The owner's team's responsibilities
The owner/engineer should establish stringent site supervision standards
that assess the contractor's site works to (a) lessen the rejects of materials,

(b) eliminate defective works, and (c) ensure proper implementation of waste management tactics. Moreover, criteria to evaluate the contractor's performance should be established (e.g. visual inspection checklists are highly recommended).

The designer's responsibilities
Good designer/contractor communication and coordination should be maintained.

The contractor's responsibilities
The contractor has to submit a detailed waste management plan by a maximum period designated by the owner/engineer, after the receipt of Notice of Award to Bid. To this end, the following should be taken into consideration while designing the complete detailed waste management plan:

Procurement

- An accurate estimate of the material quantities should be based on the detailed set of working-construction drawings to avoid overordering.
- Materials with different lengths should be ordered to meet different site requirements.
- A procurement plan should be updated periodically to ensure the on-time availability of the required materials.
- The contractor should take advantage of "supply and apply" subcontractors to assure efficient use of materials by accurate ordering and purchasing of materials.
- The contractor should purchase construction materials that have minimum packaging. Utilizing large packs can reduce the amount of packaging wastes.
- The contractor should convey the waste management objectives to the subcontractors. Contractors should select subcontractors based on their willingness to abide by and implement these goals in managing their wastes.
- The contractor should select suppliers who are willing to take back materials if ordered in excess, without additional cost.

Site layout

- A proper site layout should be planned carefully to ensure an optimum and effective use of site working areas with regard to material handling and equipment utilization.
- A defined area for the operations of each trade should be established to keep all leftovers in one area. Then the leftovers can be sorted by dimension for future use.

Storage and material handling

- Site storage time should be minimized in order to decrease the potential risk of materials loss, damage, or the possibility of theft.
- The number of times of material handling should be reduced to lessen the materials loss, damage, spillage, or breakage due to mishandling.
- Materials should be delivered according to the installation schedule and should be placed convenient to the work location. This could help in decreasing the waste generated due to excessive materials handling and misapplication.
- Materials should be stored/stacked in safe conditions that prevent any structural or finish damage. Moreover, materials and products should be kept covered, off the ground and in a dry secure area to be protected from the atmospheric conditions.
- Materials should be protected from being damaged by the activities of other trades.

Collection procedures

- Separation/segregation/sorting techniques should be implemented to the waste stream.
- Labelled containers for each waste stream and a schedule of the pick-up times of the containers should be provided.
- On-site storage areas for containers should be designated. In order to prolong the waste life and extend the reusable abilities, the storage areas should be:
  - Remote enough from the site to limit access to the stored material and hence control its contamination.
  - Labelled by large signage that describes the purpose of the area.
  - Protected from weathering conditions, such as rain and dust.

Waste Management Personnel

- A waste management team should be assigned to accomplish the tasks needed for this activity. The team could consist of a construction waste manager and a group of trained laborers from cleaning crew or specialized subcontractors. The task of the construction waste manager could be:
  - Setting up the waste management program.
  - Supervising material handling and transportation of the resources to the site.
  - Supervising the material storage areas.
  - Supervising the waste separation and sorting activities.
  - Supervising the reuse techniques of waste as per the contractor's waste plan.

- Supervising the waste preparation to be transported to recyclers.
- Supervising the legal disposal procedures of wastes.
- Informing new subcontractors about the waste management goals and techniques.
- Instructing and supervising the work of the trained laborers.
- Monitoring the wastes periodically to prevent any mixing or contamination.
- Data recording and document controlling for all the waste management procedures.

Execution

- The contractor should distribute copies of, or at least give access to, the approved waste management plan to all the project personnel.
- The contractor should attain a good site supervision and proper site management to reduce the risk of defective works that have to be redone.
- The contractor should use and select techniques that could reduce waste.
- Equipment should be selected based on its efficiency, mobility, durability, and maintenance.
- Energy and water usage during the construction phase should be minimized.
- Materials should be installed as per the manufacturer's recommendations and the accepted practice to reduce chances of material replacement.
- During the execution phase, the contractor and all his workers should respect the local inhabitants and the surrounding environment.
- The contractor should aim for lean construction.
- The contractor should provide as-built drawings that will be easily accessible after a number of years, when renovation may be required.
- The contractor should implement techniques of managing different construction waste streams.
- The contractor should compare data provided by the site engineers with that provided by the cost controllers regarding the actual waste progress within the project in order to improve the future economic analysis.
- The contractor should periodically and systematically update the data in his registers to assess the adequacy of the selected management techniques.
- The contractor should construct learning curves to update the labor force regarding separating and reusing of waste materials.
- The contractor should create a waste database based on the data generated from the previous projects.

- The contractor should submit a summary of the up-to-date waste management activities within the progress of payment application to the owner's team.

Reuse techniques

Reuse technique is defined as re-employment of materials to be reused in the same application or to be used in lower grade applications. The contractor has the major responsibly for adopting the reuse techniques in the project. Materials such as wood, earthworks, steel, concrete, masonry (e.g. blocks and bricks), tiles (e.g. ceramics, marble, granite and Terrazzo), plasterboard, insulation materials, paints, solvents, and carpets can be profitably reused on the construction site.

Recycle techniques

Recycle technique is defined as utilizing wastes as raw materials in other applications. Recycle endeavors can be successfully utilized during the construction phase. The party responsible at this stage is the contractor; the responsibilities allocated to the contractor could be as follows:

- The contractor can apply on-site/off-site waste recycling for waste materials such as earthworks, wood, concrete, masonry (e.g. blocks and bricks), asphalt, tiles, metals (e.g. steel), non-ferrous metals (e.g. aluminum), packing, plastic, glass, cardboards, and plasterboards. An example of an on-site recycling endeavor is shown in Figure 8.4.
- When recycled materials are used, the contractor should ensure that all recycled materials should be in accordance with all the quality control tests of the national specifications.

Recovery techniques

Recovery technique is a process of generating energy from waste materials that cannot be reduced, reused, nor recycled. Recovery techniques can be exhibited during the construction phase. The party responsible at this stage is the contractor who can apply various waste recovery techniques such as briquetting, incinerating, pyrolysis, gasification, and biodigestion. This recovery technique is a waste-to-energy recovery technique which is recommended universally. The best recovery technique as mentioned throughout this book is the waste-to-material recovery technique for conservation of natural resources in order to adopt a cradle-to-cradle approach as will be discussed later in the case studies.

Disposal

The last and least favorable category in the waste management hierarchy is the disposing option. Disposing of inevitable wastes in controlled dumpsites

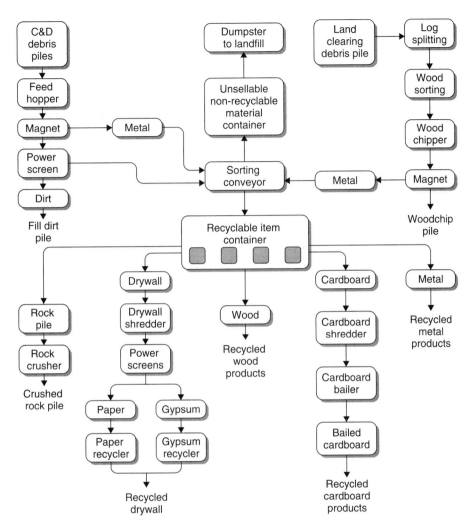

FIGURE 8.4 On-site flow pattern for recycling construction waste (Gavilan, 1994)

is the responsibly of the owner's team, designer's team and the local author-ity. It is recommended that an effective monitoring system be put in place for the legal disposal of the waste, e.g. a manifesto system. The manifesto could consist of five carbon copy certificates as summarized in Figure 8.5. The distri-bution process of the certificate could be as follows. The first one is to be signed by the contractor and the waste transporters; then forwarded to the Competent Administrative Authority (CAA). The second certificate should be kept with the contractor. The third certificate has to be signed by the dis-posal site attendant and then forwarded to the CAA. The fourth certificate should be kept in the custody of disposal site attendant. The waste transporters

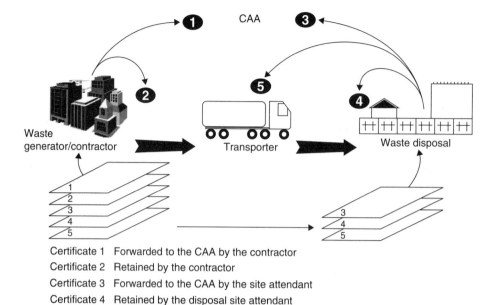

Certificate 1   Forwarded to the CAA by the contractor
Certificate 2   Retained by the contractor
Certificate 3   Forwarded to the CAA by the site attendant
Certificate 4   Retained by the disposal site attendant
Certificate 5   Retained by the waste transporter

FIGURE 8.5 Manifesto monitoring procedure

should keep the fifth certificate after being signed by the disposal site attendant. The owner's team should supervise all disposal processes stringently.

Remark
In order to apply this CWM practically in the construction industry, full cooperation between all the construction practitioners is required. In addition, more
political involvement is required to enforce the implementation of the CWM.
Moreover, it is recommended that intensive training be conducted to all parties of the construction team, especially the contractor and subcontractors, on
resource conservation and waste management strategies. In conclusion, the
integration of these proposed CWM guidelines to construction activities would
open the door to future research that could ultimately encourage the construction industry to become more environmentally sustainable.

8.4 Demolition Waste

Demolition wastes are heterogeneous mixtures of building materials such
as aggregate, concrete, wood, paper, metal, insulation, and glass that are usually contaminated with paints, fasteners, adhesives, wall coverings, insulation, and
dirt. These types of wastes are generated from the complete or selective
removal/demolishing of existing structures either by manmade processes or

by natural disasters such as earthquakes, floods, hurricanes, etc. (Al-Ansary *et al.*, 2004b), in addition to wastes generated from the renovation and remodeling works.

The composition and quantities of demolition wastes depend on the type of structure being demolished, the types of building materials used, and the age of structure being demolished. The most common types of wastes generated from demolition activities are wood, rubble, aggregates, ceramics, metals, and paper products. Although, there is no typical percentage of each waste stream generated from demolition activities, the quantity of demolition wastes from residential buildings is estimated to be 1.3 to 1.6 ton/m^2 of the ground floor area of the structure. The quantity of demolition wastes resulting from industrial structures is estimated to be from 1.5 to 2.0 ton/m^2 of the total demolished area (Al-Masha'an and Mahrous, 1999). In general, the demolition wastes are estimated to be from 1.0 to 2.0 ton/m^2 of the total ground level area.

There are many advantages for managing demolition wastes, such as reducing air borne pollutants generated from the unloading activities of waste, decreasing the possibility of heavy metals and hazardous material within the waste stream that could possibly contaminate both soil and underground water, improving the health and safety conditions by controlling the hazardous materials, broken and sharp objects and leachates from biodegradable wastes within the waste stream, and minimizing visual pollution that negatively affects the socio-economic development in any community.

Various endeavors have been attempted to manage the wastes generated from demolition activities. For example, grinding of demolition wastes has been attempted to reduce the total waste volume while the resulting powdered wastes could be landfilled (DiChristina and Henkenius, 1999). However, preventing the generation of demolition wastes is a more effective technique than managing the wastes.

8.5 Demolition Waste Management Guidelines

The proposed waste management guidelines consist of five main sections as discussed before in Figure 8.2: Reduce, Reuse, Recycle, and Recover and Disposal. Each section exhibits strategies and roles for each of the members of the project team to mitigate the generated demolition waste.

Reduce technique

Reduce is a precautionary technique aimed at minimizing the waste generated from the source before it becomes a physical problem. The reduce technique could be employed in the planning, tender, and contract formulation and execution phases as follows.

During the planning phase

During this phase the main participants are the owner and the engineer, i.e. the owner's team. It is recommended during this phase that the owner's team should choose a selective demolition technique instead of complete demolishing/removal of structures whenever possible. In that case some of the installations such as walls and ceilings can be retained while the interior systems of the structure can be renovated.

During the tender and contract formulation phase

During this phase the main participants are (1) the owner's team and (2) the contractor's team. The recommended roles for both teams are:

- The owner's team's responsibilities could be:
  - The owner has to assign an "engineer" to act as his consultant to provide the required professional and technical expertise in managing the demolition course of works.
  - The owner/engineer should avoid evaluating the contractor's bid on the lowest price but should evaluate it instead on the lowest responsible bid in which prior experience in carrying out the demolition works safely and with maximum recovery of materials is taken into consideration. The contractor should include the associated costs of implementing the waste management plan in the price quotation. If the contractor fails to submit the waste management plan within the tender/bid documents, the contractor should be held irresponsible, and should hence be disqualified.
  - The owner/engineer should assure that the waste management plan is enforceable in the contract, possibly by means of a binding clause in the contract tendering documents.
  - The content of the clause could be:

    The owner desires that as many materials as possible from this project be recovered and recycled to minimize the impact of demolition waste on the surrounding environment and to reduce the expenditures of energy and cost of fabricating new materials. To this end, the contractor shall submit a waste management plan showing the separation and mitigation actions for each material in the waste stream – generated as part or fully from demolition of the buildings, pavement, vegetation, utilities, and any other works associated with the contract scope of work – within the bidding documents. The mitigation actions should be planned to maximize the amount of reuse, recycle, and recovery of wastes and to minimize the amount of wastes to be disposed of. The waste management plan shall be part of the contractor evaluation. If the contractor is awarded the bid, he will be fully responsible to abide by his waste management plan; otherwise it will be considered as a contractual breach from the contractor side.

- The contractor's responsibilities could be:
  The contractor should prepare a draft demolition plan. The plan should include a summary/brief of the following:
  - An estimated timeframe to fulfill the goals of the waste management plan.
  - The sequences of carrying out demolition works, such as: demolition, segregation, loading, hauling, crushing, consolidation, and then stockpiling materials on site.
  - A survey of the building materials that could be reused, recycled, and recovered throughout the project – by type and quantity.
  - The quantities of disposed materials.
  - The quantities of each waste stream generated by the project. The quantities could be estimated based on either data compiled from previous projects or from experience with similar types of projects.
  - Identification of any hazardous materials and means of proper disposal.
  - The on-site separation/sorting strategies to segregate recyclables from other waste materials.
  - A list of all on-site recycling techniques.
  - Name and address of licenced disposal facilities accepting the generated waste materials.

During the execution phase

During this phase, the main participants are (1) owner's team, (2) designer's team, and (3) contractor's team. Their recommended roles and responsibilities are as follows:

- The owner's team's responsibilities could be:
  - The owner/engineer should provide very strict site supervision of the contractor's works to ensure proper implementation of waste management tactics.
  - The owner's team should supervise the contractor's performance in implementing the waste management procedures, and taking corrective action when needed.
  - Owner's team should establish criteria to evaluate the contractor's performance. Possible criteria could be visual inspection checklists.
- The contractor's responsibilities could be:
  - After the receipt of Notice of Award to Bid, by a maximum period designated by the owner, the contractor should submit a full detailed waste management plan.
  - The contractor should plan ahead the sequence of demolition that can generate the least amount of wastes while maximizing reduce, reuse, and recover endeavors.

Reuse technique

The reuse technique is defined as re-employment of materials to be reused in the same application or to be used in lower grade applications. The contractor has the major responsibly for adopting the reuse techniques in the project through the execution phase, as follows.

Collection procedures
- Separation/segregation/sorting techniques should be implemented to the waste stream.
- Labelled containers for each waste stream and scheduled pick-up times for the containers should be provided.
- On-site storage areas for containers should be designated. In order to prolong the waste life and extend the reusable abilities, the storage areas should be:
  - Remote enough from the site to limit access to the stored material and hence control its contamination.
  - Labelled with large signage that describes the purpose of the area.
  - Protected from weathering conditions, such as rain and dust.

Waste management personnel
A waste management team should be assigned to accomplish the tasks needed for this activity. The team could consist of a construction waste manager and a group of trained laborers. The task of the construction waste manager could be:

- Setting up the waste management program.
- Supervising the waste separation and sorting activities.
- Supervising the reuse of waste as per the contractor's waste plan.
- Supervising the waste preparation to be transported to recyclers.
- Supervising the legal disposal of wastes.
- Instructing and supervising the work of the trained laborers.
- Monitoring the wastes periodically to prevent any mixing or contamination.

Work activities
- The sequence of demolition activity should start by removing any valuable materials such as doors, windows, hardwoods, or flooring prior to demolition activity which could be reused, recovered or salvaged. Afterwards, the building interior should be demolished manually, followed by the core of the structure by using heavy equipment. Then excavators should be used to sort and compact recyclable and salvaged materials on site.
- Salvaged/recovered materials should be used in similar or other applications. Such materials include: wood, earthworks, plastics, vinyl,

foam, steel, concrete, masonry (e.g. blocks and bricks), tiles (e.g. ceramics, marble, and granite), plasterboard, insulation materials, paints, solvents, and carpets.
- The contractor should designate a secure and safe storage area for recovered and salvaged wastes to avoid any loss or damage that may occur to these materials.

Documentation
- The contractor should record and control all the waste management procedures.
- The contractor should periodically update the data in the registers in order to prove or disprove the adequacy of the selected management techniques during the project execution phase.
- The contractor should track costs or profits associated with various waste management methods.
- The contractor should develop learning curves to update the laborer's abilities in implementing waste management techniques.
- The contractor should document all methods and techniques of mitigating the waste, quantities and types of generated wastes experienced through to completion of the project.
- The contractor should submit within the process of payment application, a summary sheet describing all reduce, reuse, recycle, recovery and legal disposal of wastes. This summary should be supported by proper documentation. Table 8.2 is an example of a summary sheet.

Recycle technique

The recycle technique is defined as utilizing wastes as raw materials in other applications. Recycle endeavors can be successfully utilized during the execution phase by the contractor. The contractor's team's responsibilities in this stage could be as follows:

- The contractor should recycle wastes that could not be reduced nor reused. Metals such as steel, copper, and aluminium can be sold to the factories in order to be recycled to produce new metals.
- The contractor's team should assure that the recycled materials, such as recycled concrete or asphalt materials, are uniform in quality, of adequate grading, free from any contamination and meet with the Egyptian Specification and Code of Practice.
- The contractor's team should crush all materials on site such as bricks, concrete, stone, and marble in order to maximize their reuse as recycled aggregates and fill materials.
- The contractor's team should stockpile all crushed materials in separate and secured designated storage areas to avoid contamination or deterioration by weathering conditions.

TABLE 8.2
Waste Management Plan Summary Sheet (Morris, 2001)

Name of company	Contact person	Telephone No.
Project	Project type	Project size (in)

	Pre-project	Project updates	
Material	Estimated generation	Recycled/Salvaged/Disposed	Facility
TOTAL			

Signature	Title		Date

Recover technique

The recover technique is a process of generating energy from waste materials that cannot be reduced, reused, or recycled. The recover technique can be exhibited during the execution phase by the contractor. The contractor can apply various waste recovery techniques such as briquetting, incinerating, pyrolysis, gasification, and biodigestion. This recovery technique is a waste-to-energy recovery technique, which is recommended universally. The optimum recovery technique as mentioned throughout this book is the waste-to-material recovery technique for conservation of natural resources in order to adopt a cradle-to-cradle approach as will be discussed later in the case studies.

Disposal

The last category in the waste management hierarchy is the disposal technique. Disposing of wastes should be carried out in controlled landfills to prevent any contamination to water and soil. Therefore, there is a practical need to select, design, construct, and operate the landfill sites with proper environmental management systems in order to protect the environment during the whole lifespan of the landfill. The party responsible for this phase is the contractor. The major

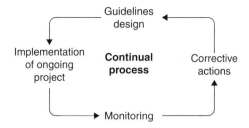

FIGURE 8.6 Continual development of guidelines

roles of the contractor are to avoid the disposal option by implementing the 4Rs Golden Rule and to manage the disposal of the inevitable wastes.

8.6 Final Remarks

It is recommended to have a continuous development process by monitoring the performance of guidelines on ongoing projects. Thus, if any discrepancy or fault in the guidelines are detected, subsequent corrective actions would be taken to rectify, improve, and upgrade the guidelines as shown Figure 8.6. Moreover, further developments are desired to calculate a scientific methodology to quantify demolition waste.

Furthermore, more political support is required to enforce the implementation of waste management schemes in the construction/building industry. This could be attained first by regulations. Second, by diligent monitoring and follow-up by municipalities and localities for illegal waste disposal. This could be attained by creating special bodies. Finally, incentives could be introduced to abide by environmental legislation such as imposing a special tax levied on wastes when exceeding a certain level.

It is also recommended to have extended research in the area of recycling techniques of building materials to induct feasibility studies including cost/benefit and payback period analysis for each technique. The research should survey the local market and seek the possibility of using waste as raw materials in manufacturing factories. The research should integrate both the construction industry and the manufacturing industry to fill the gap between the two disciplines.

8.7 Construction Waste Case Studies

Construction, Demolition and Land-clearing debris (CDL) is mainly nonhazardous solid waste resulting from construction, demolition, remodeling, alterations, repair, and land-clearing activities. CDL includes waste materials that are recycled, reused, salvaged, or disposed as municipal solid waste. Salvaged is recovery of materials for on-site reuse or donation to a third party.

As a result of thorough investigation of construction waste, it was found that the amount of recyclable waste is huge and if we pay good attention to it, it will reduce the amount of waste going to landfill and reduce costs drastically. The following examples will demonstrate the recycling capabilities of construction waste that can go all the way to 85% (Metro Solid Waste Department, 1993).

Example 1: Construction of a 5,000-square foot restaurant generated 12,344 pounds of waste, or 2.46 pounds per square foot. This waste included the following recyclable materials:

- Wood 7,440 pounds
- Cardboard 1,414 pounds
- Gypsum wallboard 500 pounds

As a result of this, the amount of construction waste that can be recycled is: [7,440 + 1,414 + 500]/12,344 = 76%

Example 2: Construction of a 17-unit apartment complex using pre-cut lumber packages resulted in 28,434 pounds of waste, or 2 pounds per square foot. This waste included the following recyclable materials:

- Wood 16,169 pounds
- Cardboard 917 pounds
- Gypsum wallboard 6,997 pounds

As a result of this, the amount of construction waste that can be recycled is: [16,169 + 917 + 6,997]/28,434 = 85%

Three case studies will be introduced for construction waste management to demonstrate the feasibility of managing the construction waste activities. One of the case studies deals with development of industrial ecology for construction waste and the way to implement the 7Rs rule for construction waste management in order to approach a cradle-to-cradle concept for conservation of natural resources. The second case study deals with construction waste from Finsbury Pavement Building located in the United Kingdom. The third case study deals with the most famous construction waste management in Sydney Olympic Village.

Development of industrial ecology and 7Rs rule for construction waste management

Construction or site waste is defined as any material that enters the construction system of any civil engineering project in its form or as an auxiliary substance, with the purpose of being utilized at any phase or in any activity towards the project's completion, and left unused. Construction wastes are heterogeneous, mixed wastes that normally consist of relatively clean building materials. The quantity and quality of the construction waste generated from any specific project will vary depending on the project's circumstances and types of materials used.

Possible sources for generating construction waste can be classified under six main categories as follows (Al-Ansary *et al.*, 2005; Gavilan, 1994): (1) "design source" which could be caused by errors in the blueprints, lack of detailing, and design change orders; (2) "procurement source" which could be caused by the inaccurate estimate of the material quantities; (3) "material handling source" which could be caused by improper storage and inappropriate on-site or off-site material handling; (4) "operational source" which could be caused by human errors, inadequate equipment or force majeure; (5) "residual source" which could be caused by poor site management and bad on-site housekeeping; and (6) "other sources" which could be due to various causes such as incompliance with manufactured dimensions of designed panels/materials, excessive material packaging, mismatching of the material quality with the required specifications, poor training in material handling, sorting and disposal methods, and incorrect use of material that requires replacement.

Industrial ecology and 7Rs rule

The industrial ecology (IE) paradigm is defined as the study of industrial systems that operate more like natural ecosystems. In a natural ecosystem there is a closed materials and energy loop, where animal and plant wastes are decomposed, by the effect of microorganisms, into useful nutrients to be consumed by plants. Afterwards, plants provide food to the animals. Finally, the loop is closed when animals die and are converted to either fossil fuels or useful nutrients to be consumed by plants. Industries can follow the same system through making one industry's waste into another's raw material, where materials and energy could be circulated in a complex web of interactions. A possible tool for the implementation of the IE concept to an industry, to implement the concept of cradle-to-cradle, could be through the full adaptation of the 7Rs Golden Rule. The 7Rs Golden Rule is one of the established waste management hierarchies, which encompasses Regulations, Reducing, Reusing, Recycling, Recovering, Rethinking and Renovation. In the following section the elements of the 7Rs Golden Rule will be illustrated.

Regulations: Regulations are one of the essential tools for enforcing laws and legislations through incentive mechanisms. There is a need for regulations that could encourage the full adaptation of the IE concept. This could be fulfilled through adding incentive articles or provisions to the existing environmental regulations.

Reducing: Reducing is a precautionary technique aimed at minimizing the waste generated from the source before it becomes a physical problem. Reducing is also called source reduction, waste minimization, and prevention (El-Madany, 1999). The reducing technique helps to minimize the waste quantity and toxicity by either decreasing the utilized quantity of raw materials or by replacing hazardous materials with more environmentally friendly ones.

Reusing: The reuse technique is defined as re-employment of materials to be reused in the same or in lower grade applications, without carrying out any further modifications or changes. Waste separation and sorting are the essential procedures for the success of this technique. Reusing endeavors could be feasible if (1) the waste materials are of proper quality and (2) the material and the waste disposal costs are greater than the reusing technique expenses.

Recycle: The recycle technique is defined as utilizing wastes as raw materials in producing other products. The recycling procedures normally depend on (1) the availability of sufficient recyclable wastes, (2) the marketing for the recyclable products, (3) the generation of a profit, (4) the landfill tipping fees, and (5) the governmental incentives and regulations regarding the recycling opportunities (Hecker, 1993). On the other hand, the most prevailing problems that could possibly hinder the recycling technique are: limited quantity of waste, space limitation, unavailability of recycling equipment, logistics and facilities, inappropriate source separation, elevated initial costs, low production rates of the recycling facility, high set-up and start-up times, inadequate staff training and equipment failure (Tchobanoglous *et al.*, 1993). In order to use the recycling techniques in construction sites, there are several methods to prepare the waste material such as (El-Madany, 1999; Hecker, 1993): (1) site separation and waste sorting, (2) site separation and processing, (3) co-mingled waste, off-site separation and processing, and (4) transfer stations.

Recovery: The recovery technique is defined as generating energy or material from waste materials that could be neither reduced, reused, nor recycled. This stage is also called "waste transformation" since physical, chemical, or biological transformation has to be applied to the wastes. There are a number of recovery techniques such as incineration, pyrolysis, gasification, and biodigestion that could produce energy in the form of steam, electricity, synthetic gases, or liquid/solid fuel. This energy recovery system might not be economically viable and therefore a material recovery system should be developed as discussed before and will be discussed under the 7th "R" Renovation.

Rethinking: After regulations, reducing, reusing, recycling and recovering, if there are inevitable, unmanageable wastes, it is important to rethink before taking action for disposal.

Renovation: Renovation is the last stage to close the loop of the industrial ecology cycle. At this stage, alternative, innovative technologies for mitigating the inevitable, unmanageable wastes are developed. The essence of this stage is to develop renewable resources by recycling, reducing, or preventing the disposal option. The development of a database is essential at this stage, which includes all types, quantity, and quality of wastes to be dumped. The most important innovative technique in this stage is to develop a simple technology to recycle the remaining waste instead of dumping it into a landfill.

It should be noted that the rethinking and the renovation techniques should be applied to the entire process starting from Regulation and Reduction to Reuse and Recycle. Rethinking and renovation techniques give room for the stakeholder to think about their waste as renewable resources throughout the entire life cycle of any construction project.

Applying 7Rs in mitigation construction material

It is recommended that the 7Rs Golden Rule tools should be applied to the whole project life cycle; from the early design phase, the planning phase, the tender and contract formulation phase, the construction and the maintenance phases. Furthermore, all the project designs, specifications, and documents should enforce the implementation of all the waste mitigation techniques. This should be developed by the project participants such as government, local authority, owner, engineer, designer, architect, planner, quantity surveyors, contractor and subcontractors, suppliers, and HSE personnel. This section will discuss the application of the 7Rs management tool in mitigating some of the construction materials.

Applying regulations techniques

Taking Egypt as an example, the Egyptian environmental legislator has regulated the dumping and treatment of solid wastes including construction wastes. In the Egyptian Environmental Law No. 4/1994, the dumping and treatment of solid wastes have been defined and regulated in many provisions. Moreover, the same law has established a system of incentives to encourage the implementation of environmental protection activities and projects. However, the industrial ecology concept has not been recommended or encouraged. Therefore, there is a vast need to add relevant provisions in the existing law to incorporate the IE paradigm into the Egyptian construction industry.

Applying reducing techniques

The reducing technique should be applied to the entire life cycle of the construction project. Reducing techniques can possibly be applied to various construction materials to prevent the transformation of such materials to become potential wastes. Table 8.3 summarizes some of the suggested techniques to reduce the generation of construction wastes.

Applying reusing techniques

The reusing technique can help in maximizing the reuse of the materials in the same or other applications within the same or different sites. Reusing techniques can be applied during the construction and maintenance phases and be adapted by the contractor team. Table 8.4 summarizes some of the suggested techniques to reuse construction waste.

Applying recycle techniques

All recycled materials should be in accordance with quality control tests and specifications in order to verify the suitability of each material for the

TABLE 8.3
Reducing Techniques

Material	Reducing technique
Wood	• The specified wood panel dimensions should correspond to the standard wood dimensions to minimize leftovers. • Efficient framing techniques should be implemented. • Detailed formwork working/shop drawings should be available to help in ordering wood lengths according to the detailed drawings. • Wooden formwork should be kept covered, off the ground, and in a dry secure area to be protected from atmospheric conditions (such as rain, sun, and humidity), physical deterioration (such as twisting), theft, or loss. • Wood cutting workshops should be centralized to optimize the use of wood scraps in other applications.
Excavated soil/earthworks	• Balance the volume of cut and fill of excavated materials. • Subject the soil, if possible, to an appropriate remediation process (e.g. solidification/stabilization) rather than excavating and clearing soil waste outside the site. • Avoid excavating unnecessary soil; for example, in case of isolated footings just excavate the place of the footings, if possible.
Concrete	• Cement bags should be ordered on time and stored properly to avoid any deterioration.
Plasterboard	• The designed plasterboard should correspond to the standard market/factory plasterboard dimensions to minimize cut-off wastes. • Plasterboard should be ordered in optimal dimensions to minimize leftovers. • Plasterboard panels should be stacked in such a way that the sequence of removal is the same as the order in which they need to be erected. Spacers or "sleepers" should be placed in between the panels to separate them. • Plasterboard panels should be kept covered, off the ground, and in a dry secure area to be protected from atmospheric conditions, deterioration, or theft. • The plasterboard should be utilized in the most efficient way and with skilled installation that could reduce leftovers.
Insulation materials	• Use the cold method of applying bitumen rather than the hot method to avoid carbon dioxide emissions. • If hot bitumen is to be used, a good combustion system with complete emission control should be applied.
Packaging and shipment	• Avoid excessive packaged materials. • Assure packaging is adequate to prevent material damage.

(Continued)

TABLE 8.3 (Continued)

Material	Reducing technique
	• Choose the supplier who is willing to recycle packaging. • Reuse shipment containers in further shipments, if possible.
Cardboard and paper	• Avoid unnecessary reproduction and copying of drawings and sketches. • Segregate cardboard and paper from other waste streams and store them in a separate area.
Hazardous wastes	• Employ materials that produce minimum hazardous/ toxic effects. • Prepare well-documented Material Safety Data (MSD) sheets that identify all the hazardous wastes that could be generated from the project.

TABLE 8.4
Reusing Techniques

Material	Reusing techniques
Wood	• Increase the number of times for reusing wooden formwork. • Clean wood waste could be reused in many applications such as: superstructure and substructure formwork, blocking wall and floor cavities, surfacing material, and landfill cover.
Excavated soil/ earthworks	• Excavated soil could be used for landscaping and as noise reducing embankment. • If excavated soil is clean and complies with the quality specifications and code of practice, it could be used in backfilling between tie beams, under plain concrete and slabs, under foundations, and under structural walls.
Steel	• Steel reinforcement wastes could be straightened and reused in the reinforcement of any pavement, sidewalks and curbs, concrete lintels, and openings. • Steel reinforcement wastes could be used as spacers between the main reinforcement grids, mainly in stairs.
Concrete	• Wastes of green concrete could be used in manufacturing on-site curbs and concrete blocks and in architectural decoration (e.g. hard landscape) applications. • Concrete wastes could be used in non-structural works such as in windows and door openings and in road construction applications. • Spoiled cement can be used in non-structural purposes such as masonry works, cement paint, plain concrete, and tile mortars.

(Continued)

TABLE 8.4 (Continued)

Material	Reusing techniques
Masonry (blocks and bricks)	• Masonry wastes can be used in non-structural applications such as capping layers for road embankment, landscape cover, and sub-base fill and backfill. • Block and brick waste can be used for producing light-weight concrete. • All brick leftovers in the different floors could be collected on the top floor where it could be used as temperature insulating material instead of using foam or other chemicals.
Plasterboard	• Plasterboard wastes could be used to insulate wall cavities, doors, and windows to improve both the sound and thermal proofing. • Large pieces of plasterboard waste could be used as fillers.
Insulation materials	• Bituminous waste could be used to improve the surfaces of footpaths. • Leftovers from insulation materials could be used in filling the interior wall cavities or on the top of installed insulation to enhance thermal performance. • Large pieces of hard insulation materials could be used under concrete floors.
Carpet	• Carpets if left in good condition could be used in areas where aesthetics play fewer roles such as in basements. • Carpet leftovers could be used to make mats for hallways and entryways.
Others	• Leftovers of materials such as paints and sealants could be used on other areas.

intended purpose. Table 8.5 summarizes some of the suggested techniques to recycle some of the construction wastes.

Applying recovery techniques

Recovery techniques can be classified into two main groups: a waste-to-energy (waste recovery) technique and a waste-to-material (material recovery) technique. Organic wastes can be used to recover energy according to a waste-to-energy recovery technique. For example, wood wastes that could not be reduced, reused, or recycled, could be incinerated to produce energy in the form of fuel or collected in a digester (container) for anaerobic biological reaction to produce gaseous fuel (biogas). Proper segregation of construction waste will help waste-to-material recovery techniques and reuse or recycle material as discussed in Tables 8.4 and 8.5.

TABLE 8.5
Recycle Techniques

Material	Recycle technique
Concrete	• In order to recycle concrete waste, the following procedures should be executed. First, concrete waste should be crushed, and then ferrous metals (such as steel bars and bolts) should be separated from concrete either manually or by specialized scissors. Afterwards, screening for sizing should be performed to separate different sizes to meet with the quality control specifications. • Recycled concrete could be used as a fill material (e.g. backfill, general fill, and base or sub-base course) and/or as secondary aggregate.
Wood	• In order to recycle wood to be used in further applications, the following procedures should be followed. First, wood grinders should shred the wood wastes. Second, ferrous metals should be separated magnetically from the other wood particles. Then, the shredded parts should be passed through trammel or mechanical screens to separate between the different sizes. Oversized wood pieces could be reused in other applications. Normal sized wood particle could be used in landscaping or could be burnt to produce energy/fuel. Undersize or fine particles could be used in landscaping or as animal bedding. • Wood waste can be reduced to fibers to be used in producing processed wood products such as composite panels (e.g. particleboard) and reconstituted boards. • Wood waste could be used in the manufacture of pulp and paper since the secondary wood fiber could provide the required longer and stronger fibers. • Clean sawdust (untreated and unpainted) could be composted to produce soil amendments, mulch, and soil fertilizers and conditioners. It could also be mixed with yard mulch to produce boiler fuel. • Wood waste could be mixed with cement to produce cement-wood composite and used in structural applications, in the manufacture of construction panels and in the construction of highway sound barriers.
Steel	• Steel reinforcement wastes could be recycled into new steel bars.
Masonry	• Masonry wastes could be used in manufacturing new blocks and bricks.
Tiles	• Tiles wastes could be crushed and used in the manufacture of agglomerated marble (i.e. tiles with polished marble/granite pieces) or the manufacture of mosaic tiles.
Asphalt	• In order to recycle asphalt waste, the following steps should be fulfilled. First, asphalt waste (e.g. from pavement or roofing

(Continued)

TABLE 8.5 (Continued)

Material	Recycle technique
	applications) should be crushed followed by the magnetic removal of ferrous metals. Then, the crushed asphalt should be screened and graded. Afterwards, the graded material could either be used as: (1) road base with other crushed and screened aggregates, (2) new paving material, and (3) asphalt products by mixing it with new asphalt binders. • Asphalt waste could be used as an additive in the production of hot asphalt mix, after being ground to 0.5 inch. • If asphalt waste is mixed with rock and gravel, it could be used as a groundcover that is used in rural roadways or temporary construction surfaces to control the generated dust from the construction activities. • Asphalt wastes could be utilized in the manufacture of new types of composite roofing shingles.
Packaging	• Corrugated cardboard could be recycled to be used in the manufacture of outside skin layers and internal rolled layers for new containers. • Paper packaging, in general, could be recycled into paper products.
Plastic	• Recycle plastic wastes into further plastic products.
Glass	• Glass could be recycled to produce new containers and bottles. The presence of glass helps in reducing the required furnace temperature, thus saving energy and prolonging the furnace lifetime.
Plasterboard	• Plasterboard wastes could be shredded and pulverized to 1 to 0.25-inch size, after removing the paper faces. The powdered gypsum could be utilized as soil conditioner to improve plant growth due to its high calcium and sulphur contents. • Plasterboard wastes could be utilized as raw materials in the manufacture of new gypsum products. Generally, the new plasterboard products could utilize about 10–20% of the recycled gypsum content and 100% of the recycled liner paper. • Plasterboard wastes could be used as animal bedding after being crushed and mixed up with wood chips. • Plasterboard wastes could be processed and blended with gypsum rock and then utilized as a granular gypsum product. • Plasterboard wastes could be used in the manufacture of non-structural building such as lightweight interior walls, sound barrier, walls, textured wall sprays, acoustical coatings, fire barriers, and absorbent products.

Applying rethinking and renovation techniques

Construction wastes that could not be reduced, reused, recycled, or recovered should not be sent to disposal facilities before applying rethinking and renovation techniques. In this stage, these kinds of waste could be used to

TABLE 8.6
Renovation Techniques

Material	Renovation technique
Steel	• Coupling technique could be used for the reinforcement connection instead of the overlapping technique (i.e. with effective length) to provide the required strength.
Tiles	• Tile wastes (ceramics, mosaic, terrazzo, marble, granite, etc.) could be used in architectural and decorative applications.
Glass	• Glass could be used in manufacturing glasswool, fiberglass insulation and glasphalt (paving material) and in the production of bricks, ceramics, terrazzo tiles, and lightweight foamed concrete.

develop some innovative technologies in order to mitigate them. Table 8.6 shows some of the recommended innovative techniques.

Remarks

A possible means for fulfilling the "sustainability triple bottom line" of attaining environmental benefits, economical development, and social enhancement is to incorporate the industrial ecology paradigm. The adaptation of the IE concept could be possibly fulfilled by the full implementation of the 7Rs Golden Rule. The 7Rs Golden Rule is a sustainable waste management hierarchy that consist of Regulations, Reducing, Reusing, Recycling, Recovering, Rethinking, and Renovation.

The application of the IE concept to the construction industry can help it become a more sustainable business. This could lead to many benefits such as attaining resource optimization and waste management enhancement as well as complying with the environmental protection regulations. The case study illustrated some examples of the mitigation techniques that could optimize the use of the construction materials by utilizing the IE concept. It could be concluded that if the 7Rs Golden Rule is well implemented throughout the whole life cycle of a construction project, ultimately zero waste could be reached and hence there would be no need for landfills. Furthermore, this approach could fulfil the cradle-to-cradle concept, save the finite materials and hence redefine the construction waste as renewable resources.

Construction waste from Finsbury Pavement Building, UK

The construction waste of a 50 m high building was monitored during a 9 month period to investigate the amount of daily waste generated from the construction activities. The Finsbury Pavement Building was to be constructed on 114,000 sq. ft (10,500 m²) costing approximately £14.3 million sterling (Bovis Lend Lease, 2001). The major sources of wastage on site were plasterboard, M&E metal, and timber respectively as shown in Table 8.7.

TABLE 8.7
Breakdown of Waste Materials on a Finsbury Pavement Site (Bovis
Lend Lease, 2001)

Item	Percentage of project waste
Concrete	4.3
Metal decking	1.3
Conlit fire protection	2
Timber	11.1
Pallets	5.2
Plasterboard	16.7
Blockwork	9.4
Stonework	2.6
Roofing	2.6
Ceiling tiles	3.4
Flooring tiles	3.8
M&E metal	13.8
Cable drums	3.4
Plastic packaging	3
Cardboard packaging	8.5
Miscellaneous	8.9

Analysis

- One of the major findings of this case study was the fact that around 70–75% of the volume of waste disposed was actually air voids. This was discovered by comparing the volume of compacted waste recorded on site and the volume of waste removal that was paid for as follows:

Volume of compacted waste	$2{,}009 \, m^3$
Volume of waste removal paid for	$6{,}960 \, m^3$
Volume of air voids in the waste disposed	$6{,}960 - 2{,}009 = 4{,}951 \, m^3$
Percentage of air voids	71%

The highest percentage of construction waste is plasterboard. Plasterboard constitutes 16.7% of the overall waste generated on site because of the following:

- The original estimated amount of plasterboard for the project was $9{,}688 \, m^2$.
- Due to design modifications, the quantity increased to $11{,}587 \, m^2$.
- Due to poor deliveries on site that resulted in overordering of plasterboard causing a wastage of $1900 \, m^2$ (20%) of the amount originally planned.

The main reasons behind plasterboard waste were:

- An excessive amount of off-cuts because the design included some awkward angles and constricted cuts. This led to wasting considerable amounts in the process of reformatting the boards to fit the project needs.
- The need to rework poorly executed work because of poor quality of workers and the damage that occurred because of poor storage on site, etc.
- The misconception that plasterboard is regarded as a cheap material.

Possibility of recycling plasterboard: Plasterboards are mainly made from gypsum. In fact, in the United Kingdom, the major manufacturers of plasterboards recycle their own factory waste. Hence, technically there isn't any obstacle that prevents the recycling of plasterboard; it might require some management and regulations.

The ceiling tiles waste as indicated in Table 8.7 represents 3.4% of total construction waste materials generated on the Finsbury Pavement site because of the following:

- The expected area to be covered on the project was 4,800 m^2 of ceiling. This would need around 19,200 tiles with an extra 10% as a safeguard against waste.
- Even though the contractor assumed a 10% waste, the actual waste of ceiling tiles on site exceeded that figure by 580 tile (around 3% more).
- The financial loss incurred due to the loss of ceiling tiles was about £16,000.
- Reasons for emergence of this high wastage rate of ceiling tiles are defects during the manufacturing process and the need to rework some badly installed tiles.

Financial loss: There are two major sources for financial loss in this project. The first loss is due to the disposal cost. The second loss is due to depleting the natural loss or the cost of raw material that will be disposed of. The disposal cost of construction waste is as follows:

- Total estimated cost of waste disposal   £331,093
- Total estimated cost of waste disposal/unit   £31.26/sq. m
  area of site

Construction waste from Sydney Olympic Village

One of the most famous construction projects for its sound management and utilization of waste generated from construction activities is the Sydney Olympic Village. This project was considered to be environmentally friendly when it came to construction waste due to the fact that 90% of waste materials generated from construction activities were reused or recycled. Highlights of the environmental awareness that were reflected in this project include the following:

- Building insulation came from recycled paper.
- Recycled concrete, steel, and timber were reused as structural and architectural components.
- Rainwater was collected into underground tanks to be used in irrigation.
- Solar energy was the major source of energy for the whole village. This reduced the total energy consumption by 50%.

Obstacles for recycling construction material waste:

- Economic feasibility: Transportation to recycling facility is a concern.
- Construction site managers are typically more occupied with the hassle of a pressurized construction program and the risk of delays.
- Segregating items to be recycled and collecting them is not an appealing activity on construction sites as it consumes much time and money in terms of labor and disposal costs.
- Possible scarcity of space on constricted construction sites.
- Fluctuating rates of waste generation of a certain waste material on construction sites may not justify regular collection processes from recycling firms.

Questions

1. Develop a comparative analysis between construction waste and demolition waste.
2. What is the 7Rs rule for industrial ecology and list one example for implementing such a rule in:
   - Construction waste.
   - Demolition waste.
3. What are the major barriers of implementing industrial ecology in construction waste?
4. What is the current situation of C&D waste in your area? Recommend some corrective action to approach a cradle-to-cradle concept.
5. What activities in your area can you change to make your community more sustainable for C&D waste?
6. What regulations should you recommend to your government to make C&D more sustainable?
7. What indicator(s) would you select to measure sustainability in your community and your country?

Chapter 9

Sustainability of Clinical Solid Waste Management

9.1 Introduction

Clinical waste is defined as the waste generated from healthcare facilities. It includes sharp and non-sharp objects, blood, body parts, chemicals, pharmaceuticals, clinical devices, radioactive materials, etc. Clinical waste can be found in municipal solid waste originating from domestic clinical activities, expired pharmaceuticals, etc.

The main objective of managing clinical wastes produced from hospitals or any clinical activity is to constrict the spread of disease by disinfecting the waste. The management of clinical waste generally poses a vital problem for countries around the globe, especially for the hazards it can cause to the surrounding environment. Technological advances in recent years have made the problem of clinical waste management even more complex with the introduction of disposable needles, syringes, and similar items.

Most of the available clinical waste treatment methods are not cost effective or environmentally acceptable and do not deal with clinical waste in a safe manner. The mismanagement of clinical waste can cause catastrophic effects not only to the people operating in clinical facilities, but also to the people in the surrounding areas. The infectious wastes can cause diseases to spread, and these diseases are some of the most fatal on earth, such as hepatitis A, B and C, AIDS, typhoid, boils, and many others. If waste is not disposed of properly, there is the potential for ordinary individuals to collect disposable clinical equipment (particularly syringes and plastic containers) and to sell these materials for reuse/recycling without sterilization. This reuse/recycling of unsterilized clinical waste material is currently the common practice in most developing countries and is among the significant causes for the diseases that develop due to poor clinical waste management.

It is very difficult to estimate the amount of clinical waste generated in developing countries. But it is recommended to use the average indicator for the amount of medical waste generated per bed of 1 kg per day to estimate the amount of clinical waste for planning and management. It is worth noting that almost 80% of the total waste generated by healthcare facilities is municipal waste, while the remaining 20% of waste is considered to be hazardous (clinical waste) and may be infectious, toxic, or radioactive. Thus, mixing non-hazardous waste with hazardous waste or infectious waste will produce hazardous waste. So, it is always recommended to sort hazardous waste from non-hazardous waste and manage each group separately.

Types of clinical wastes

The World Health Organization classified clinical waste into the following categories:

Infectious: These are materials containing pathogens – bacteria or fungal activities that can cause the spread of diseases. Such activities include surgeries and autopsies conducted on patients carrying a contagious disease.

Sharps: This includes items such as disposable needles, syringes, saws, blades, and any other tool that can cause a cut, and consequently transmit diseases. An item defined as a sharp does not necessarily have to have been in contact with human blood, body fluids, or an infectious agent.

Pathological: Anything related to human or animal body parts, such as tissues, organs, human flesh, fetuses, blood, and body fluids.

Pharmaceuticals: This type of clinical waste includes the drugs and chemicals that are no longer needed, or that have spilled, become contaminated, or expired.

Radioactive: Such as any radioactive material in the form of a solid, gas, or liquid that may have been used in the diagnosis and treatment of toxic goiters.

Others: Wastes generated from rooms, including bed linens, offices, kitchens, etc.

9.2 Methodology

The cradle-to-grave process would involve industries extracting resources, processing them, using the processed products, and finally dumping the waste into the environment. This scenario has been long adopted by industries around the world, until awareness of the harm caused to our planet forced such acts to change. This was due to the fact that treating and dealing with the waste meant money, and in reality money is the most important factor not the environment or natural resources. Hence, the environment was sacrificed for years to save money. Most countries have shown that about 75–85% of the products that are consumed are discarded after the first usage in a landfill depleting natural resources. The cradle-to-grave concept is perceived as a closed loop that would eventually lead natural resources to total depletion and thus should not be adopted by any country nowadays as discussed in Chapter 1.

The natural world around us behaves in a manner different from the artificially created industrial world. In nature, a predator can limit the number of its prey in the region; however, it does not deplete it, and no wastes are often left. The waste left by the predator is picked up by scavengers, and the leftovers of the bones are decomposed as nutrients to the plants. This concept applies to all the ecosystems in the environment; they act in a closed loop – the waste of one organism is used by another; which is again the basic concept of industrial "artificial" ecology as discussed in Chapter 3. Thus establishing an industrial ecology would be the solution to avoid landfills, incineration, and treatment. This would result in a transfer from the cradle-to-grave concept toward a cradle-to-cradle concept, with maximum utilization of the available resources.

Generally, connecting individual firms into industrial ecosystems is attained through reuse and recycling, maximizing efficient use of materials and energy use, minimizing waste generation, and defining all wastes as potential products and finding markets for them. Furthermore, balancing inputs and outputs to natural ecosystem capacities is attained through reducing the environmental burden created by the release of energy and material into the natural environment, designing the industrial interface to match the sensitivities and characteristics of the natural receiving environment, and avoiding or minimizing creation and transportation of hazardous materials. In addition, the re-engineering of industrial use of energy and materials is attained through redesigning processes to reduce energy usage, substituting technologies to reduce the use of materials or energies that disperse waste beyond the environment's ability to recapture it, and manufacturing more products using fewer resources.

Thus, it is our objective in this chapter to apply the "cradle-to-cradle" concept with respect to clinical wastes produced and collected from hospitals and clinical facilities. This approach can be achieved if complete disinfection of clinical waste is implemented through the usage of the electron beam accelerator technology for disinfection of the said wastes, and the usage of the disinfected waste as municipal solid waste to produce new products.

9.3 Clinical Waste Management

As clinical waste touches upon an important element, which is human health, there has been a universally defined strict process for managing it to ensure the safety of the environment. Thus, the Environmental Agency of Queensland (EcoAccess, 2006b) has indicated that, prior to final disposal, an effective clinical waste management system should include the following:

- Waste minimization
- Sorting

- Handling
- Interim storage
- Treatment for disinfection
- Disposal

Clinical waste minimization techniques: To begin with, certain initiatives should be taken to minimize the generation of clinical waste. In order to achieve any significant and sustained reduction in clinical waste, it is essential for those in charge of purchasing clinical items to fully comprehend and take into consideration the quality of each item, its date of expiration, what and how much waste it would generate (i.e. which parts of it would be discarded after usage, and which parts could be reused), and their medical facility's need. The most important initiative is to develop a rewarding mechanism for all employees within the clinical facilities for their successful waste reduction ideas. Studies of clinical waste minimization techniques – of how to reduce the amount of waste generated daily in hospitals and clinical facilities – offer very helpful suggestions that advise hospitals and clinics to be selective when purchasing their supplies:

- Buy in bulk whenever possible; it saves packaging.
- Consider switching from disposable to reusable medical instruments (e.g. stainless steel trays, laparoscopic instruments).
- Purchase washable surgical and isolation gowns and sterilization trays.
- Improve ordering practices so perishable products don't become outdated and unusable.
- Purchase reusable pillows and washable linens and bed pads.

Sorting clinical wastes: The sorting of wastes produced by hospitals and clinical facilities must be done in an ordered manner. Hazardous waste should be sorted from the non-hazardous in a designated color-coded bag. The sorting should be done at the source, at the stage where the waste is generated – this will make the process of sorting easier and faster. Sorting the waste at this early stage would separate and highlight the wastes that would require special handling – the "clinical waste" – as opposed to the general waste that would be handled according to Chapters 5 and 6.

Handling clinical wastes: The handling of clinical waste is considered another important stage in the waste management process because it involves collecting and transporting the waste within the clinical facility. Waste should not be transported within a hospital except through special areas, in order to avoid harming the staff, clients, and visitors. Moreover, the waste containers should be dealt with when they are three quarters full or at least once daily or after each shift, regardless of how many times this would be done,

to minimize the risk of the plastic bags containing the waste splitting open and spilling the waste over the area.

Interim storage: Interim storage of clinical waste is storing waste within the clinical facility until it can be transported for treatment or final disposal. The waste should be stored in an area of limited access to people, and only a few staff members should be allowed to be in contact with the interim storage room. This storage room should be air conditioned.

Treatment or disinfection of clinical wastes: It is not recommended to dispose of clinical waste in a sanitary landfill before treatment according to international regulation because clinical waste can be hazardous to the environment. Clinical waste must be disinfected prior to its final disposal by one of the following technologies:

- Incineration
- Autoclaving
- Chemical disinfection
- Microwaving
- Screw-feeding
- Pyrolysis
- Gasification
- Plasma-based systems and irradiation

Evaluating any of the above treatment methods, according to Health Care Without Harm Organization (Emmanuel *et al.*, 2006), should include examining how well it performs in the following criteria:

- Throughput capacity
- Types of waste treated
- Microbial inactivation efficacy
- Environmental emissions and waste residues
- Capital and operating cost
- Regulatory acceptance
- Space requirements
- Reduction of waste volume and mass
- Occupational safety and health
- Noise and odor
- Automation
- Reliability
- Level of commercialization
- Community and staff acceptance

Disposal of clinical wastes: While landfills may seem like the optimum solution due to their relatively low cost, they nonetheless come with a number of disadvantages that have led this type of activity to be abandoned by developing and developed countries. First of all, landfill construction requires a high capital cost, for the installation, operation, and management as discussed in Chapter 1. Moreover, a poor management of landfills can lead to soil and water contamination as well as to air pollution, thus making it a definite source of pollution to the entire surrounding environment.

9.4 Disinfection of Clinical Wastes

There are a number of technologies used to disinfect clinical waste as described above such as incineration, autoclaving, microwaving, gasification, etc.

Incineration: According to the Environmental Agency of Queensland (EcoAccess, 2006a), "Incineration involves the high temperature destruction of wastes." Incineration is a suitable process for disinfecting clinical, cytotoxic, pharmaceutical, and may be even chemical wastes, but it must not be used for the destruction of radioactive wastes.

Incineration is a method for dealing with solid and liquid wastes as discussed in Chapter 1. Incinerators are very expensive and some countries cannot afford to buy or maintain them. Conventional incinerators burn wastes in a refractory-lined primary chamber, the exhaust gases are then burned in a secondary combustion chamber to guarantee high temperature destruction efficiency of toxic emissions. One must keep in mind that the air pollution control system, which is very expensive, is a very essential part of the incineration process; otherwise the emission will be harmful to the environment. The fly ash as well as the bottom ash should be treated very carefully to protect the environment. A fly ash collection system should be installed within the exhaust gas path. However, the bottom ash should be collected as a solid hazardous waste to be landfilled in sanitary landfills. The advantages and disadvantages of the incineration process were discussed in Chapter 1.

Autoclave (steam sterilization): One of the most efficient and common technologies for dealing with clinical waste is the steam autoclaving technology. The most common autoclaves are meant to deal with biohazard wastes as well as with basic hospital wastes simultaneously. They were not, however, meant to deal with pathological, chemotherapy, and radioactive wastes. Autoclaves that are designed to deal with medical waste often operate along with a shredder and a compactor in order to minimize the volume of waste produced. The minimum temperature for a medical autoclave is 140°C, for a time period of half an hour according to the Environmental Agency for Queensland (EcoAccess, 2006a). However, it is very difficult for the steam to penetrate sharps, and in order to promote effective treatment, sharps must be treated for 40 minutes in order for the process to be effective. Autoclave technology operates as follows: the heat from the saturated steam combined with high pressure decontaminates the medical waste by demolishing the microorganisms present. According to Hong Kong's Advisory Council on the Environment (The Government of Hong Kong, 2004), autoclaving is a relatively cheap disinfection method; however, residual chemicals in the waste that cannot be destroyed would be vaporized and would escape, thereby contaminating the environment. Therefore, the advantages and disadvantages are:

Advantages:

- Proven technology that has been adopted for many years.
- Technology is simple and easily implemented.
- Autoclaves are manufactured in varying sizes, thus the desired size could be obtained.
- Capital costs are relatively low.

Disadvantages:

- Offensive odors are usually generated and thus require special ventilation systems.
- Wastes should not contain chemical content which would contaminate the environment.
- Large metal objects can damage the shredders.
- A shredder and a compactor must be adopted to reduce the volume of the waste.
- Wastes are heavier because of condensed steam if there is no drying mechanism.

Chemical disinfection: Chemical disinfection systems are suitable for treating clinical waste and human body parts; however, they must not be used for the treatment of cytotoxic, pharmaceutical, radioactive, and chemical wastes (EcoAccess, 2006b). Chemicals can be categorized as follows:

- Using hydrogen peroxide, lime, grinding, and shredding: The system involves shredding of the waste together with spraying of hydrogen peroxide. The spraying serves to destroy any airborne pathogens that may be released during shredding. After passing through a screen, the shredded waste is mixed with burnt lime that has a high pH value. Moreover, bentonite is added at the end of the treatment process to absorb excess liquid from the waste. Immediately prior to discharge onto a conveyor belt, sodium silicate is added to the waste. This reacts with the free lime to form calcium silicate (cement). The treated material must then be stored for at least 48 hours in order for the temperature of the waste to rise to at least 70°C to complete the waste disinfection process
- Using sodium hypochlorite, grinding, and shredding: The clinical waste is first passed through the shredder, the bags are cut open and

any fibrous materials or lengths of tubing are cut into short pieces. The material is then drawn into a grinder that cuts and grinds the waste, in the presence of a fine mist of sodium hypochlorite. Then, it is passes through a sieve into another grinder, which further reduces the size of the waste. This process also ensures that any sharps are unable to puncture the skin. Again, the waste is then soaked with sodium hypochlorite as it passes into an air classifier. The solid particles and the fluid are then mixed with sodium hypochlorite for a period of not less than 15 minutes. The material is then de-watered and removed for disposal in sanitary landfill.

According to Hong Kong's Advisory Council on the Environment (The Government of Hong Kong, 2004), though chemical disinfection methods are relatively economical means, they can introduce an additional chemical burden on the environment. Thus, it is concluded that chemical disinfection has the following advantages and disadvantages:

Advantages:

- Proven technology with a long and successful track record.
- Well-understood technology.
- Liquid effluents are easily discharged.
- No combustion by products are produced.
- Automated technology.

Disadvantages:

- Possible toxic byproducts in the wastewater from chemicals.
- Possible chemical hazards.
- Noise from shredder is very high.
- Offensive odors are generated that require special air handling equipment.
- If waste includes chemical materials, it will contaminate the environment.

Microwave disinfection: The microwave disinfection process treats clinical waste by wetting the waste, which is entered in batches or in a continuous manner, with steam or water and then heated through microwave radiation. The microwave technology involves the use of high intensity radiation to heat the moisture inside the waste. Microwave disinfection technology is suitable for the treatment of clinical waste; it must not be used for the treatment of cytotoxic waste, pharmaceutical waste, radioactive waste, chemical waste, and human body parts (EcoAccess, 2006a).

Similar to autoclaving, residual chemicals in the clinical waste that cannot be destroyed under low temperature would be vaporized and escape into the environment (The Government of Hong Kong, 2004). Thus, it is concluded that microwaves have the following advantages and disadvantages:

Advantages:

- Proven technology with long and successful track record.
- Easily understood technology.
- No liquid effluents.
- Reduces volume by 80%.
- Automated technology.

Disadvantages:

- Large metal objects may damage the shredder.
- The shredder used is very noisy.
- Offensive odors are generated that require special air handling equipment.
- If waste includes chemical materials, they will contaminate the environment.
- Capital cost is very high.

Screw feeding: This is another form of thermal treatment process, where the waste is heated up to 100°C by a rotating screw. Shredding is required before the treatment through screw feeding. It is not appropriate for dealing with pathological, pharmaceutical, and chemical wastes. Screw feeding has the same drawback and inadequacies as autoclaves (The Government of Hong Kong, 2004). The advantages and disadvantages of screw feeding are as follows:

Advantages:

- Simple design.
- The varying moisture content can treat wastes including blood fluids.
- Treated waste is unrecognizable and dry.
- Reduces volume by 80%.
- No liquid effluents.
- Automated technology.

Disadvantages:

- Relatively new technology.
- Large metal objects may damage the shredder.
- Offensive odors are generated near the compactor.
- If waste includes chemical materials, they will contaminate the environment.

Pyrolysis: Pyrolysis is a thermal–chemical decomposition process where waste is heated to very high temperatures (2,500°C) in the absence of oxygen to chemically decompose waste. This process is suitable for dealing with pharmaceutical wastes and toxic chemicals in an effective manner. However, the pyrolysis process is not yet commonly used in the clinical waste management sector.

Gasification: Gasification is a thermal process with a limited amount of oxygen to heat wastes that have a high carbon content to about 1,300°C. Gasification is still in the research and pilot stage.

Plasma system and irradiation: This type of disinfection process uses very high temperature (10,000°C) ionized gas to convert waste to a vitrified substance with separation of molten metal. Also, irradiation is the use of electron beam or other high energy particles emitted from radioisotopes to disinfect waste. Both technologies are deemed able to kill all microorganisms and make clinical waste unrecognizable. This is a reliable process that disposes of any type of clinical waste and eliminates all possible types of microorganisms (The Government of Hong Kong, 2004).

9.5 Current Experience of Clinical Wastes

In the past decades, Europe has started an initiative of moving towards non-incineration technologies such as:

- In France, starting 1993, Centre Hospitalier in Roubaix closed its incinerator and used hot steam technology.
- In Portugal, starting 1996, the incinerators were being closed and an autoclaving process was adopted.
- Ireland and Slovenia are currently adopting shredding and steam technologies for clinical wastes.

China has a population of 1.3 billion and over 310,000 medical institutions where there is a high demand for healthcare products. It is one of the ten biggest markets for healthcare products in the world, ranking just behind Japan and Asia. In 2000, the market for medical devices was worth 22.7 billion RMB (RMB = $0.125), corresponding to 3% of the global medical device market and showing an average annual growth rate of 15%. A significant proportion of China's products are exported and 90% of single use medical devices are sterilized using electron beam technology.

In most developing countries, there is difficulty in assessing the way that clinical wastes are managed. However, it is very much apparent that there is no clear pattern for clinical waste management. The standard of management in hospitals varies greatly, which makes the uniform disposal of waste difficult to monitor. It was observed that some developing countries dictate that every hospital should incinerate its wastes, others might opt for landfill of clinical waste without disinfection because of lack of awareness and cost.

9.6 Electron Beam Technology

Analyzing the above-mentioned treatment methods, it is evident that though most of them are efficient tools for disinfecting clinical wastes, they are harming the environment with different degrees of emissions and follow a cradle-to-grave concept. However, it is our main objective in this chapter to follow a cradle-to-cradle concept that reaches an industrial ecology where a sustainable environment is dominant. This can be achieved through electron beam technology where clinical wastes are completely disinfected without harming the environment and then can be recycled as an ordinary waste.

According to Dr Uwe Gohs, Professor at Fraunhofer-Institut Elektronenstrahl und Plasmatechnik, electron beam technology has been well established for about 50 years. Its uses the energy input of accelerated electrons not only for local and temporal precise heat generation (thermal applications) but also for generation of ions and excited particles (non-thermal applications). The behavior of electrons in matter is a function of:

- Penetration depth that depends on:
  - Acceleration voltage
  - Product density.
- Absorbed energy that depends on:
  - Penetration depth
  - Generation of dose gradient.

It is worth noting that the units associated with electron beam technology are concerned with:

- Dose: Absorbed energy per mass measured in units of Gy.
- Energy: Measured in MeV.
- Square weight: Total area of unit weight of product measured in units of g/m^2.
- Mass per hours: Measured per dose.

The phases for electron beam treatment usually comprise:

- Physical phase: Characterized by energy absorption and depends upon density, atomic weight, atomic number, and energy.
- Physical–chemical phase: Characterized by energy transfer and rearrangement of atoms.
- Chemical phase: Characterized by energy transfer and depends upon pressure, temperature, molecule structure, and state of aggregate.

Electron beam technology has a wide range of applications that include, inter alia, the following:

- Cross linking: Transformation of linear polymers into three-dimensional C-C systems with enhanced chemical, mechanical, and thermal properties. It is used for cables, isolated wires, pipes, and tubes.
- Electron beam curing: Transforming reactive organic liquids into solid polymer networks with enhanced chemical, mechanical, and thermal properties. It is used for façade cladding as well as production of industrial flooring.
- Food treatment: Killing of germs, bacteria, and parasites ensuring safer food with higher storage duration.
- Sterilization: Inactivation of microbial contamination and transformation of non-sterile products into sterile ones. It is used for clinical wastes.
- In addition to these known applications, electron beam technology is used also for the treatment of seeds, modification of polymers, decontamination of exhaust gases/wastewater/groundwater/sewage water.

Thus, it is concluded that electron beam technology has the following advantages:

- Used for more than two decades especially for cancer treatment.
- No toxic emissions except for ozone.
- No ionizing radiation after machine is turned off.
- No liquid effluents.
- Nothing is added to the waste, room temperature process.
- Low operating cost.
- Automated technology.

On the other hand, electron beam technology has the following precautions:

- Personnel should be protected from radiations.
- If shielding is not part of the design, a concrete shield is needed which adds to the capital cost.
- Ozone gas should be removed.
- Low radioactivity levels are observed in food treatment.
- Does not reduce volume or weight without usage of shredders.

In summary, electron beam technology is very promising as it is environmentally friendly, powerful, and relatively economic.

9.7 Electron Beam for Sterilization of Clinical Wastes

As previously mentioned, electron beam technology is used for clinical wastes through sterilization, that is inactivation of microbial contamination and transformation of non-sterile products into sterile ones. The process of radiation sterilization is based on the exposure of a product to ionizing radiation. This means that the clinical waste is exposed to ionizing radiation in specially designed irradiation containers moving on a specified way through the treatment area. The sterilization process is dependent upon:

- Number and species of microorganisms.
- Environmental conditions prior to and during treatment.

However, it is worth noting that when dealing with clinical wastes, sterilization is not enough, thus a process of radial inactivation of possible viruses is of great importance. At this point, the sterilized materials:

- Are free of sterilization agent.
- Have a high and reproducible sterilization assurance level.
- Available for immediate use due to absence of quarantine.

Cost analysis for electron beam technology

A market survey was conducted by Dr Uwe Goh, Professor at Fraunhofer-Institut Elektronenstrahl und Plasmatechnik (2005), and revealed the cost of electron beam accelerators as shown in Table 9.1.

Moreover, according to a relative cost estimate study carried out in 2004 (El-Haggar, 2004c) at The American University in Cairo, Table 9.2 portrays the investment cost analysis for the usage of different clinical waste treatment technologies relative to electron beam technology.

Usage of disinfected clinical wastes

Now, the clinical wastes have been completely disinfected and can be treated as municipal solid waste as discussed in Chapters 5 and 6. Thus, in order to

TABLE 9.1
Estimated Cost of Electron Beam Accelerator

Energy range	Beam power (MeV)	Beam current (kW)	Reference (mA)	Price (USD)
ELV–0.5	0.4–0.7	25	40	375,000
ELV–1	0.4–0.8	25	40	400,000
ELV–2	0.8–1.5	20	25	520,000
ELV–3	0.5–0.7	50	100	450,000
ELV–4	1.0–1.5	50	100	650,000
ELV–6	0.8–1.2	100	100	690,000
ELV–8	1.0–2.5	100	50	1,000,000

TABLE 9.2
Relative Cost Analysis of Clinical Waste Treatment Technologies

Used technology	Capacity (kg/hr)	Investment cost/kg
Incineration	100–200	3.4
Autoclaving	10–50	2.6
Microwave	25–450	1.1–2.6
Gamma radiation	500	7.6–9.3
Electron beam	400	1.0

apply a cradle-to-cradle concept, these disinfected materials (waste) should be used as sources for new products.

Even after disinfection, it is preferred that products made from clinical wastes do not come into contact with humans or food. Even separating the wastes into plastics, bones, glass, aluminum, and the like, might not be a cost beneficial idea. Accordingly, it is advisable to treat the lump sum of clinical wastes as rejects that are recycled as explained in detail in Chapter 6. Knowing that electron beam treated material enjoys better thermal/chemical/mechanical properties, we can use the wastes as rejects to produce:

- Bricks
- Interlocks
- Table toppings
- Wheels
- Breakwaters for wave barriers in seas or oceans
- Manholes
- Etc.

Questions

1. Is disinfection of clinical waste a must before landfill? Why?
2. Explain the difference between sterilization and disinfection of clinical waste.
3. Draw a process flow diagram for recycling of disinfected clinical waste.
4. Can an electron beam accelerator disinfect clinical waste completely? Explain the precautions for such a system.
5. What is the recommended clinical waste management in a small clinic?
6. Why do most countries try to develop regulations to stop using incinerators for clinical waste?
7. What is the current situation of clinical waste management in your country? Recommend some corrective action to approach sustainable clinical waste management.

Chapter 10

Sustainability of Industrial Waste Management

10.1 Introduction

An industry can be defined as any organized manmade activity that might generate wastes and/or pollution such as the tourism industry, construction industry, petroleum industry, recreation industry, planning industry (urban, rural, industrial, agricultural, desert, development or any other type), etc. Industrial processes are also encompassed by this definition such as the textile industry, food industry, cement industry, etc.

Since the beginning of human history, industry has been an open system of materials flow. People transformed natural materials such as plants, animals, and minerals into tools, clothing, and other products. When these materials were worn out they were dumped or discarded, and when the refuse buildup became a problems, the refuse was simply relocated, which was easy to do at the time due to the small number of habitants and the vast areas of land.

The goals of any industry must include the preservation and improvement of the environment as well as the conservation of natural resources. With industrial activities increasing all over the world today, new ways must be developed to make large improvements to our industrial interactions with the environment.

An open industrial system – one that takes in materials and energy, creates products and waste materials and then discards most of these – will probably not continue indefinitely and will have to be replaced by a different system. This system would involve, among other things, paying more attention to where materials end up, and choosing materials and manufacturing processes that generate a more circular flow as already discussed. Until quite recently, industrial societies have attempted to deal with pollution and other forms of waste largely through pollution control and regulations. Although this strategy has been partially successful, it is nonetheless unsustainable, since it still involves the depletion of natural resources. Consequently, a life-cycle assessment based on cradle-to-cradle concepts discussed in Chapter 1 has been recommended and implemented in the different industries. The case studies in this chapter will take a closer look at successful implementation of cradle-to-cradle concepts in the following industries:

- Cement
- Iron and steel
- Aluminum
- Marble
- Drilling cuttings of petroleum/gas wells
- Sugarcane
- Tourism

Other industrial activities were studied in detail in previous chapters such as the textile industry, the oil and soap industry, the food industry, and the wood industry.

10.2 Cement Industry Case Study

The cement industry is an essential industry, necessary for sustainable development in any country. It can be considered the backbone for development. The main pollution source generated from cement industry is the solid waste called cement bypass dust, which is collected from the bottom of the dust filter. It represents a major pollution problem, producing around 300 tons a day from each production line. The fine dust generated can be easily inhaled and thus poses an air pollution problem if it is not properly collected and disposed of or utilized.

Limestone and clay are the main raw materials necessary for producing cement. Other additives may be included according to the cement properties desired. Heating the raw materials with additives in a rotating kiln produces clinker, which is mixed with a small amount of gypsum and ground to produce the fine powder which we know as cement. Crushing limestone into smaller rocks in a crusher and then grinding it takes place before preheating the raw materials to about 800°C. Entering the rotating kiln, the mix is heated to nearly 1,450°C to form the clinker which is then ground to yield the fine powder of cement. The bypass dust is collected below the stacks in the cement factories by means of filters and then transported in closed or open container trucks outside the factory to open disposal sites (landfills) nearby the factories. The probability of air pollution created by this dust increases in and around the open dumping site where there are no wind barriers to prevent the wind from carrying back the bypass dust to nearby residential areas. This results in serious health problems to those residents as well as to the surrounding environment.

Thus cement bypass dust is considered the most dangerous industrial pollutant because the particle size is very small ranging between 1 and 10 microns and is highly alkaline with a pH factor of 11.5. The emissions produced from the stacks of factories in the different processes of the cement industry are also considered another environmental problem. These emissions, however, are being controlled through air pollution control units like electrostatic precipitators, bag house filters, etc. But cement bypass dust dumped near residential areas still remains a major problem. This problem has further increased since the use of filters on the stacks has increased the dust accumulated down the sides of the stacks.

During the manufacturing process, the stack emissions are dust, sulfur oxides, organic compounds, nitrogen oxides, carbon oxides, chlorine compounds, fluorine and its compounds, etc. depending on the type of fuel and quality of raw material used in the manufacturing process. These emissions can be controlled through a combustion process and air pollution control devices such as bag filters, cyclone separators, and electrostatic precipitators. In spite of the variability of these emissions and their concentrations, what is known as cement bypass dust remains the greatest – if not the only significant – one of them all. According to estimates, a single production line of cement in any factory produces a minimum of 300 tons of bypass dust per day. Most cement factories consist of three or more production lines. Therefore, the estimated amount of bypass dust from each factory is 1,000 ton/day or 0.35 million ton/year, which is reason enough to deal with this effectively.

The main problems of this cement bypass dust – other than its large quantities – are its size (1–10 microns), which is a problem for humans – as particulates of such small size affect the lungs when inhaled, which causes serious health problems, and its alkalinity (pH = 11.5) which can be highly corrosive and may cause damage to machines or buildings.

The most popular way of disposal of this bypass dust is landfilling which – to be done properly with all the required lining and covering – costs a lot of money and still pollutes the environment. That's why proper and effective disposal or reuse of cement bypass dust is always one of the main concerns of both the cement industry and environmental protection. The chemical analysis for the bypass dust is shown in Table 10.1.

Methodology

Treatment or proper disposal of wastes to be able to comply with environmental protection regulations has always been considered as an additional cost that has no return to industry or the community. This cost sometimes represents a significant portion of the total cost of the product, which is passed on to consumers or deducted from the profits of the industry resulting in an indirect waste of money.

The core element of cleaner production as discussed in Chapter 2 is prevention vs clean-up or end-of-pipe solutions to environmental problems.

TABLE 10.1
Chemical Analysis of Cement Bypass Dust

Chemical formula	Percentage
SiO_2	9.0–13.0
Al_2O_3	3.0–4.0
Fe_2O_3	2.0–2.5
CaO	45.0–48.0
MgO	1.7–1.9
SO_3	4.0–11.0
Na_2O	3.0–8.0
K_2O	2.0–6.0
Cl	4.0–13.0

In other words, the establishment of a safe sustainable environment using methods in the most efficient way to avoid any unnecessary cost. It is considered a new and creative approach towards products and production processes.

Cleaner production focuses on reduction of use of natural resources, thus minimizing the waste generated from the process. It also stresses how to prevent these wastes at the source by the use of cleaner technologies. This does not mean changing processes as a process may be made "cleaner" without necessarily replacing process equipment with cleaner components, it can simply be done by changing the way a process is operated. Simply cleaner production's only prerequisites are willingness, commitment, an open mind, team work, and a structured methodology as explained in Chapter 2. Some people might suggest changing the technology or process from dry to wet to reduce the amount of cement bypass dust as one of the cleaner production techniques discussed in Chapter 2. This is true from the bypass dust point of view only, but a wet process will increase the energy consumption and the amount of fuel consumed. This will lead to more consumption of natural resources, more air pollution, more air pollution control devices to be installed and more money spent in terms of capital and running costs. Others may think of recycling the bypass dust as one of the cleaner production technique options. This option might be cost effective, if the product from the recycling process is competitive with similar products in the market from quality and price points of view.

Therefore, cleaner production has many benefits if managed properly; it can reduce waste disposal and raw material costs, improve profitability and worker safety, reduce the environmental impact of the business and improve public relations and image thus also improving local and international market competitiveness. It is simply a win–win concept and the challenge that rises here is how to achieve cleaner production in an appropriate manner. The proper approach to tackle this challenge is: (1) source inventory (where are wastes and emissions generated?), (2) cause evaluation (why are wastes and emissions generated?), and (3) option generation (how can these causes be eliminated?).

Knowing the answer to these questions and achieving a proper cleaner production can be made using the six-step organized approach, which is:

- Simply getting started and taking action towards the issue.
- Analyzing process steps, inputs, outputs (flows and emissions).
- Generating cleaner production opportunities.
- Selecting cleaner production solutions that fit the environment and that are capable of being done.
- Implementing cleaner production solutions.
- Sustaining cleaner production.

Following this simple logical approach one can simply help alleviate the problems of environmental pollution with significant economical benefits.

Approach

Under what we may call the same umbrella of cleaner production, a philosophy/strategy/framework of industrial ecology came to life (Cote and Cohen-Rosenthal, 1996). However, most definitions of industrial ecology discussed in Chapter 3 comprise similar attributes with different emphases, and these attributes include the following:

- A systems view of the interactions between industrial and ecological systems.
- The study of material and energy flows and transformations.
- A multidisciplinary approach.
- An orientation toward the future.
- An effort to reduce the industrial systems' environmental impacts on ecological systems.
- An emphasis on harmoniously integrating industrial activity into ecological systems.
- The idea of making industrial systems emulate more efficient and sustainable natural systems.

The idea behind the concept of industrial ecology beyond already existing practices attempting to reduce negative impacts of industrial waste is the systems-oriented approach and the link with natural ecosystems in a twofold manner:

- Analogy: Natural systems are seen as an ideal model for management of natural resources including waste and energy, which industrial systems should try to adopt, which is the same idea that had been introduced and discussed in the cleaner production approach.
- Integration: Industrial systems are viewed as only one part of the surrounding systems with which they have to be in concert (Jelinski *et al.*, 1992), which is – to a limit – a new point that can be taken more into consideration and studied.

Using the industrial ecology concept, many benefits would be provided other than the same gains that are provided by cleaner production, which are reduction in the use of virgin materials, reduction in pollution, increased energy efficiency, and reduction in the volume of waste products requiring disposal:

> As discussed before, cleaner production allows for a reduction in the use of virgin materials; reduction in pollution, increased energy efficiency, and reduction in the volume of waste products requiring disposal. In addition to these, other benefits of great importance are to be gained from implementing the industrial ecology concept such as increasing the amount and types of process outputs that have market value and the birth of new industries which means more work opportunities and consequently, higher economic benefits. Furthermore, industrial ecologies will produce hidden resource productivity gains through the implementation of cradle-to-cradle concepts and the creation of eco–industrial parks with inter-firm relations.

Eco-industrial parks actually represent the form that every industrial community should be trying to achieve according to industrial ecology. In these parks every single waste is either reused or retreated in a certain way that sustainability of resources and the environment is assured. Most industrial ecologists believe that Kalundborg, a small city on the island of Seeland, 75 miles west of Copenhagen, was the first-ever recycling network (Gertler, 1995) – four main industries and a few smaller businesses feed on each other's wastes and transform them into useful inputs. As Gertler (1995) explains, the basis for the Kalundborg system is "creative business sense and deep-seated environmental awareness," and "while the participating companies herald the environmental benefits of the symbiosis, it is economics that drives or thwarts its development". For more detail regarding the contribution of the cement industry in Kalundborg, see Chapter 3 case studies.

The same idea and approach to reach a sustainable environment is what the cradle-to-cradle concept calls for. The cradle-to-grave concept means that the raw materials of a certain industry are being used only once then dumped in a landfill, which is a one-way stream of materials. This is happening all over the world ever since the industrial revolution. But to realize environmental reform in the cement industry, the cradle-to-cradle concept has to be implemented, where every material is considered as a nutrient either to the environment or to the product itself. In order to form a community to cope with such a concept and action, according to Michael Braungart, there are four phases to be made:

Phase 1 – Creating community: Identification of willing industrial partners with a common interest in replacing hazardous chemicals with natural or less hazardous ones, targeting of toxic chemicals for replacement.

Phase 2 – Utilizing market strength: Sharing a list of materials targeted for elimination, development of a positive purchasing and procurement list of preferred intelligent chemicals.

Phase 3 – Defining material flows: Development of specifications and designs for preferred materials, creation of a common materials bank, design of a technical metabolism for preferred materials.

Phase 4 – Ongoing support: Preferred business partner agreements among community members, sharing of information gained from research and material use, co-branding strategies.

Cleaner production opportunities

According to cleaner production techniques and the industrial ecology concept, a number of alternatives can be demonstrated to utilize the cement bypass dust as a raw material in another industry or another process such as:

- On-site recycling in cement production process (most efficient in wet process) and more research required to optimize the percentage of bypass dust to be recycled without affecting the cement properties.
- Production of tiles/bricks/interlocks blended cements.
- Enhancing the production of road pavement layers.
- Production of safe organic compost (soil conditioner) by stabilizing municipal waste water sludge.
- Production of glass and ceramic glass.

On-site recycling of bypass dust

On-site recycling of bypass dust within the industrial process can be done as a raw material to produce clinker. The wet process was found to be efficient and economic where 87% of the alkaline salts were removed by this method. On the other hand, the dry process requires treating the dust to remove the salts before mixing with the raw materials. The dust treatment process can be done by one of two ways:

- Heating the dust in furnaces to 1,450°C to extract the salts but this method was found not to be economic.
- Adding the dust to water basins with mechanical stirring and sometimes with heating using gases from the kiln; the solution is then left for a specific residence time before filtering the excess water.

The treated dust can then be recycled in the production line by 16% of weight from the raw materials used or if dried can be added to clinker with 8% by weight to be able to have the same standards of the produced cement. However, recycling the cement bypass dust will not utilize all the quantity of dust generated daily in cement factories.

Off-site recycling through tiles/bricks/interlocks

By pressing the cement bypass dust in molds under a certain pressure, bricks/ interlock/tiles can be formed with a breaking strength directly proportional to the pressure used to form them; sometimes the breaking strength is even higher than the pressure used. In the Turah cement factory in Egypt, experiments were conducted on the following:

- Using 100% cement bypass dust with a pressure of 200 kg to form cylindrical cross-section bricks of 50 cm^2 area were the breaking pressure of these bricks reached 120 kg/cm^2. In addition, chemical treatment of these bricks during the hydraulic molding can achieve a breaking pressure of 360–460 kg/cm^2 for the 100% cement dust bricks.
- Using 15–20% cement dust with clay and sand along with pressure thermal treatment to reach breaking pressure of 530–940 kg/cm^2.
- Using 50% cement dust with sand along with thermal treatment only to reach a breaking pressure of 1,300 kg/cm^2.

Using cement bypass dust with clay to produce bricks has proven to reduce the weight of the bricks along with reducing the total linear drying shrinkage. In addition, this opportunity can utilize very high percentages of the bypass dust. However, this will still depend on the market needs and the availability of easily transported bypass dust.

Off-site recycling through glass and ceramic glass

Using bypass dust as a main raw material (45–50%) along with silica and sandstone and melting the mix at temperatures ranging between 1,250 and 1,450°C, glass materials were obtained. The glass product has a dark green color with high durability due to the high calcium oxide (CaO) content in cement bypass dust. It can be used for bottle production for chemical containers. This step was then followed by treatment for 15–30 minutes at temperatures ranging between 750 and 900°C to form what is known as ceramic glass. This new product, unlike glass, has a very high strength and looks like marble. The produced ceramic glass is highly durable and can resist chemical and atmospheric effects. Consequently, this new product opens the way for utilizing huge quantities of bypass dust in producing architectural fronts for buildings, prefabricated walls, interlocks for sidewalks, and many other engineering applications.

Sewage sludge stabilization using bypass dust

Because of the high alkalinity and pH value of cement kiln dust, it can be used in stabilizing municipal sewage sludge. Municipal sludge contains bacteria, parasites, and heavy metals from industrial wastes. Therefore, if used directly as a soil conditioner, it will cause severe contamination to the soil and the environment and may be very hazardous to health.

Two types of sludge from sewage treatment plants can be used; the first one from a rural area where no heavy metals were included and the second from an urban area where heavy metals might exist depending on the level of awareness and compliance.

Due to the high alkalinity of cement bypass dust, when it is mixed with municipal sludge, it enhances the quality of sludge by killing the bacteria and viruses (e.g. Ascaris) in the sludge. Also, it will fix the heavy metals (if they exist) in the compost and convert them into insoluble metal hydroxides thus reducing flowing of metals in the leachate. Agricultural waste, which is considered a major environmental problem in developing countries as discussed in Chapter 7, can be added to the compost to adjust the carbon to nitrogen ratio and enhance the fermentation process. Agricultural waste will also act as a bulking agent to improve the chemical and physical characteristics of the compost and help reduce the heavy metals from the sludge.

The uniqueness of this process is related to the treatment of municipal solid waste sludge which is heavily polluted with Ascaris eggs (a most persistent species of parasite) using a passive composting technique. This technique is very powerful, very efficient with much less cost (capital and running) than other techniques as explained in Chapter 5. First, primary sludge is mixed with 5% cement dust for 24 hours. Second, agricultural waste as a bulking agent is mixed for passive composting treatment. Passive composting piles are formed from sludge mixed with agricultural waste (bulking agent) and cement dust with continuous monitoring of the temperatures and CO_2 generated within the pile. Both parameters are good indicators of the performance and digestion process undertaken within the pile.

Passive composting technology has shown very promising results, especially by adding cement dust and agricultural wastes. Results show that Ascaris has not been detected after 24 hours of composting mainly due to the high temperature elevations reaching 70 to 75°C for prolonged periods, as well as the high pH from cement dust. Also, the heavy metal contents were way beyond the allowable limits for both urban sludge as well as rural sludge.

As a result of previous discussion, three major wastes (cement bypass dust, municipal wastewater sludge (MWWS) from sewage treatment plants as well as agricultural waste) can be used as byproducts to produce a valuable material instead of dumping them in landfills or burning them in the field. This technique will protect the environment and establish a new business. If cement bypass dust does not exist, quick lime can be used to treat the MWWS. Sludge has a very high nutritional value but is heavily polluted with Ascaris and other pollutants depending on location. Direct application of sludge for land reclamation has negative environmental impacts and health hazards. Cement bypass dust is always considered a hazardous waste because of high alkalinity. The safe disposal of cement dust costs a lot of money and still pollutes the environment because it is a very fine dust with a high pH (above 11) and has no cementing action. Agricultural waste has no heavy metals and contains some nutrients, which will be used as a bulking agent. The bulking agent can influence the physical and chemical characteristics of the final product. It will also reduce the heavy metal content of the sludge and control C/N ratio for composting.

Off-site recycling through road pavement layers

Three applications can be used to utilize cement bypass dust through road pavement layers. The first application deals with the subgrade layer, the second application deals with the base layer, while the third application deals with asphalt mixture as will be explained in this section.

Subgrade layer: Adding 5–10% of cement bypass dust to the soil improves its characteristics and makes it more homogeneous and hard wearing to maintain loads.

Base layer: It is quite well known that limestone is used in the base layer (which is located right below the asphalt layer) for road paving. Also good binding and absence of voids in this layer is crucial to maintain strength and to prevent settlement and cracking. Therefore, due to its softness adding cement bypass dust as filler material to the base layer fills the voids formed between rocks. This helps increase the density (weight/volume) of this layer due to increase of weight and fixation of volume improving the overall characteristics of binding especially if base layers of thickness more than 25 cm are required. Also, the absence of voids in the base layer prevents the negative impact of acidic sewage water and underground water which work on cracking and settling the base layer.

Asphalt mixtures: Asphalt is a mix of sand, gravel, broken stones, soft materials, and asphalt. In Marshall's Standard Test* for designing asphalt mixtures, it was found that the percentage of asphalt required can be reduced as the density of the mixture increases. Therefore, adding cement bypass dust which has very fine and soft particles improves the mixture efficiency by filling the voids. Also, the bypass dust contains high percentages of dry limestone powder and some basic salts which in nature decrease the creeping percent of the asphalt concrete, enhance the binding process, and reduce the asphalt material required, which is very desirable in hot climates.

This process was implemented in the road joining the stone mill of Helwan Portland Cement Company and the company's factory in Egypt. The results of binding the base and subgrade layers assured that adding the cement bypass dust to the layers improved the overall characteristics of the road. The road is still operating and is in perfect condition even though the trucks using the road have load capacities not less than 100 tons.

---

* http://www.dupont.com/asphalt/mix.html

Final remarks

The environmental and social benefits as a result of utilizing cement bypass dust are:

- Reduced air pollution problems.
- High percentage of bypass dust used.
- Improved pavement layers with very low cost.
- Reduced percentage of asphalt required for the same performance asphalt mixture.
- More job opportunities.
- Lower price product with high quality and strength.
- A new business opportunity.
- Prevention of biomass field burning to get rid of the agricultural waste in the sludge treatment process.
- Reduced greenhouse gas emissions.
- Reduced microbes and parasites in sludge forming a high quality soil conditioner as well as improving land reclamation and public health.

10.3 Iron and Steel Industry Case Study

The steel industry is one of the essential industries for the development of any community. In fact, it is really the basis for numerous other industries that could not have been established without the steel industry. The European industrial revolution was actually founded on this industry. There are three basic routes to obtain finished steel products: (1) integrated steel production, (2) secondary processing, and (3) direct reduction.

Integrated steel production involves transforming coal to coke in coke ovens, while iron ore is sintered or pelletized prior to being fed into the blast furnace (BF). The ore is reduced in the blast furnace to obtain hot metal containing some 4% carbon and smaller quantities of other alloying elements. Next the hot metal is converted to steel in the basic oxygen furnace (BOF). Then it is continuously cast to obtain semi-finished products, such as blooms, bars, or slabs. These semi-finished products are rolled to the finished shapes of bars, sheet, rail, H or I beams.

Secondary processing, often called minimills, starts with steel scrap which is melted in an electrical arc furnace (EAF). The molten steel produced is possibly treated in a ladle furnace and then continuously cast and finished in a rolling operation. Originally, minimills provided only lower grade products, especially reinforcing bars. But recently it has been able to capture a growing segment of the steel market.

An alternative mode of steel production is the direct reduction method. In this method, production starts with high grade iron ore pellets which are reduced with natural gas to sponge pellets. Then the sponge iron pellets are fed into an electrical arc furnace. The resulting steel is continuously cast and rolled into a final shape.

The problem of the solid waste generated from iron and steel industry "slag" is not only preventing the use of millions of square meters of land for more useful purposes but is also contaminating it. Many of these waste materials contain some heavy metals such as barium, titanium and lead. Also, it is well known that toxic substances tend to concentrate in slag. The health hazards of heavy metals and toxic substances are well known. Based on the concentration levels, some slags may be classified as hazardous waste materials. Furthermore, ground water is susceptible to serious pollution problems due to the likely leaching of these waste materials. This section describes the different types of solid waste materials generated from the iron and steel industry and the associated environmental problems. Different techniques of managing these waste materials are presented with a focus on utilizing slag and dust in civil engineering applications. Test results of many research efforts in this area are summarized. In addition, numerous ideas to mitigate the environmental impacts of this problem are suggested.

Environmental problems worldwide

Iron making in the BF produces a slag that amounts to 20–40% of hot metal production. BF slag is considered environmentally unfriendly when fresh because it gives off sulfur dioxide and, in the presence of water, hydrogen sulfide and sulphoric acid are generated. These are at least a nuisance and at worst potentially dangerous. Fortunately, the material stabilizes rapidly when cooled and the potential for obnoxious leachate diminishes rapidly. However, the generation of sulfuric acid causes considerable corrosion damage in the vicinity of blast furnaces. In western Europe and in Japan, virtually all slag produced is utilized either in cement production or as road filling. In Egypt, almost two-thirds of the BF slag generated is utilized in cement production. Some 50 to 220 kg of BOF slag is produced for every metric ton of steel made in the basic oxygen furnace, with an average value of 120 kg/metric ton (Szekely, 1995). At present, about 50% of BOF slag is being utilized worldwide, particularly for road construction and as an addition to cement kilns (Szekely, 1995).

Recycling of slags has become common only since the early 1900s. The first documented use of BF slag was in England in 1903 (Featherstone and Holliday, 1998); slag aggregates were used in making asphalt concrete. Today, almost all BF slag is used either as aggregate or in cement production. Steel-making slag is generally considered unstable for use in concrete but has been commercially used as a road aggregate for over 90 years and as an asphalt aggregate since 1937. Steel-making slag can contain valuable metal and typical processing plants are designed to recover this metal electromagnetically. These plants often include crushing units that can increase the metal recovery yield and also produce materials suitable as construction aggregates. Although

BF slag is known to be widely used in different civil engineering purposes, the use of steel slag has been given much less encouragement.

BF dust and sludge are generated as the offgases from the blast furnace are cleaned, either by wet or dry means. The dust and sludge typically are 1 to 4% of hot metal production (Lankford *et al.*, 1985). These materials are less effectively utilized than BF slag. In some cases, they are recycled through the sinter plant, but in most cases they are dumped and landfilled. Finding better solutions for the effective utilization of BF dust and sludge is an important problem that has not yet been fully solved. BOF dust and sludge are generated during the cleaning of gases emitted from the basic oxygen furnace. The actual production rate depends on the operation circumstances. It may range from about 4 to 31 kg/metric ton of steel produced, and has a mean value of about 18 kg/metric ton (Szekely, 1995). The disposal or utilization of BOF dust and sludge is one of the critical environmental problems needing solution.

Electrical arc furnaces produce about 116 kg of slag for every metric ton of molten steel. Worldwide, about 77% of the slag produced in EAFs is reused or recycled (Szekely, 1995). The remainder is landfilled or dumped. Due to the relatively high iron content in EAF slag, screens and electromagnetic conveyors are used to separate the iron to be reused as raw material. The remaining EAF slag is normally aged for at least 6 months before being reused or recycled in different applications such as road building. All efforts in Egypt have focused on separating the iron from EAF slag without paying enough attention to the slag itself. However, pilot research work conducted at Alexandria University has investigated the possibility of utilizing such slag (El-Raey, 1997). The test results proved that slag asphalt concrete could in general fulfill the requirements of the road-paving design criteria.

EAF dust contains appreciable quantities of zinc, typically 10 to 36%. In addition, EAF dust holds much smaller quantities of lead, cadmium, and chromium. EAF dust has been classified as a hazardous waste (K061) by the United States Environmental Protection Agency, and therefore its safe disposal represents a major problem. Although there are several technologies available for processing this dust, they are all quite expensive, on the order of US$150 to $250 per metric ton of dust.

The problem in Egypt

Until recently, there was a lack of consciousness in Egypt either to the environmental impacts or to the economic importance of waste materials generated from the iron and steel industry. Apart from the granulated BF slag used in producing slag cement, all types of waste from any steel plant were simply dumped in the neighboring desert. Based on the production figures of the major steel plants in Egypt and the generation rates of different waste materials per each ton of steel produced, the annual waste materials generated in Egypt was estimated and summarized in Table 10.2 over and above the

TABLE 10.2
Waste Materials Generated from Iron and Steel Industry in Egypt

Type of waste	Annual amount generated (tons)
Blast furnace slag	600,000
Basic oxygen furnace slag	200,000
Electrical arc furnace slag	300,000
Blast furnace dust	20,000
Basic oxygen furnace dust	Not collected
Electrical arc furnace dust	15,000
Rolling mill scales and sludge	25,000

stockpile. There are stockpiles of 10 million tons of air-cooled BF slag and another 10 million tons of BOF slag lying in the desert.

Obviously, waste generators will be required to pay a certain fee per ton disposed. In 1993, the cost of disposing one ton of steel plant wastes in the United States was $15 (Foster, 1996). Large tonnages of iron and steel slags are increasingly produced in Egypt and the huge space needed to dump them has become a real challenge. To have an idea of the considerable area needed for disposal, it is quite enough to know that the 20 million tons of BF and BOF slags currently available are estimated to occupy 2.5 million square meters.

Potential utilization of slag

Many of the environmental problems of solid waste materials generated from the iron and steel industry have been known for some time and attempts have been made to tackle them with varying degrees of success. During the past few years, the iron and steel industry has been able to produce some creative solutions to some of these environmental problems. It is highly probable that many other creative solutions could also emerge as a result of a well-thought out and well-supported research programs.

Processing of slag is a very important step in managing such waste material. Proper processing can provide slag with a high market value and open new fields of application. Cooper *et al.* (1986) discussed the recent technologies of slag granulation. The main steps of the granulation process were addressed with schematic drawings including verification, filtering, and denaturing systems. The most recent continuous granulation technology at that time was introduced in detail with the help of many illustrative figures as shown in Figure 10.1.

Foster (1996) addressed the high cost of disposing of wastes generated from the steel industry and discusses an innovative idea from South Africa to manage BOF slag which has a limited usage. He came up with a new idea for processing BOF slag. This process starts with preparing the slag by grinding it, mixing it with a reductant such as sawdust or charcoal, and feeding it into a

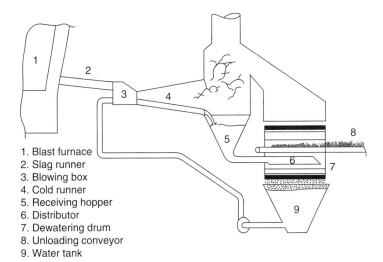

1. Blast furnace
2. Slag runner
3. Blowing box
4. Cold runner
5. Receiving hopper
6. Distributor
7. Dewatering drum
8. Unloading conveyor
9. Water tank

FIGURE 10.1 Continuous slag granulation system (Cooper *et al.*, 1986)

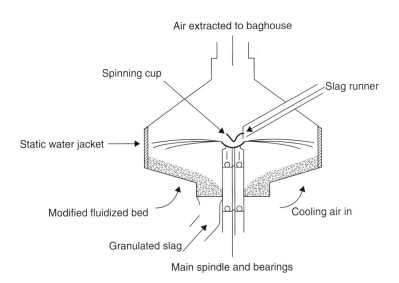

FIGURE 10.2 Dry slag granulation (Featherstone and Holliday, 1998)

modified cyclone type preheater. This reduction process removes iron oxide from iron. The slag is then passed over a magnet which removes the iron particles. The low iron slag is then mixed with other materials, such as clay, to produce an acceptable type of cement kiln feed.

Featherstone and Holliday (1998) introduced the idea of dry slag granulation shown in Figure 10.2. The existing slag treatment methods, the new dry granulation method, and the value of granulated slag products were reviewed. The development, application, and advantages of the dry method of granulating molten slags were described. The dry granulated slag was proved to have many environmental advantages over conventional processes while generating a product of equal quality in addition to its low cost and simplicity.

Swamy (1993) presented an extensive and critical examination of the use of ground granulated BF slag in concrete. It was shown that the use of BF slag can lead to concrete that combines high strength and excellent durability. Apart from its ability to reduce the temperature rise due to hydration, test results showed that BF slag has a hidden potential to contribute high early age strength, excellent durability, and very good chemical resistance. A mix proportioning method was advanced which assured the development of early strength for slags of normal surface area of 350–450 m^2/kg. Table 10.3 summarizes the compressive strength development up to 180 days age for mixes with 50% (A) and 65% (B) slag replacement of coarse aggregates. Curing was shown to be a critical factor which affects early age strength, continued strength development, and fine pore structure responsible for durability. It was also shown that with a well-defined curing period, the mineralogy and chemistry of slag could be mobilized to develop a very fine pore structure which is far superior to that of Portland cement concrete. Such a fine pore structure can impart a very high resistance to concrete to the transport of sulfate and chloride ions and water.

Nagao *et al.* (1989) proposed a new composite pavement base material made of steel-making slag and BF slag. When the new composite base material was prepared by mixing steel-making slag, air-cooled slag, and granulated

TABLE 10.3
Compressive Strength Development of Slag Concrete
(Swamy, 1993)

Mix	Age (days)	Compressive strength (MPa)
A	1	7.20
B	1	4.10
A	3	28.90
B	3	19.00
A	7	39.00
B	7	27.40
A	28	46.80
B	28	34.20
A	90	54.90
B	90	38.20
A	180	57.10
B	180	36.47

blast furnace slag in proportions of 65%, 20% and 15%, respectively, it was found to have material properties and placeability similar to those of conventional hydraulic and mechanically stabilized slags. Also, it was found feasible to quickly and economically suppress the swelling of steel-making slag by the hot water immersion that involves hydration reaction at 70–90°C, under which slag can be stabilized in 24 hours at an expansion coefficient of 1.5% or less as proposed by the Japan Iron and Steel Federation.

One of the interesting research efforts in Egypt to find fields of application for iron and steel slags among other waste materials is the project carried out by Morsy and Saleh (1996) during the period from 1994 to 1996. This project was funded by the Scientific Research and Technology Academy to investigate the technically sound and feasible utilization of two solid waste materials, iron and steel slags, and cement dust, in addition to some other different liquid wastes. The study dealt with BF slag, both air-cooled and water-cooled, and BOF slag. The use of such slags in road paving as a base, sub-base, and surface layers was examined through laboratory and pilot field tests. The results proved that these slags are suitable for use in all paving layers. Better performance and higher California bearing ratio were obtained for slags compared to conventional stones. The Egyptian standards for ballast require that the weight of a cubic meter of ballast be not less than 1.1 ton and that the Los Angeles abrasion ratio does not exceed 30%. Test results showed that the properties of iron and steel slags surpassed the requirements of these standards and can be used as ballast provided that suitable grading is maintained.

Another study performed in Egypt in 1997 (El-Raey, 1997) by the Institute of Graduate Studies and Research at Alexandria University in conjunction with Alexandria National Iron and Steel Company was aimed at investigating the use of EAF slag as road-paving base material and as coarse aggregate for producing concrete suitable for applications such as wave breakers, sidewalk blocks and profiles, and manhole covers. The physical and chemical properties of EAF slags of different ages were determined. EAF slag was crushed to the desired size and the applicability of slag in producing asphalt concrete was tested in the laboratory using the Marshall test. Table 10.4 shows some of the results obtained for asphalt concrete containing slag. The test results proved

TABLE 10.4
Marshall Test for Asphalt Concrete Containing Slag

Asphalt content (%)	Unit weight (lb/ft^3)	Stability (lb)	Flow/ inch	Voids in agg. (%)	Voids in total mix
3.5	2.62	2279	10.2	14.8	9.2
4.5	2.72	2726	11.7	14.3	5.9
5.5	2.77	2550	12.4	14.5	3.5
6.5	2.73	2100	15.9	15.6	1.5
Design criteria	—	>1800	8–18	>13	3–8

that slag asphalt concrete could in general fulfill the requirements of the road-paving design criteria. EAF slag was successfully used as a coarse material for the base layer in a field-scale test.

The study also covered the use of EAF slag as coarse aggregate for concrete and the obtained results revealed that slump values for slag concrete were lower than those of gravel concrete by 33% for the same water/cement ratio. The unit weight of slag concrete was found to be 2.6–2.7 t/m^3 while it ranged from 2.35 to 2.38 t/m^3 for gravel concrete. The higher unit weight is attributed to the higher specific gravity of slag compared to gravel, 3.5 and 2.65, respectively. For the same cement content and water/cement ratio, and at both early and later ages, slag concrete exhibited higher compressive strength than gravel concrete, with an average increase of 20% at 7 days age and 10% at 28 days age. The same classical effects of water/cement ratio, aggregate/cement ratio, and curing on gravel concrete were observed for slag concrete. From a durability point of view, no sign of self-deterioration was noticed for slag concrete. Slag concrete has been successfully applied in manufacturing sidewalk blocks and profiles, manhole covers, and balance weights.

Kourany and El-Haggar (2000) investigated the utilization of different slag types as coarse aggregate replacements in producing building materials such as cement masonry units and paving stone interlock. Cement masonry specimens were tested for density, water absorption, and compression and flexural strengths, while the paving stone interlocks were tested for bulk density, water absorption, compressive strength, and abrasion resistance. They also studied the likely health hazards of the proposed applications. The test results proved in general the technical soundness and suitability of the introduced ideas. Most of the slag solid brick units showed lower bulk density values than the commercial bricks used for comparison. All slag units exhibited absorption percentages well below the ASTM limit of 13%. Substantially higher compressive strength results were reached for all masonry groups at 28 day age compared to the control and commercial bricks as seen in Figure 10.3. All test groups showed higher compressive strength than the ASTM limit of 4.14 MPa for non-load-bearing units. At slag replacement levels higher than 67%, all groups resulted in compressive strength higher than the ASTM requirement of 13.1 MPa for load-bearing units. All slag types resulted in paving stone interlocks having water absorption values far below the ASTM limit as shown in Figure 10.4. All slag paving stone interlocks showed higher compressive strength and abrasion resistance than the control specimens made of dolomite. Moreover, the proposed fields of application were found to be safe to the environment and have no drawbacks based on the heavy metal content and water leaching test results.

Szekely (1995) proposed a comprehensive research program to reduce fume formation in the BOF and EAF, find an effective approach to reduce and utilize steel-making slag, and effectively use the oily sludge produced in rolling mills. Related environmental problem areas were discussed and preliminary solutions were identified. From Szekely's viewpoint, although

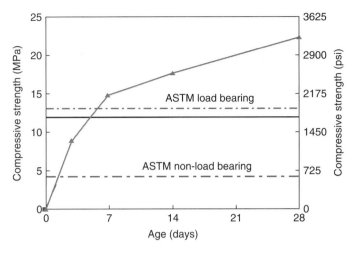

FIGURE 10.3 Development of compressive strength with age for solid brick groups at 100% slag replacement levels (Kourany and El-Haggar, 2000)

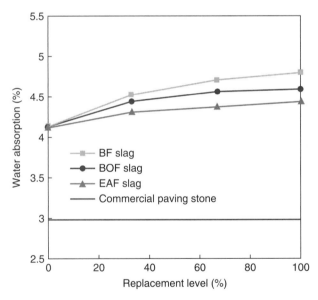

FIGURE 10.4 Comparison between absorption ratios of the different slag types with replacement level (Kourany and El-Haggar, 2000)

several technologies are available for treating EAF dust, they are quite expensive and satisfactory solutions for the EAF dust problem have not yet been produced. He suggested some possible solutions worthy of exploration such as modifying the charging, blowing, and waste gas exhaustion system to minimize dust formation. Another proposed solution is to examine the composition of the dust produced during different phases of furnace operation and, if appropriate, segregate the recovered dust.

Some of the methods used to turn steel plant dust into a valuable raw material were described by one of the solid waste processing companies in its article published in 1997 (Heckett MultiServ, 1997). One of the commonly used methods is micro-pelletizing where dust is mixed with lime as a binder and pelletized to produce a fine granular form, the major proportion being in the size range of 2–3 mm with a total size range of 1–10 mm. The water content is adjusted to 12% during mixing and the pellets are air-cured for a minimum of 3 days before charging to the sinter plant where they account for 3% of the total charge. The article also addressed the direct injection process which is currently on trial in Germany and the UK where injection is used to pass fine dust into the liquid metal in the furnace. Fine dust is blended with hydrating dusts such as burnt lime and carbon. The metal oxide content of the material is about 70% and has a particle size range of 0–8 mm making it ideal for direct injection into a range of furnaces.

Final remarks

Besides the economic and technical importance of utilizing waste materials generated from the iron and steel industry, this activity is of great importance from the environmental protection point of view. The first environmental impact is the useful consumption of the huge stockpiles of these waste materials. When these waste materials are used as replacements for other products such as cement, the natural resources which serve as raw materials are preserved. Also, air pollution levels will be reduced due to the reduction in fuel consumption. All slag types should be treated as byproducts rather than waste materials. All existing and new steel plants should have a slag processing unit within the factory premises to extract steel from slag and to crush the slag and sieve it to the desired grading for ease of promotion.

10.4 Aluminum Foundries Case Study

Aluminum has been recycled (The Secondary Aluminum Industry) since the days it was first commercially produced and today recycled aluminum accounts for one-third of global aluminum consumption worldwide. The growth of the market for recycled aluminum is due in large measure to economics. Today, it is cheaper, faster, and more energy efficient to recycle aluminum than ever before. For instance, only about 5–8% of the energy required to produce primary aluminum ingot is needed to produce recycled aluminum ingot. This roughly translates into an annual energy saving of 13 million gigajoules, the equivalent of 2.1 million barrels of oil (Natural Resources – Canada 2002). In addition, to achieve a given output of ingot, recycled aluminum requires only about 10% of the capital equipment compared with primary

- Collect the slag in bowls to facilitate handling for transport to the slag department as well as keeping the area reasonably clean.
- Arrange the mold in such a way to be close to the smelter and away from traffic.
- Stack the products loosely to facilitate the heat transfer and subsequently reduce the annealing time.
- Reduce the size of the slabs during the pouring phase in order to reduce the scrap (which has to be melted again and reduce unnecessary stretching).
- Change the square slabs into round slabs.
- Cutting scrap should be collected and compacted to a suitable size for the smelters.
- Install temperature measurement and temperature control for quality and energy management in both smelter and annealing furnace.
- A lighting system should be installed for better illumination and of course better product quality.
- A good ventilation system is needed to improve working conditions in the finishing department.
- Designate an area for slag waste (powder) collection.
- Upgrade several processes: smelter oven, annealing furnace.
- Safety protective equipment for workers is required such as gloves, boots, masks, etc.

Implementation of cleaner production techniques

The cleaner production measures which have been identified for implementation are briefly outlined below. Particular attention was paid to the reduction of costs to the factory.

Good housekeeping

- Space management:
  - Better scrap management in the storage area.
  - Arrange molds near the smelters away from traffic.
  - Stack the products more loosely.
  - Cutting scrap should be collected and compacted.
- Shape and dimensions alteration:
  - Cut the big blocks into smaller pieces before smelting.
  - Reduce the size of the slabs during the pouring phase in order to reduce the scrap.
  - Change the square slabs into round slabs to reduce scrap.

Process modification

Upgrading the smelting furnace: Smelting is the most important process in an aluminum foundry to mix and melt aluminum scrap and aluminum ingots in order to control the pure aluminum content according to specification.

A huge amount of energy and material loss were found in all smelters in small and medium sized enterprises. The design, manufacturing, and operation of an aluminum smelter will be discussed in detail throughout this case study taking the following into consideration:

- Improvement of the combustion process.
- Improvement to temperature distribution.
- Installation of temperature control equipment.

Upgrading the annealing oven: The annealing oven has no temperature measurement and control with no homogeneous distribution of heat within the furnace. Furthermore the heat transfer is obstructed by the fact that the products are tightly stacked on top of each other. The rapid annealing of the top products forces the management to open the oven at specific times to remove the top products or add more products at the top. This consumes extra energy, as the furnace door is opened four times during one charge. The upgrading of the annealing oven will be discussed in detail throughout this case study from design, manufacturing and operation points of view, taking the following into consideration:

- Improvement to the distribution of heat within the furnace.
- Installation of temperature control equipment.

Upgrading the rolling machine: Noise and vibration generated from the rolling machine are very high because of the transmission system. It is recommended to replace one drive for two rollers with two independent drives for two rollers.

Upgrading the spinning machine: Noise and vibration generated from the spinning machine are very high because of the transmission system (pulley and belt system). Large physical effort is needed to operate the machine which could cause health hazards if operated for a long period. It is recommended to change the pulley and belt system for a motor driven system.

Upgrading the polishing processes: Polishing is an important process for finishing aluminum products prior to market. Cloth brushes are used to clean aluminum products from oil, grease, and aluminum oxides. Poor polishing leads to high suspended particulates in the atmosphere, bad working environment, and high heat losses as a result of friction causing high temperature in the working environment. Poor polishing requires frequent maintenance and repair of machines.

The most problematic environmental aspect of the polishing system is the particulate emission. A good polishing design should be based on dust collection and ventilation concepts. All particulates generated from the polishing processes will be collected through a hood, blower, and duct system to ensure a good working environment. The water scrubbing mechanism shown in

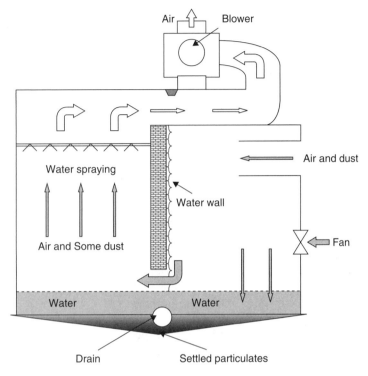

FIGURE 10.5 Water settling scrubbing system

Figure 10.5 will collect all dust and ensure a good surrounding environment. Providing a good working environment can guarantee a long lifespan for machines and good occupational health for the worker force, which increases their productivities and reduces time off work due to sickness.

Aluminum smelter

Smelting is an important process to mix and melt aluminum scrap and aluminum ingots in order to control the pure aluminum content according to specification. Degassing chemicals are used to improve the quality of aluminum. The aluminum scrap is stacked either loosely or in large blocks near the smelter. The modified crucible oven is connected to a fuel supply (LPG bottles and/or diesel fuel) to provide the oven with the required energy for melting. The combustion process can be controlled by adjusting the air/fuel ratio and installing a chimney to ensure a good working environment. The gaseous emission from degassing chemicals and aluminum dust from the aluminum scrap can be controlled by installing a hood system above the smelting oven. All these emissions will be collected through a chimney and hood system to ensure a good working environment.

A poor smelting oven leads to high combustion emission, bad working environment and high heat losses. A poor smelter requires frequent maintenance, repairs, and replacement of the smelter. A good smelter design should be based on conservation of energy and materials, and preventive maintenance concepts.

Poor smelter design leads to excessive slag as a result of the formation of aluminum oxides from direct burning at the surface of the crucible. Because of the indirect combustion process with the new furnace, the amount of slag reduces and the raw material increases. The slag is continuously removed from the top of the melt. Because of the closed smelter with the new furnace, the use of degassing chemicals can be handled very easily in order to improve the quality of the melt.

Smelter design
The heart of the smelting furnace is the combustion process. Attention should be given to the burner, combustion chamber design, and the insulation to insure minimum heat loss and ensure that desired temperature is attained with minimum fuel consumption. An indirect flame should always be used to guarantee less slag and more material produced.

The smelter should be equipped with a chimney to ensure removal of all combustion gases from the working environment. An insulation cover should be used to cover the crucible and protect the working environment from the emission produced as a result of using degassing chemicals and aluminum scrap.

A new closed indirect flame smelter was designed, with the same capacity as that used in the Technical foundry in Met Ghamr, Egypt, to replace the open smelters operating with very poor energy efficiency as shown in Figure 10.6. The smelter was designed to accommodate all types of scrap, and was equipped with a burner, chimney, and hood. The burner was located in a tangential position to the combustion chamber in order to allow the flame to enter with a swirling motion away from the crucible. This facilitated a better combustion and heating process as well as less damage to the crucible and less maintenance to the smelter.

The combustion chamber was located between the crucible and outer casing. At the end of the combustion chamber, all exhaust gases were sent through the chimney. The chimney was designed to allow air entrainment through four rectangular slots at the entrance to dilute and control the exhaust gases before going to the atmosphere. The crucible was covered with an insulated cover, to decrease the heat loss and the material's emissions (particulate matter). In addition, the smelter was covered with a hood to collect any emissions and exhaust gases to protect the working environment. The specification of the smelter is as follows:

- Crucible size = 250–500 kg aluminum per charge
- Smelter height = 130 cm

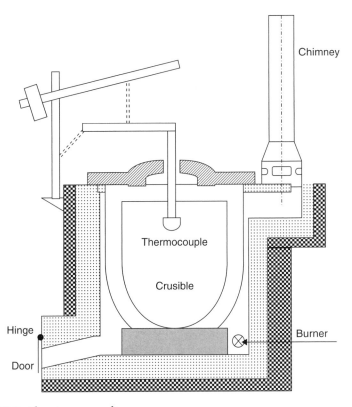

FIGURE 10.6 Aluminum smelter

- Smelter diameter = 130 cm
- Lid: insulated with ceramic fibers (80 cm diameter and 5 cm thickness)
- Air blower capacity = 2 hp
- Chimney height = 180 cm
- Insulation: 5 cm ceramic fibers followed by 5 cm light bricks and 10 cm firebricks

Benefits of improved smelter design
In the case of the Technical foundry where an improved smelter design was introduced, indirect burning prevented metal oxidation and decreased the slag by 3% thereby increasing production by 3%. This led to an income of $18,950/y assuming an average of 2 tons of aluminium production each day.

In the same company, the new smelter could afford a longer duration of maintenance-free operation of up to 45 days because of the improved material specifications and design, instead of an earlier maintenance schedule of 15 days. This led to reduced maintenance, repairs, and replacement costs resulting in total savings of $1,400/y.

In the old smelter, the crucible had to be replaced more frequently. In the new smelter, the lifetime of the crucible increased by 50% and led to savings of $4,385/y.

These benefits led to a net saving of at least $24,735 on an annual basis. The investment made was paid back in less than 3 months. There were several other non-quantifiable benefits including:

- Reduction in slag formation, coupled with less maintenance duration, led to an increase in production capacity of more than 3%.
- Reduction in slag resulted in less release of waste to the environment.
- Reduction in risks to workers, as in the open smelter there used to be frequent incidences of flying ash.
- Reduction of emissions of total suspended particulates in both the working and surrounding environments, due to closed operation of the smelter and because the stack provided better pollutant dispersal. The concentration of particulates near the old smelter was 15 ppm and higher. This concentration was reduced to less than 10 ppm, which is the allowable limit according to the Egyptian environmental regulations.
- Environmental compliance achieved.

Implementation and operation advice:

- Good insulation should be used to prevent heat loss.
- Exhaust emission (Total Suspended Particulate) control should be implemented to protect the working environment.
- The crucible should be covered with a lid to prevent heat and emissions generation.
- Burner location should be tangential to the oven wall and away from the crucible.
- Crucible base should be used to have enough room for the burner.
- It is recommended to have rectangular slots in the bottom of the chimney.
- The degassing chemicals can be safely used, as the smelter functions under closed operation. If degassing chemicals are used, the quality of the molten metal will further improve and this would lead to a further reduction in the generation of the slag, leading to even higher profits.
- The lid should always be used during operation and compact scrap blocks used instead of loose scrap.

Annealing furnaces

The annealing process is performed to relieve the internal stresses that have occurred during the rolling process. The temperature inside the annealing furnace reaches 400–600°C, which is sustained for 1.5 hours using LPG bottles or diesel as fuel. The annealing process is done two or three times before reaching the cutting stage, and after longitudinal stretching, transversal stretching, and cutting into circular discs. The number of times the annealing process is repeated depends on production and furnace capacities.

The annealing furnace implemented at Tiba Company is equipped with temperature measurement and control devices. The homogeneous distribution of heat inside the furnace is guaranteed by the circulation of the hot gases from the combustion chamber all the way up in the furnace and down into the product before flowing into the chimney.

Design of annealing furnace
A new annealing furnace was designed, manufactured, and installed at Tiba Company in Met Ghamr, Egypt. The new design had 25% higher capacity than the old furnace to allow extending of the annealing time from 1.25 hours to 1.75 hours. The hot gases of the new furnace flow all the way up from both sides of the furnace wall and move down through the aluminum products, then exit through the chimney as shown in Figures 10.7 and 10.8. This led to better temperature control and distribution of heat.

The furnace consisted of two burners, control unit, carriage, main furnace body, and door as shown in Figures 10.9 and 10.10. The burner consisted of: two high pressure 2 hp fuel pumps, 0.75 inch gas filter (250 mesh size), 0.75 inch safety valve 10 bar pressure, 0.75 inch solenoid valve, two 4 hp 60 cm blowers, 4 gallons per hour diesel fuel atomizers, and air–fuel mixing chamber as shown in Figure 10.11. A control unit with on/off signals to fuel pumps was provided to maintain optimum temperature. The main body of

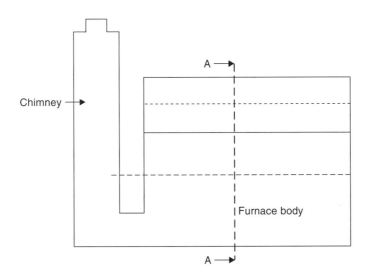

FIGURE 10.7 Annealing furnace – side view

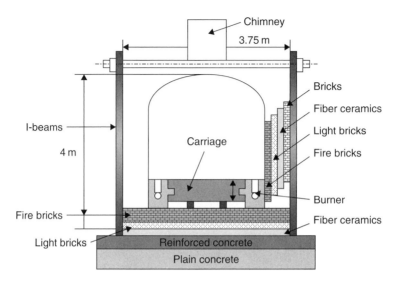

FIGURE 10.8 Annealing furnace – sectional front view A-A

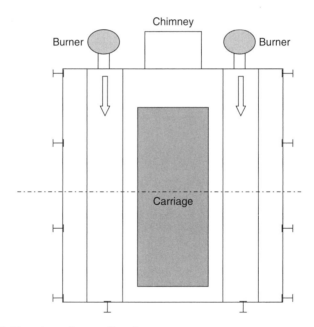

FIGURE 10.9 Top view of annealing furnace

the furnace was well insulated with light bricks, ceramic fibers, and clay hollow bricks.

The oven door as well as the carriage was heat insulated to reduce the amount of heat loss. The furnace door was insulated with ceramic fibers and

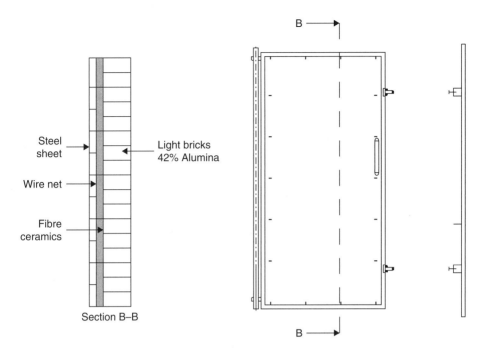

FIGURE 10.10 Annealing furnace door

light bricks as shown in Figure 10.10, while the carriage was insulated only with light bricks. The specification of the annealing furnace is as follows:

- Furnace dimensions: length = 5 m, width = 3.75 m, height = 4 m
- Carriage dimensions: length = 3.6 m, width = 1.2 m, height = 1.7 m
- Chimney: clay bricks 25 cm thickness followed by 25 cm firebricks. Additional sheet metal chimney is added at the top to move the exhaust gases outside the factory. The internal cross-section = 40 cm × 40 cm with 5.5 m height.
- Furnace insulation: 5 cm fiber ceramics followed by light bricks of thickness 6 cm for the floor and 12.5 cm for all walls with outside wall made of clay hollow bricks of thickness 40 cm. The fire bricks of thickness 15 cm for the floor and 25 cm for all walls were added inside the furnace
- Door dimensions: width = 1.2 m, height = 2.2 m, thickness = 22 cm
- Dimension of combustion chamber: length = 4.2 m, width = 25 cm, height = 25 cm.

Benefits of improvement in the annealing process:

- Heat losses from the furnace were reduced by 80%. This reduction, coupled with better temperature control, led to a drop in fuel consumption

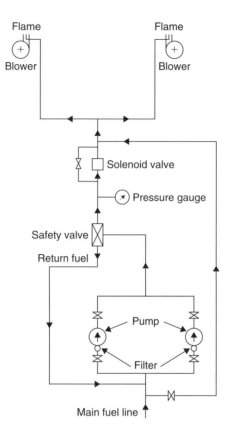

FIGURE 10.11 Combustion system and control

with a total saving of \$4,210/y. In this process, the electricity consumption, however, increased by \$350 as a result of using a fuel pump for fuel atomization and better combustion performance.

- The material wastage or rejects after the rolling process were reduced from 4% to 3% of the production, as a result of the homogeneous temperature distribution within the aluminum sheets. This led to a total saving of \$7,000/y as more material was now available to sell.
- Again, due to the homogeneous temperature distribution within the annealed discs, the trimming losses of the forming process in the products were reduced by 20% with a total saving of \$2,800/y.
- For the old furnace, maintenance was required once a month. The new furnace required maintenance only once every 4 months. The maintenance costs of the annealing furnace therefore decreased leading to a total saving of \$395/y.

These benefits led to a net saving of \$14,055 on an annual basis. The investment made was thus paid back in less than 14 months. Further, since the capacity of the furnace was increased by 25%, the production was increased by 25% and this could lead to a potential benefit of another $8,770/y. There were several non-quantifiable benefits as well, including:

- It is expected that the lifetime of the new annealing furnace would be double that of the old furnace.
- Because of the homogeneous distribution of hardness in the annealing process, the instances of disintegration of metal reduced by 50%, which reduced injury to the workers in the rolling and forming areas.
- Heat and air emissions in the workspace and neighborhood were considerably reduced due to improved operation of the annealing process and because the stack provided for better pollutant dispersal. The temperature at the immediate workspace was reduced from 36°C to 27°C and the air pollutants were maintained within the regulatory limits.

Implementation and operational advice:

- Good insulation should be used to prevent heat loss not only on the walls but also for the floor.
- Temperature control in different locations should be maintained to guarantee homogeneous distribution of heat.
- Exhaust gases enter the chimney from the bottom to increase the gas circulation within the furnace and guarantee the heat distribution. The air-to-fuel ratio should be adjusted to control the exhaust gas emissions.
- The I-beam should be tightened after warming up the furnace to allow the furnace body to expand during the commissioning phase.
- The railway should be protected from heat exposure to avoid distortion, this can be achieved with proper insulation of the carriage.
- Heat losses around the door frame should be minimized by using a proper lock system and extending the fiber ceramic insulation layer all around the boundaries.
- The combustion process should be gradually changed from a traditional to a full control system in order for workers to adapt to the management of change criteria.
- The fuel pump should be turned off during the exit of carriage from the annealing furnace.
- The fuel filter should be cleaned continuously (if liquid fuel is to be used).

10.5 Drill Cuttings, Petroleum Sector Case Study

The oil and gas industry is growing every day with a lot of investments and returns. Although this industry returns a lot of income worldwide, it is associated with several wastes that have great impacts on our environment. These wastes relate to the different activities done in the exploration stage all the way to the production stage such as:

- Drilling operations: Drilling activities are the essential activity in the oil and gas production processes. This operation involves drilling a well in the ground (on-shore or off-shore). Several chemicals for oil-based, water-based, and synthetic-based mud are used for the drilling operation. The cuttings that come out of the ground are usually contaminated with chemicals, mud, and oil. The wastes associated with the drilling operation include the following: drill cuttings, solids, hydraulic fluids, used oil, rig wash, spilled fuel, spent and unused solvents, and paints, scrap metal, solid wastes and garbage. The drilling fluid ("mud") that is used in drilling and responsible for getting the drilling cuttings out of the well include chemical contaminants as well as the above-mentioned wastes. The mud waste includes soda ash, calcium carbonate, caustic soda (NaOH), magnesium hydroxide, acids, bactericides, defoamers, and shale control inhibitors along with other chemicals.
- Production operations: Production operations include well operation and piping facilities to produce and separate oil from gas, water, and other impurities. This is a wide range of activities, which result in a wide variety of wastes that include: produced sand and shale, bottom sludge, paraffin, slop oil, oil and produced water-contaminated soils, treating chemicals, gas, hydrogen sulfide, hydrocarbon condensates, and naturally occurring radioactive materials.

Waste generated

This section will look closely at the wastes associated with the oil industry. It is evident that the drilling fluids and drill cuttings constitute a great portion of these wastes and they are associated with the drilling stage. Drill cuttings are the cuttings that come out of the well during the drilling operation. The cuttings are usually mixed with the drilling fluid which is used to maintain hydrostatic pressure during well drilling and at the same time flushes the cuttings back to the surface. There are different types of drilling fluids used in the drilling operations:

- Water-based mud (WBM) in which the cuttings are suspended in water.
- Oil-based mud (OBM) in which the drill cuttings are suspended in a hydrocarbon distillate rather than water. The oil base includes diesel oil and more recently low toxicity oil-based mud.
- Synthetic-based mud (SBM) in which the cuttings are suspended in synthetic oil rather than distillates of mineral oil. SBMs share the desirable drilling properties of OBMs but have lower toxicity, faster biodegradability, and lower bioaccumulation potential.

The objective of this case study is to develop a cradle-to-cradle solution for two of the oil industry wastes: the drilling fluids and drilling cuttings. Several alternative solutions are to be explored. One of the recommended alternatives for implementation will be justified by simple economic analysis.

Methodology

The methodology followed in this case study could be summarized in the following steps:

- Identify problem(s) through focusing on one of the oil industry's wastes and figure out its environmental impact and current ways of handling the waste.
- Identify cleaner production opportunities for this waste.
- Prepare a quick analysis for each opportunity.
- Research some of the opportunities where required.
- Develop a cradle-to-cradle solution for the selected oil industry waste during the drilling operation and assess this solution with simple economic analysis.

Approach

Industrial ecology is an environmental concept developed by researchers to improve environmental management as discussed before. Industrial ecology attempts to induce balance and cooperation between industrial processes and environmental sustainability, such that neither violates the other. This approach, thus, aims to develop industrial processes that minimize material waste and pollutants in materials, according to the cradle-to-cradle concept. The cradle-to-cradle concept can be considered an extension to the industrial ecology concept. L.W. Jelenski defines industrial ecology as: "a new approach to the industrial design of products and processes and the implementation of sustainable manufacturing strategies. It is a concept in which an industrial system is viewed not in isolation from its surrounding systems but in concert with them. Industrial ecology seeks to optimize the total materials cycle from virgin material to finished material to component, to product, to waste products, and to ultimate disposal" (Garner, 1999). Under this definition, industrial ecology would involve the optimization of the whole material life cycle from the raw material extraction to the final product including components and the wastes that result from each stage of the material production and process.

Cleaner production opportunity assessment

The first cleaner production opportunity is to recycle the mud (with water base, oil base, or synthetic base) in a cement kiln as an alternative fuel and raw (AFR) material. This will reduce the amount of fuel used and will replace some raw material provided that the end products will not be affected technically (i.e. the specifications of the cement produced will be maintained or enhanced, not worsened). Thus, it is essential to thoroughly test the AFR material before employing it in the cement kiln to ensure that the cement specifications are not negatively affected.

The mud coming out of the well is usually mixed with cuttings and some chemical additives, and the selection of which base type depends on the well formation. Therefore, there are three stages involved in the cradle-to-cradle scheme that would present reuse/recycle options for the drill cuttings:

- Stage one: Mud-cutting separation.
- Stage two: Treatment.
- Stage three: Reuse/recycle opportunities.

Mud-cutting separation
The cycle of the drilling fluid is considered to be a closed cycle, where the fluid is pumped to the bore hole and upon returning to the surface it is filtered to separate the cuttings before repumping it to the well. Below is a brief description of the filtration process which consists of two stages: the first involving shale shakers; the second desilters and desanders.

- Shale shakers are the main components in the mud filtration cycle. The main function of the shale shakers is to remove the cuttings from the circulating fluid. These devices are mounted on the tanks in order to purify the mud before storage and recirculation. The cuttings are then transferred to the cutting pits or land storages. This process is commonly implemented in most drilling rigs.
- Desilters and desanders are used to purify the mud from the fine particles that could increase the weight of the mud and erode the drilling stem if they are not removed.

Treatment
In addition to the oil covering the rock particles, the drill cuttings contain other contaminants even after the separation process. And those contaminants are in the form of chemicals from the drilling, fluid and suspended toxics, and heavy metals. As a result, any plans for reusing the cuttings or proper disposal should be done after treatment, where the purpose of the treatment is to get rid of the toxic and liquid fluids (water or oil). There are different methods of treatment that could be used in this regard, such as:

- On-site indirect thermal desorption: This process depends on evaporating the hydrocarbon oils and contaminated soils.
- Distillation: A process that enables the solids and liquid mixtures to separate each constituent through evaporating at different temperatures.

It is usually done by thermo-mechanical conversion and cracking, or by thermal stripping.

- Solvent extraction: Involves mixing the cuttings with specific solvents that allow for distillation and separation without causing damage to the oil constituents.
- Combustion: An incineration process that gets rid of contents that are toxic, flammable, and resistive to biological breakdown. It produces solid waste and ashes which are easily disposed of.
- Stabilization: Modifying the cuttings into more usable form or less hazardous waste. This is done by encapsulating the wastes into a large solid mass to minimize leaching. Organic or inorganic additives can be used to stabilize the mass.
- Biological treatment: Depends on the process of microbial breakdown of the waste. Techniques employed here are the same ones employed for organic wastes such as the aerated, anaerobic, and composting techniques. The process speeds up the natural decomposition of organic liquids in the cuttings. The advantage of this process is that the resultant compost could be used as a soil nutrient and in sand reclamation.

Reuse/Recycle opportunities (Smith et al., 1999)
Mud cuttings are considered a waste that could have many reuse/recycle opportunities. This is due to the fact that they are available in large quantities at any oil or gas exploration site. Therefore they are an important candidate for environmental researchers to create possible reuse options in the manner of achieving a cradle-to-cradle cycle.

From the previous discussions, the first step in reusing/recycling mud cuttings is treatment to get rid of oils and toxic constituents mixed with the cuttings. Therefore the first reuse opportunity is recovering the oil in the treatment process such as the distillation process or solvent extraction process as discussed before. In fact mud cuttings have several reuse options that could be implemented depending on the type of cuttings (formation) and the amount of minerals available. The reuse techniques could be classified under the category of construction and landscaping.

Cuttings could be reused in construction and landscaping, for example in concrete products, coastal defenses, and land reclamation, pipe beddings, landfill cells, roads and pavements, top soil admix, and finally fill materials as follows:

- Concrete products: It was found that treated cuttings could be used as aggregates in concrete mix. However, this is not always applicable due to different cutting piles with different types of rocks and minerals that depend on the type of earth formation. Further research confirmed that the cuttings could be used as "binder filler with Portland cement as an activator or as an aggregate" (Smith *et al.*, 1999).

Other applications in the construction industry are brick manufacturing. However, this issue is still impractical and needs more research and testing to overcome the problem of different rock types in the cuttings.

- Coastal defense: Due to the fact that some cutting types have high salinity, they are possible candidates for costal and sea defenses in addition to marine structures as an aggregate in the concrete mixture.
- Land reclamation: Drill cuttings could be reused in filling of costal reclaimed land. This is done by filling pre-isolated cells with cuttings and fill materials.
- Pipe bedding: Pipe beddings are the material used to support pipes. General construction standards require 10 mm particle size aggregates; this creates a possible reuse for drill cuttings.
- Landfill cell construction: Drill cuttings could be used to construct cell walls or cell covers. However, this is not considered a reuse but a suitable disposal of the cuttings.
- Roads and pavements: One use of cuttings is to stabilize surfaces that are subject to erosion, such as roads or drilling pads. Oily cuttings serve the same function as traditional tar-and-chip road surfacing. Application rates should be controlled so that no free oil appears on the road surface. However, in most cases drill cuttings are not reused until they go through a treatment process.
- Topsoil admix: Admix is a low grade aggregate material used as topsoil in landscaping. Studies encourage the reuse of drill cuttings as a mixture with different topsoil aggregates. Although it doesn't seem to be economically encouraging it is considered a reuse option.

Constructing roads requires very large amounts of aggregate materials in addition to asphalt. As a result drill cuttings could be reused in the filling process. The process of reusing the drill cuttings in road filling and construction depends on filling the land with cuttings and spreading it along the roadway.

However, the cutting aggregates need to be stabilized or in other words reinforced with heavier aggregates. The stabilization is done by spreading stone rejects above the cuttings. Finally, the road is surfaced with asphalt to complete the road construction. However, as the reuse of drill cuttings in road construction seems encouraging, it is limited to low and medium quality roads, as the roads' service life is still under examination.

- Fill material: Drill cuttings could be used as a fill material in landscaping, as requires the use of aggregate materials.

Cradle-to-cradle approach

To achieve a cradle-to-cradle approach, analysis should be done to assess how useful this approach is in the process. The analysis should also include a technical–economic assessment of the process (i.e. an assessment of whether

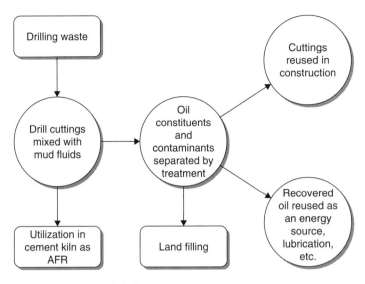

FIGURE 10.12 Utilization of drilling waste

it is technically applicable and economically feasible). When carrying out an assessment according to the cradle-to-cradle concept, it is important to keep in mind that this concept seeks to reach a closed loop cycle for the product under consideration.

In the previous discussion it was shown that the cuttings could be reused/recycled in different applications after treatment. In addition some treatment processes are able to recover a considerable amount of the oil coating, the cuttings, and other mud fluids.

Figure 10.12 demonstrates different alternatives to utilize the drilling waste in order to approach a cradle-to-cradle concept by creating, from a waste material, different products that enter into a closed life cycle. Thus, in the case of drilling fluids/drill cuttings, one cannot say that a cradle-to-cradle concept is 100% fulfilled as the cuttings are only partially reused and the treatment process involved produces some wastes that are disposed of in landfills.

Economic assessment
An economic assessment of the developed cradle-to-cradle solution for the drill cuttings reveals the feasibility of such a solution. The following estimates demonstrate this.

First an estimation is made of the amount of cuttings being extracted from the ground during drilling. Average well depths vary from 6,000 ft–14,000 ft with an average bore diameter of 10 in. Using these estimates, the volume of the drill cuttings per well would be 3,275–7,650 ft^3. This volume could generate a road construction area of 305–710 m^2. Average market prices for paving a road are around \$5.30/m^2 of road. Thus, it is expected that this method would yield a profit or savings depending on the use of about $1,600–$3,740 per well. It is important to note that an oil field will have a number of wells, producing a large amount of drilling fluids and drill cuttings.

10.6 Marble and Granite Industry Case Study

The marble and granite industry is one of the largest industries worldwide. Marble and granite have been used in Egypt since the time of the ancient Egyptians. We can see them in the Egyptian as well as Greek temples in Alexandria, Egypt. Marble is a calcium carbonate base formed by sedimentation, while granite is formed from volcanic lava. These two materials differ in their physical and chemical characteristics. Granite is stronger and more strongly bonded than marble, which is why it is stronger and harder to break than marble. Granite is, furthermore, more chemically stable due to the strong bonds between particles. In contrast, marble reacts vigorously with H_2SO_4.

Egypt is abundant with mines of granite and some types of marble. Famous Egyptian granite mines include Ahmaer Asswan, Brichia, and Zaaferna; famous Egyptian marble mines are Glala and Fleto. Egypt produces nearly 400,000 tons of granite and 100,000 tons of marble every year. Marble and granite blocks are cut into 20 mm thick sheets by gang saws having 5 mm thick blades using water as a coolant. As a result of the blade used, for each 20 mm sheet of marble or granite, 5 mm are crushed into powder during the cutting process. This powder flows along with the water forming marble slurry. In other words, 20% of marble/granite produced results in powder in the form of slurry. The marble/granite slurry has an approximate 35–40% water content.

Industrial process

Mining
Marble and granite mines can be found all over mountainous regions, where blocks are cut using giant saws. Special care must be exercised when determining where and how to seize and cut the blocks as this process involves a careful study of the internal cracks within each block and the amount of veins and selling sizes required in the finished product. Handling the blocks is also an important issue which ultimately affects the size of each block. The total waste generated from the entire mining process, through the manufacturing process, and ending at the finished product is in the range of 50–60% of the mineral.

Cutting (sawing)
- The blocks are carried by cranes.
- The blocks are placed according to the dimensions of the slices needed (nearly 2 × 1 m).

- The blocks are cemented and held by wooden ribs so as not to slide or fall.
- A water jet is fixed in the machine as a coolant agent for the saws.
- The saws have a width of around 5 mm each and move in a continuous forward/backward motion.
- Each saw can slice two blocks each day (assuming 8 working hours per day).

Grinding and polishing
After the marble is sliced, the slices are grinded and polished.

- The sheets are first cured in case of cracks or veins using kola (oily filler benzene byproducts that are very transparent and have a highly adhesive effect).
- The marble sheets are polished with rocks of three degrees that abrade any protrusions that might occur throughout each sheet. Salt and oils are added to avoid capillarity and to remove any organic material.
- The sheets are glazed.

Trimming and cutting
The glazed sheets are cut according to the desired shapes and sizes – usually 40 × 40 cm tiles with a 2 cm thickness.

Waste generated and impacts

Slurry is the main byproduct (waste) generated in the manufacturing process of marble and granite. This is due to the fact that slurry is generated at almost all the stages of the process. However, the largest amount is generated at the cutting stage: for each 20 mm thick sheet cut, a 5 mm thick sheet is crushed into slurry. This forms 20% of the block volume. Around 35–45% of the resulting slurry is water. The chemical analysis of marble slurry is shown in Table 10.5.

TABLE 10.5
Chemical Properties of Marble Slurry (Singh and Vijayalakshmi 2006)

Test carried out	Test value (%)
Loss on ignition	43.46
Silica	1.69
Alumina	1.04
Iron oxide	0.21
Lime	49.07
Magnesia	4.47
Soda	Less than .01
Potash	Less than .01

Another byproduct (waste) is marble/granite scrapes (small pieces) which represent nearly 10–15%. These are usually dumped into landfills without any precaution. There are two types of scrapes: bulk stone and polished scrapes. Bulk stone scrapes are the huge masses produced at the mine when the large blocks are cut out of their natural environment. These masses are usually crushed to small aggregates by the use of a crusher, and ultimately used in tiles. Polished scrapes are normally 2–4 cm thickness. Both types of scrapes have the following in common:

- They contain slurry and can negatively affect the soil, water and air.
- They consume space.
- When piled up, they form hills of scrape that can house harmful animals like snakes and rats.
- They are not self-degrading materials.

A third byproduct (waste) is the amount of wastewater generated during all the processes of marble/granite cutting. This water can be circulated as a coolant from the settling tank to the saw nozzle after collecting the slurry for possible utilization as tiles, bricks, or to be used in the cement industry as road paving, as will be explained later.

Impacts

The environmental problems that result from marble/granite slurry are as follows:

- The slurry is collected in a wet form and is discharged in a nearby desert causing soil pollution and possible contamination to surface/ground water.
- Dry slurry diffuses in the atmosphere causing air pollution and possible pollution to nearby water and agricultural lands because of its high alkalinity. In addition, it has a high pH which makes it a corrosive material that is harmful to the lungs and may cause eye sores.
- It may corrode any nearby machinery.
- It depletes natural resources in terms of wastewater, marble/granite powder, and small pieces (solid waste) of marble and granite.

Cleaner production opportunity assessment

Slurry/Scrape recycling
Slurry can be used as a building material by mixing slurry with cement or resins. Marble/granite slurry was successfully used in the following applications:

- Tiles made with marble/granite dust up to 90% bound with resins.
- Tiles made with marble/granite dust up to 40% with small pieces (scrape) up to 50% bound with resins.

- Tiles made with marble/granite dust up to 50% with small pieces of marble/granite scrape up to 40% bound with resins.
- Building bricks have been developed using marble slurry mixed with cement.
- Other recommended applications of marble dust include:
  - Can be used in the cement industry/ceramic industry.
  - Can be used for road-making purposes.
  - Can be used as water injecting material to trap underground water during construction due to its low porosity.
  - Can be used in fire fighting wall panels or as powder in fire extinguishers due to self-extinguishing and light weight properties.
  - Can be used as a material to hold the soil during excavation, like bentonite.
  - Can be used as aggregates in concrete.

Properties of slurry polymer composite resin bound are shown in Table 10.6 and slurry mixed with cement base tiles with marble slurry are shown in Table 10.7.

Wastewater recycling
Over time the slurry's heavy particles settle at the bottom (of the tank, and the remaining water flows to another tank). However, this system still has one disadvantage. It did not cure the water from the alkalinity it gets due to

TABLE 10.6
Slurry Polymer Composite Resin Bound (Singh and Vijayalakshmi 2006)

Property	Result
Density	$1.96\,g/cm^3$
Moisture content	0.2–0.40%
Modulus of rupture	21–26 MPa
Tensile strength	23–25 MPa
Compressive strength	77–96 MPa
Water absorption # 2 hours	0.15–0.40%
Fire retarding tendency	Self-extinguishing
Exposure to boiling water	No change in dimensions
Chemical action	Appears chemically inactive

TABLE 10.7
Slurry Cement Bound (Singh and Vijayalakshmi 2006)

Property	Result
Compressive strength	$80–120\,kg/cm^2$
Water absorption	10%

the presence of lime in slurry. Therefore, it whitens the sheets of the marble and granite while polishing, and it cannot be used in certain automatic polishing systems because it corrodes the machinery.

Two samples were taken from one of the workshops and analyzed chemically. One of the samples was taken from under the polishing machine and the other was taken from the physical settling tank. The following results were generated:

- Regarding alkalinity we can see that the longer the water tubes from the tanks to the machine the less alkaline the water becomes as it drops from 280 mg/l to 120 mg/l. Therefore the physical treatment with time can help the water alkalinity.
- The pH is within normal range as the canal water.
- The very high percentage of dissolved salts was observed and it is not recommended to use this water directly in concrete work or mortar.
- The water contains a very high percentage of sulfates which neutralize the lime effect on water and decrease the pH.

As a conclusion of the above analysis, one can reach the following:

- The physical treatment (settling) is not enough as the reused water whitens the marble due to the soda, potash, and sulfur.
- The high amounts of salts and sulfates will prevent the direct use of this water without proper treatment according to quality of water produced from the physical treatment tank and the quality of water required.

10.7 Sugarcane Industry Case Study

Sugarcane is cultivated in tropical and subtropical areas, as it requires at least 60 inches of irrigation or rainfall per year. The major world producing countries of sugarcane are Brazil, India, and China. Their combined total production exceeds half of the world's production.

Egypt ranks second after South Africa among African sugar producing countries. Over the past decade Egypt's sugar production contributed 13–16% to Africa's total sugar production (Lichts, 2004). The amount of total sugar production and cane sugar production in Egypt are shown in Tables 10.8 and 10.9 respectively.

The sugar produced in Egypt is either from cane or beet; however, cane sugar constitutes 70–90% of Egypt's annual total production (Lichts, 2004).

There are eight cane sugar producing factories in Egypt, most of them located in upper Egypt, and there are three sugar refineries located in lower Egypt. Egypt's average annual refined sugar production is 1.03 million tons, which is 10% by weight of cane production. The average proportions of the

TABLE 10.8
Egypt's Sugar Production (Lichts, 2004)

	95/96	96/97	97/98	98/99	99/00	00/01	01/02	02/03	03/04	04/05
1,000 tons	1,222	1,230	1,171	1,266	1,476	1,586	1,555	1,397	1,488	1,544

TABLE 10.9
Egypt's Cane Sugar Production (Lichts, 2004)

	95/96	96/97	97/98	98/99	99/00	00/01	01/02	02/03	03/04	04/05
1,000 tons	1,108	1,083	1,016	1,006	1,128	1,097	1,061	1,020	1,089	1,090

TABLE 10.10
Average Annual Cane Consumption, Solid Wastes Generation, and Refined Sugar Production of the Sugar Mills in Egypt

Location of sugar factories in Egypt	Average annual cane consumption (×1,000 tons)	Bagasse (×1,000 tons)	Mudcake (×1,000 tons)	Refined sugar (×1,000 tons)
Komombo, Aswan	1,800	630	72	180
Edfu, Aswan	1,150	402.5	46	115
Naga Hamady, Quena	1,750	612.5	70	175
Armant, Quena	1,250	437.5	50	125
Uose, Quena	1,650	577.5	66	165
Deshna, Quena	1,000	350	40	100
Gerga, Sohag	900	315	36	90
Abu Uoras, Menya	800	280	32	80
TOTAL	10,300	3,605	412	1,030

resulting bagasse and cachaza are 35% and 4%, respectively, by weight as shown in Table 10.10. The bagasse has 50% moisture content and the cachaza's moisture content ranges from 50 to 60%. The average annual cane consumption, solid waste generation, and refined sugar production of the sugar mills in Egypt are shown in Table 10.10.

Traditional energy form

The sugar production process is very energy intensive; it requires steam and electrical power at different stages. Sugar mills worldwide often resort to using bagasse, a byproduct of the cane sugar production process, in its loose bulky form as a boiler fuel. Bagasse has a gross calorific value of 19,250 kJ/kg

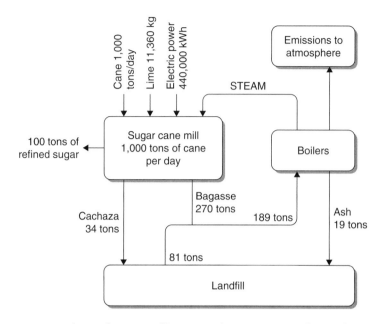

FIGURE 10.13 Traditional sugar mill in Egypt (EL Haggar, *et al.*, 2005)

at zero moisture and 9,950 kJ/kg at 48% moisture (Deepchand, 2001). The net calorific value at 48% moisture is around 8,000 kJ/kg (Deepchand, 2001). Burning bagasse in its loose bulky form is an inefficient process, a proportion of the bagasse remains unburnt and is dumped with the cachaza in a landfill. This alternative is adopted by sugar mills as it helps to reduce production costs. Figure 10.13 illustrates the operation of a typical sugar mill in Egypt such as Komombo Sugarmill located in Aswan.

Impacts of the traditional energy form
The traditional energy form, which is adopted by all sugar mills in Egypt and other developing countries, has major negative impacts:

- Contamination of the surrounding environment: Burning of bagasse in its loose bulky form generates fly ash that causes air pollution. The sugar mills will require expensive scrubbers and filters to purify the emissions.
- Loss of resources: The ash generated at burning is lost due to the bulkiness of bagasse and the lack of control over the burning process. The ash lost to the atmosphere is a resource that should be preserved and used efficiently. The chemical composition of the ash as shown in Table 10.11 reveals its rich nutrient content, which qualifies it for use as a fertilizer.
- Energy inefficiency: The bulkiness of bagasse causes it to have a low energy per unit volume and leads to a low burning efficiency of 60%.

TABLE 10.11
Chemical Composition of Ash (Dasgupta, 1983)

Composition (%)	Ash
Organic matter	17.13
Total potassium	12.5
Total phosphorus	1.24
Iron	0.181
Manganese	0.006
Copper	0.0121
Zinc	0.009
Calcium	0.4133
Magnesium	0.0127

TABLE 10.12
Chemical and Physical Composition of Bagasse and Mudcake
(Dasgupta, 1983)

Constituent (%)	Bagasse	Mudcake
Cellulose	46	8.9
Hemicelluloses	24.5	2.4
Lignin	19.9	1.2
Fats and wax	3.5	9.5
Carbon	48.7	32.5
Hydrogen	4.9	2.2
Nitrogen	1.3	2.2
Phosphorus	1.1	2.4
Silica	—	7.0
Ash	2.4	14.5
Fiber	40.8	15.0

In addition, due to the uncontrolled burning approximately 30% of the bagasse by weight does not burn in sugar mills, which is then used as a fuel for brick manufacturers or dumped in a dumpsite. Other countries dump bagasse and cachaza in a landfill (Nemerow, 1995).

Currently, sugar mills in Egypt use cachaza and the remaining bagasse in a very inefficient way. However, the chemical compositions shown in Table 10.12 reveal a high cellulose content for bagasse and a high organic content for cachaza, which qualifies them as possible energy sources. The high nutritional value of both bagasse and cachaza qualifies them as a good candidate for fertilizer.

The sugar production season lasts for 5 months. For example, in Egypt, the season starts in December and ends in May. The mills operate 24 hours a day 7 days a week throughout the 5 months. Due to the tight schedule and the high costs of production, high efficiency is a vital issue that needs to be maintained and constantly improved. The current process of production in sugar mills is highly inefficient. Resources such as bagasse, cachaza and ash should be utilized by the sugar mills. The traditional energy form is one of the major problems in sugar mills. In addition to its negative impacts, it also causes the loss of these rich resources. Therefore, an alternative energy form should be adopted that utilizes the excess bagasse, cachaza, and ash.

Solution

The possible alternatives for increasing production efficiency and overcoming the negative impacts of the traditional energy form are threefold: bagasse briquetting, biogas, and natural gas or oil fuel. Adoption of any of these alternatives will require modifications to the sugar mills. A description of each of the possible alternatives is provided below.

Alternative 1: Solid fuel using briquetting technology

This alternative involves compressing the bagasse and cachaza at high pressure into densely packed briquettes. The cachaza acts as a binding agent due to its fat and wax content. The resulting briquettes are a possible fuel for boilers in the sugar production process as they have a calorific value of 15,000 kJ/kg.

High compression pressure increases density of the briquettes, which improves their handling and storage properties. In addition, higher density leads to higher energy per unit volume, which is more economical.

Residue size changes density; however, high pressure leads to densities $\geq 0.7\,g/cm^3$ for all residue sizes (fine, coarse, stalk), which is sufficient to provide a high energy per unit volume (Adeleke, 2003). The optimum process conditions are (Adeleke, 2003):

- Pressure applied: 100–120 Mpa.
- Residue moisture content should range between 9 and 12%.
- Cachaza inclusion should not exceed 10%.

Briquetting increases combustion efficiency from 60 to 80%, this increase in efficiency reduces the amount of harmful pollutants to the atmosphere, leads to a more controlled burning process, increases time efficiency, and reduces the amount of bagasse required for burning. The use of briquettes allows the ash, which is rich in nutrients, to precipitate in the boiler, therefore reducing harmful emissions to the environment. The ash can also be easily collected after burning and used as a fertilizer.

The implementation of this process requires the establishment of a briquetting unit. The sugar mill will be more efficient by combining the briquetting unit with the sugar mill to form an environmentally balanced industrial complex (EBIC). Figure 10.14 illustrates a model of an EBIC for a

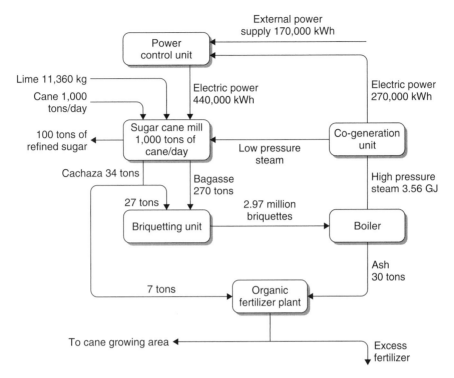

FIGURE 10.14 Environmentally balanced sugar cane industrial complex using briquetting technology (EL Haggar *et al.*, 2005)

sugar mill that processes 1,000 tons of cane per day, utilizing the briquetting technology. The quantities of bagasse and cachaza generated from processing 1,000 tons of cane are 270 and 34, tons respectively (Nemerow, 1995). The amount of cachaza used in the production of briquettes should not exceed 10%, therefore 27 tons are sent to the briquetting unit with the bagasse and the remaining 7 tons will be used in the production of fertilizer. Assuming the mass of a briquette is 100 g, the number of briquettes generated from 270 tons of bagasse and 27 tons of cachaza is 2.97 million briquettes. The entire quantity of briquettes produced will be used as boiler fuel. At 80% combustion efficiency, the boiler produces 3.56 GJ of steam. The steam produces 270,735 kWh of electricity. The electric power generated will be used to supply a proportion of the power requirements at the sugar mill and the remaining amount of power, 169,265 kWh, will be purchased from the grid.

The amount of ash precipitating in boilers is 29.7 tons per 1,000 tons of cane. The ash is collected and mixed with the excess cachaza in the organic fertilizer unit to produce 18.5 tons of fertilizer that can be either used in the cane growing area or sold to local consumers depending on requirements. The price of 1 ton of fertilizer is $44.

Economic evaluation

Although the amount of electric power generated due to the combustion of briquettes is quite high, it is insufficient for the mill requirements and the remaining amount of electric power is bought from the grid at a price of $0.03 per kWh. The cost of electric power from the grid is $4,750 per 1,000 tons of cane (per day). For the briquetting unit, the total labor cost is $42 per 1,000 tons of cane (per day). In the organic fertilizer the total labor cost is $5.30 per 1,000 tons of cane (per day) and the revenue from selling of fertilizer is $811 per 1,000 tons of cane (per day).

Processing of 1,000 tons of cane using briquetting technology costs $3,987. Although in the EBIC the revenue from selling the fertilizer does not cover the costs of processing 1,000 tons of cane, it has to be compared to the traditional alternative, where processing 1,000 tons of cane cost $12,350 since all of the power requirements of the mill are bought from the grid.

Alternative 2: Gasification of bagasse–cachaza, biogas

In sugar mills the processing of cane generates a mixture of bagasse and cachaza with an average 8:1 ratio, respectively. Traditionally, 70% of the bagasse is inefficiently burnt in boilers. The remaining mixture of bagasse and cachaza, which is usually dumped, has an average ratio of 2.4:1, respectively. Dasgupta (1983) investigated the possibility of utilizing the bagasse–cachaza mixture in the generation of biogas (70% methane and 30% carbon dioxide gas) through anaerobic fermentation. Anaerobic digestion is performed by a microbial culture that is developed for this substrate.

Trials were performed using both mixture ratios of bagasse to cachaza and the resulting gas yield was measured. Results revealed that the 2.4:1 ratio of bagasse to cachaza led to a higher gas yield (Dasgupta, 1983). The optimum process parameters for biogas generation are (Dasgupta, 1983):

- Organic loading of 1.0 g Volatile Solids/l/day.
- Detention time of 30 days.
- 100% circulation of the filtrate.
- 6 ml of nutrient solution per liter per day.

Given the above optimum conditions the gas yield is 0.33 l/g V.S. added resulting in a methane yield of 0.24 l/g V.S. added. Volatile solid reduction under this condition is 41%.

The complex proposed in Figure 10.15 illustrates the concept of biogas generation in a sugar mill. For illustrative purposes the estimated mass balances are based on processing 1,000 ton of cane per day. The corresponding amounts of bagasse and cachaza generated are 270 and 35 tons, respectively (Nemerow, 1985).

Most of the bagasse produced, 189 tons (70%), is used in the production of animal fodder and the remaining amount, 81 tons, is mixed with the resulting cachaza, 34 tons, and used in the production of biogas. Anaerobic digestion of

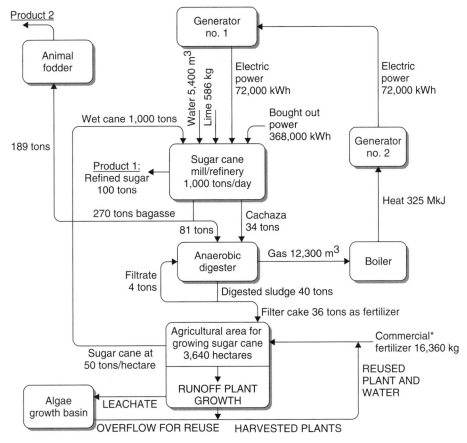

FIGURE 10.15 Environmentally balanced sugar cane industrial complex using biogas technology (Nemerow, 1995)

the bagasse/cachaza mixture generates about 12,300 cubic meters of gas, 70% of which is methane, 36 tons of filter cake and 4 tons of filtrate. The filtrate is recirculated to the digester to enhance the digestion process. The gas burnt in the boiler produces 325,000,000 kJ of steam, which is used in energy (heat and electricity) production. The generator uses the steam to produce 72,215 kWh of the 440,000 kWh of electric power required by the mill. The residual amount of power can be bought from the national grid.

Approximately 3,640 hectares of land are required for harvesting 180,000 tons of cane as feedstock for the refinery at a rate of 1,000 tons per day, for a 180 day growing and harvesting season. The filter cake produced, 36 tons, is used as fertilizer; however, the lack of essential nutrients in the filter cake necessitates the addition of and mixing with commercial fertilizer to guarantee healthy cane growth. Fertilizer and pesticide residues are washed off to a runoff collection basin that drains excess water to the algae growth basin. Plant growth from the runoff basin and algae from the algae growth basin are mixed together and reused with excess water from the growth basin in the sugarcane growing area as fertilizer.

Animal fodder
Bagasse is a suitable animal fodder due to its high fiber and carbohydrate content, as seen in Table 10.13. There are several processes for treatment of

TABLE 10.13
Comparative Analysis of Alternative Fuel Technologies

Traditional (loose bulky bagasse)

Sources of energy

Environmental	Benefits	None.
	Drawbacks	Pollution of the surrounding environment due to fly ash and smoke.
Economical	Benefits	No purchasing cost of fuel.
	Drawbacks	• Requires construction of special furnaces. • Requires the installation of expensive scrubbers and filters. • Waste of renewable natural resources (bagasse, cachaza, and ash).
Other	Benefits	No time delay, fuel always available.
	Drawbacks	• Low energy gain per unit volume due to high moisture content. • Health risks due to polluted environment. • Highly flammable during storage.

Alternative (1) Briquettes

Environmental	Benefits	• Reduced emissions to atmosphere. • Almost no waste generation.
	Drawbacks	Very small amounts of emissions.
Economical	Benefits	• Lower cost of processing 1,000 tons of cane. • Income generated from selling fertilizer. • Work opportunities due to briquetting unit and organic fertilizer plant.
	Drawbacks	Requires additional costs for carrying out modifications to the mill.
Other	Benefits	• Higher combustion efficiency. • High energy per unit volume. • Controlled moisture content.

(Continued)

TABLE 10.13 (Continued)

Traditional (loose bulky bagasse)

		• Compact form that allows for better storage and handling. • Simple production process so no time delays.
	Drawbacks	Flammable.
Alternative (2) Biogas		
Environmental	Benefits	• Reduced emissions to atmosphere. • Negligible amounts of waste generated.
	Drawbacks	Emission of carbon dioxide.
Economical	Benefits	• Lower cost of processing 1,000 tons of cane. • Revenue from selling animal fodder. • Work opportunities for fermentation process and production of animal fodder, and cane growing area.
	Drawbacks	Additional costs due to modifications to mill.
Other	Benefits	Growing own supply of cane.
	Drawbacks	• Anaerobic digestion of organic materials is a complicated chemical process that involves many intermediate compounds and reactions each catalyzed with a specific enzyme or catalyst. • Strict preservation techniques for anaerobic digester are essential. Anaerobic bacteria are highly sensitive to environmental conditions.
Alternative (3a) Fossil fuel: natural gas		
Environmental	Benefits	• Reduced harmful emissions. • Minimum waste generation.
	Drawbacks	• Emissions of carbon dioxide. • Availability of work opportunities.
Economical	Benefits	Profits from selling fertilizer.
	Drawbacks	None.
Other	Benefits	
	Drawbacks	
Alternative (3b): Fossil fuel: heavy oil		
Environmental	Benefits	None.
	Drawbacks	Harmful emissions of sulfur oxides.
Economical	Benefits	• Profits from selling fertilizer. • Availability of work opportunities.
	Drawbacks	
Other	Benefits	
	Drawbacks	Health risks.

bagasse and making it suitable for animal fodder. The following is a brief description of the processes.

- The mechanical process involves shredding the bagasse and soaking it in steam under high pressure and temperature. This process accelerates the digestibility of the fodder without giving it much time for complete digestion. The main drawback of this process is its high cost.
- The chemical process involves the shredding of bagasse into fine sizes and adding chemicals such as urea or ammonia. The chemicals increase the nutritional value of bagasse by increasing its protein content and digestibility. Treatment of bagasse with chemicals lasts for 2 to 3 weeks depending on surrounding temperature. This procedure is inexpensive due to the cheap price of urea and can be easily applied.
- Biological process: Bagasse is buried in the soil, with no aeration for a period of 2 to 3 months, after which it becomes suitable for feeding to animals. The bagasse can be kept in the soil and used for as long as 18 months. This process is inexpensive and is simple to apply. The produced fodder has a high nutritional value and is easily digested.

Economic evaluation In this evaluation it is assumed that the sugar mill will use the chemical process for treating bagasse and making it suitable for animal fodder. The price of urea and the labor required for processing of bagasse to animal fodder is low and can be evaluated at $5.30 per ton. The selling price of 1 ton of animal fodder ranges from $35 to $70. At the given prices and quantities, revenues from selling animal fodder will cover the cost of buying electric power for the mill and the cost of urea and labor if sold at a price of $61.40. This price may be considered high in some areas and the mill might have to sell the fodder at a lower price. Yet, the mill is still running more efficiently than in the traditional form because the revenues compensate for some of the costs of buying electric power. Initially, the mill paid for its entire power supply (444,700 kWh) which cost $12,456.

The amount of animal fodder is 189 tons per 1,000 tons of cane. The cost of processing 1 ton of bagasse (urea and labor) is $1,000. The revenue from selling animal fodder at a price of $44 is $8,290. The cost of buying electric power (at $0.03 per kWh) is $10,526 per 1,000 tons of cane.

Alternative 3: Traditional fossil fuels
Natural gas
Natural gas is a strong candidate for boiler fuel in the sugar mill if there is a natural gas network in the area. Combustion of natural gas produces mainly carbon dioxide and water, which do not cause serious pollution problems to the surrounding environment. Natural gas causes the least pollution as opposed to other fossil fuels and has a high calorific value, 1 ton of natural

gas is equivalent to 1.1 TOE (ton oil equivalent). One ton of natural gas generates 12,795 kWh of electric power.

In this alternative the sugar mill will supply all of its natural gas requirements from the grid. The power requirement in a sugar mill for the processing of 1,000 tons of cane is 440,000 kWh, which requires 35 tons of natural gas.

The bagasse and cachaza produced during the sugar mill process will be used in the production of organic fertilizer. The fertilizer is sold at a price of $44 per ton.

Economic evaluation The price of 1,000 ft^3 of natural gas is US$0.8. At an exchange rate of US$1 = £E5.7 the cost of 35 tons of natural gas is $1,280 per 1,000 tons of cane, as opposed to $12,350 in the traditional alternative. In addition to the low price of fuel, revenue is generated from the production of organic fertilizer. The bagasse and cachaza generated during the processing of cane, 270 and 34 tons respectively, produce 152 tons of organic fertilizer. The cost of producing 1 ton of fertilizer is $5.30, therefore producing 152 tons of fertilizer costs $800 and generates a revenue of $6,667. Therefore, processing 1,000 tons of cane generates a profit of $4,586.

Heavy oil
Heavy oil is a high density, highly viscous petroleum product from petrochemical refining. Its high content of sulfur, heavy metals, wax, and carbon residues makes it unsuitable for combustion. Although major pollution and health problems arise due to the combustion of heavy oil, sugar mills in Egypt are using it as a boiler fuel. Heavy oil has a high calorific value, not as high as natural gas, 0.972 TOE. One ton of heavy oil generates 11,306 kWh of electric power.

In this alternative the mill purchases sufficient heavy oil to supply all of its electrical power requirements. The sugar mill will require 40 tons of heavy oil per 1,000 tons of cane. The bagasse and cachaza generated will be used in the production of organic fertilizer.

Economic evaluation The price of heavy oil is $44 per ton. The cost of purchasing heavy oil is $1.754 per 1,000 tons of cane, as opposed to $12,350 in the traditional alternative. The revenue generated from the production and selling of fertilizer covers the cost of heavy oil, generating profits for the mill. During processing bagasse and cachaza produce 152 tons of organic fertilizer at a cost of $800, which generates a revenue of $6,667. Therefore, processing 1,000 tons of cane generates a profit of $4,112.

Final remarks

A comparative analysis of alternative fuel technologies for a sugar mill in Egypt is shown in Table 10.13. Each technology was presented and theoretically applied to a mill that processes 1,000 tons of cane per day, for the purposes of illustration. It was found that all alternatives were far more efficient than the traditional technology, which burns 70% of the resulting bagasse from cane processing. In addition, it was shown that bagasse can have other uses, other than as a boiler fuel, such as animal fodder and production of fertilizer.

According to the economic evaluations of the different alternatives the use of natural gas and heavy oil would yield profits from the selling of fertilizer. However, economic assessment of the damage costs due to environmental degradation was not included in the economic analysis, which if included would not justify heavy oil as a feasible alternative for boiler fuel.

10.8 Tourist Industry Case Study

The need for natural resources has grown inversely proportional to their availability in the past. Day after day the natural resources are depleting and becoming scarcer, because we use them without generating more resources since the process is irreversible. This has led not only to the scarcity of these resources but also to the contamination of the environment. According to this strategy, sooner or later the world will run out of natural resources, and we would be incapable of providing ourselves with the daily products we take for granted today. Even the environment we live in would end up by being too small to live in since more and more land is becoming contaminated because of our unsustainable strategy of waste management. For that reason, our strategy needs to change, a new methodology of dealing with our waste should be adapted according to a cradle-to-cradle approach in order not to discover one day that we have run out of natural resources and that the piece of land on which we live is now uninhabitable. This strategy should be implemented from the smallest to the largest activities in the community. It is restricted not only to factories manufacturing products that we use; but should be implemented at all levels of industry and society such as tourism.

An eco-touristic resort is the result of using industrial ecology within the tourism resort. This will encourage national and international communities to enjoy their holidays in a safe environment. The resort might consist of hotel(s), housing units, clinic, marina, shopping center(s), water purification plant, wastewater treatment plant, desalination plant, etc. The wastewater treatment plant, water purification plant, desalination plant, solid waste management system, etc. should always be located away from the hotels and housing units so as not to disrupt the residents and tourists.

Although the industrial ecology concept should be applied to all communities, it should be particularly applied to tourist villages, where the environment is the main resource that makes this resort extraordinary and makes it attract tourists from all parts of the world. This strategy will not only protect the environment in the resort, but it will also increase the profits of resort's owners, provide more work opportunities, and ultimately lead to a more sustainable community.

Approach and methodology

There is no doubt that raw materials are becoming scarcer and one day we will find ourselves incapable of making products that we take for granted now. Not only that, but the environment is also becoming more polluted every day because of the way we dump, landfill or incinerate our waste. A new approach should be used to change this dull future. This approach is to design factories, resorts, and communities (rural communities, industrial communities, remote communities, desert communities, urban communities, etc.) in a way that eliminates the waste and pollution totally and enables the efficient use of materials and waste.

Touristic resorts should put the industrial ecology concept at the top of their environmental policy's priorities to be implemented during the construction phase. All liquid and solid wastes generated from the resort activities should be reused or recycled in a closed loop where the natural resources can be reutilized according to life cycle assessment using a cradle-to-cradle not a cradle-to-grave concept. Approaching "cradle-to-cradle" effectively within the resort community might require the use of the 7Rs Golden Rule to identify the wastes and develop innovative techniques to utilize such waste. This of course is not the case for all items; however, the environmental manager should in fact attempt to minimize the amount of waste at the source through an environmental awareness program.

After identifying the flow of materials in the touristic resort, and the flows that cause the loop to be open, we need to find ideas to close this loop in order to approach a cradle-to-cradle concept as shown in Figure 10.18. In order to do that a material and energy balance should be first identified as well as the wasteful processes. The cleaner production technologies should be first implemented within a good environmental management system.

Desalination plant

Most of the resorts worldwide are located next to the sea where there is no fresh water available and the cost of constructing pipelines to transport the fresh water is too expensive. To counter this owners of touristic resorts can construct a desalination plant on the site as shown schematically in Figure 10.16. The desalination plant utilizes water from ground wells. Usually the water from the ground wells close to the sea has a high salinity of 6,000–12,000 ppm. This water is taken to the plant and the fresh water that is produced is taken to the hotels and housing units. The daily volume of the byproduct reject water (brine) from the desalination plant is large. The salinity and temperature of the reject water can be very high depending on the incoming salinity and the technology used. The mechanism of disposing of this reject water (brine) is a continuous problem. The reject water could not be dumped in the sea at such high salinity and temperatures, because it would disrupt the existing ecosystem. The other option is to dump the reject water in a nearby area. However, this was also found to be disruptive because of land use and the highly saline waters will create swamps.

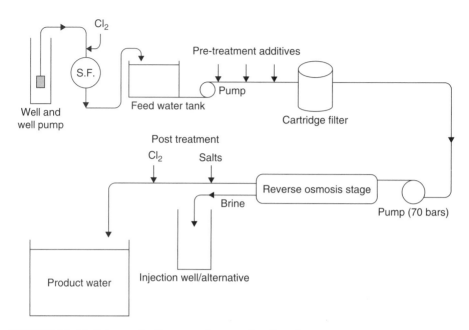

FIGURE 10.16 Schematic diagram of water desalination system

Brine disposal/utilization alternatives

Deep well injection: This method is already adopted at different locations where the brine is injected in wells many meters underground and the salt seeps through the soil and stabilizes itself. This disposal method is very expensive (capital and running cost) and depleting natural resources.

Evaporation ponds: The evaporation ponds are large shallow concrete ponds where the brine is pumped to let the water evaporate and the residue can be collected and turned into commercial products that can be resold to the salt refinery plant.

Solar ponds: The solar ponds are large concrete ponds where the brine is pumped. Then, brine is heated and pumped into the bottom of the pond. This creates a convective current and a small amount of energy can be produced from this process. This alternative requires some experience and solar energy all year round.

Brine shrimp harvesting: Brine shrimps live in salt ponds or salty lakes. They can be used to feed small fish and can also be used as a food supplement for humans. They have a very high nutritional value with vitamins and proteins, can be crushed and given as pills or sometimes added as a gel on the food. Their easy and short production cycle makes them easy to grow and profitable.

Fish farms: A chemical analysis of the reject water should be done to realize the suitability of raising table fish such as grey mullet, sea bass, sea bream and groupers. Thus the idea of a fish farm project can be established in order to dispense the reject water efficiently. El-Guna, Hurghada, located along the Red Sea in Egypt, was constructed in the desert land close to the desalination plant. Ponds were made in the desert where they could raise the fish. The topography of the desert land helped to divide the ponds into three levels to separate the herbivorous and carnivorous species while keeping all the ponds connected. The daily water loss by evaporation and seepage in the fishponds was measured and it was found that the total pond area should be around 9.5 acres for the daily reject water to compensate for the water loss, and through this there will be no extra reject water to get rid of. The only problem at this point was that the reject water from the desalination plant passing through filters and membranes contained no dissolved oxygen or any type of live food such as phyto and zooplanktons. This was necessary for the survival of the fish in the ponds. To solve this problem a waterfall was constructed over the collecting tank where the reject water passes over, increasing the dissolved oxygen to the maximum and finally reaching the ponds for the fish. The fish were then raised in these ponds and taken to the hotels for food or sold to the local markets.

Irrigation of salt tolerable plants: Jojoba is a plant that can grow in many semi-arid regions of the world. It requires little water and maintenance and yields a crop of seeds that have many uses. The seed-oil has been used in many parts of the world as lubricants, cosmetics, pharmaceuticals and as a replacement for sperm oil in manufacturing inks, varnishes, waxes, detergents, resins, and plastics (Tremper, 1996).

Wastewater treatment plant: All wastewater discharged from the hotels and the housing units and any other domestic activity is taken to the water treatment plant where a treatment process reuses the water in the hotels, or to be used to irrigate the crop farms. Out of the treatment process, the byproduct of sludge is collected and used as a fertilizer in the compost fields to create better compost. The sludge contains nutrients that are essential fertilizer for the compost fields and thus can be utilized and not wasted. The water is very sacred, since it is a scarce resource. Thus it must be cautiously and efficiently used.

The wastewater treatment process commonly used in touristic resorts is the activated sludge process. As shown in Figure 10.17 the wastewater is pumped from the sewage network to the homogenization basins where it is homogenized primarily and the bacteria and wastes settle down. Air

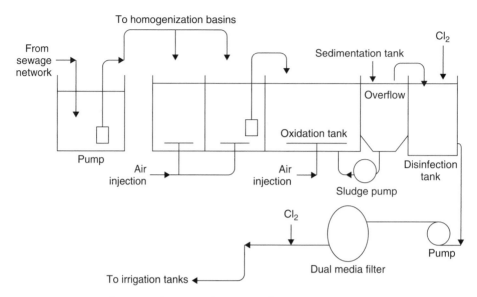

FIGURE 10.17 Schematic diagram of municipal waste water treatment system

diffusers are distributed along the length of the basins to introduce oxygen and enhance the degradation process. A bar screen is located prior to the basin that holds up any big wastes like plastic bags or carton paper so as not to clog the filter. Wastewater is pumped to the oxidation tank where more oxygen is pumped in; this activates the bacteria which start eating up the wastes. The wastewater then goes to the sedimentation tank where the sludge settles down and the relatively clean water overflows to the disinfection tank. Part of this sludge is sent back to the oxidation tank to activate the bacteria digestion cycle. The wastewater is pumped to a dual media filter consisting of sand and gravel. The wastewater will finally be chlorinated to disinfect it and stored in tanks before irrigation.

Animal farming: There is a large amount of fresh food waste that comes out of the hotels and housing units. The best way to recycle fresh food is to raise animal and poultry breeding. The food waste was collected and divided according to category to feed the animals their desired diet. For example, vegetable waste like carrots, lettuce, tomatoes, potatoes, cucumbers, cabbage, etc. went directly to feed the sheep, goats, and ducks. Bread and similar carbohydrate byproducts need to be sun-dried. They are then crushed together with maize bought from the local market to feed the turkeys and ostriches. The animal manure is taken as fertilizer to the compost fields. The animals are bred and either taken to the hotels as a source of meat or sold to the local markets for profit. The sheep also provide wool which is sold to the markets as a raw material.

Organic farming: All food products served in touristic resorts should be grown using organic farming since there is enough compost to be converted into organic fertilizer by adding natural rocks as discussed above. There is no need for chemicals or pesticides to be used in the plantations at all. The fertilizer that is used comes from compost fields. Manure from the animal farms goes into the compost as a means of fertilizer. Plant residue from the golf courses is thrashed and placed into the compost. Also, the sludge byproduct from the wastewater treatment plant is used in the compost as a fertilizer. By this mechanism the compost fields are left to aerate, by means of aerobic aeration. This enables them to control the carbon to nitrogen (C/N) ratio within the compost to ensure the effectiveness of the aerobic fermentation process.

Garbage collection: The garbage collection system might change from one resort to another according to the size of the resort. If the resort is large enough a collection (transfer) station may be developed as discussed in Chapter 5. First, the garbage is collected into garbage rooms near the hotels. In these rooms, the garbage is separated according to material. These rooms should be air conditioned to provide good ventilation to eliminate the bad odors emanating from the garbage. Once this is complete, the garbage is transported by trucks to the large garbage unit outside the resort where the recycling plant is located. On the recycling plant, the workers commence further separation of the garbage on a conveyor belt where each worker is designated a certain item or material like glass, or plastic, etc.

Once the garbage is separated according to material, a process of reuse or recycling occurs depending on the material as discussed above in Chapter 5. Wet food products are taken to the animal farms to feed the animals. Any remaining food rejects might be taken to the biogas unit where it is used to generate biogas and fertilizer.

Medical waste from the clinic is collected separately and placed in an air conditioned room to be collected and sent to the central facility as explained in Chapter 9. This is done to prevent any sort of contamination or infection from the medical utensils.

The flow chart in Figure 10.18 shows how everything in a touristic resort is connected and how the concept of industrial ecology is prevalent in eco-tourism. It is noticed that, all raw materials and wastes are fully utilized in a closed loop. The loop of industrial ecology can successfully prove that an eco-touristic resort can be self-sufficient with its resources utilizing and reusing all the wastes generated in order to minimize/prevent the waste and ultimately save themselves the costs incurred. The flow diagram shown in Figure 10.18 has proven to be a role model for a truly environmentally friendly eco-touristic resort.

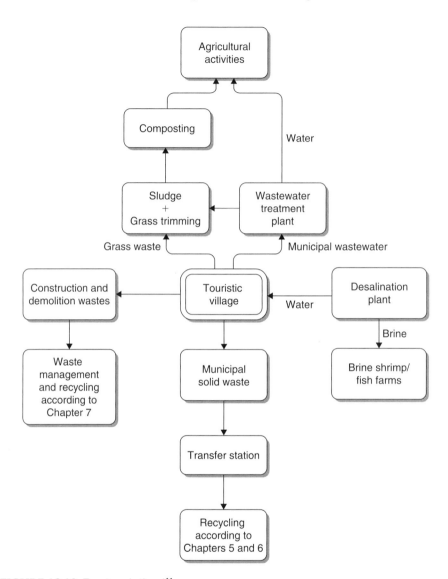

FIGURE 10.18 Eco-touristic village

Questions

1. Can we develop a strategy and action plan to stop new industrial waste landfill? How? Explain with justification.
2. Discuss different alternatives to utilize the bypass dust in cement industry. Can cement industry approach cradle-to-cradle system? How?
3. Discuss the wastes generated from tanning industry. Draw the process flow diagram for tanning industry showing all wastes generated. Develop

a case study to group all tanning industries in one industrial estate with the possibility of having an Eco-Industrial Park within the tanning industrial estate.

4. What is the use of Aluminum slag produced from Aluminum smelters as a result of recycling aluminum scrap?

5. What is the current situation of hazardous and non-hazardous industrial waste management in your country?

6. What is the recommended waste management hierarchy to approach cradle-to-cradle for an existing industrial estate in your country? Propose a short-range plan and long-range plan to approach cradle-to-cradle concept for such recommendations.

7. Most countries try to develop regulations to stop constructing any more incinerators and/or landfills to manage the waste generated for different industries. Do you recommend developing any incentive mechanism to encourage or enforce such regulations? What?

References

ABC Hansen (2001), *Wood Waste Briquetting Plant*, Denmark: ABC Hansen A/S.

Abou Khatwa, M.K., H. Salem, and S.M. El-Haggar (2005), "Building Material from Waste", *Canadian Metallurgical Quarterly*, Vol. 44, No. 3.

Alcoa Aluminio Inc 2006, "Itapissuma", http://www.alcoa.com/locations/brazil_itapissuma/en/home.asp

Adeleke, I.O. (2003), "Durability and Stability of Briquetting from Solid Agricultural Wastes", M.Sc. Thesis, American University in Cairo, Egypt.

Akaranta, O. (2000), "Production of particle boards from bioresources", *Bioresource Technology*, **75**, 87–89.

Al-Ansary, M.S., S.M. El-Haggar, and M.A. Taha (2004a), "Proposed Guidelines for Construction Waste Management in Egypt for Sustainability of Construction Industry", Proceedings of the International Conference on Sustainable Construction Waste Management, Singapore, 10–12 June.

Al-Ansary, M.S., S.M. El-Haggar, and M.A. Taha (2004b), "Sustainable Guidelines for Managing Demolition Waste in Egypt", Proceedings of the International RILEM Conference on the Use of Recycled Materials in Building and Structures, Barcelona, Spain, 9–11 November.

Al-Ansary, M.S. and S.M. El-Haggar (2005), "Construction Wastes Management Using the 7Rs Golden Rule for Industrial Ecology", Proceedings of the Global Construction: Ultimate Concrete Opportunities, Dundee University, 5–7 July.

Al Masha'an, M.A. and F. Mahrous (1999), "Environmental Strategy for Managing the Kuwait's Solid Wastes until 2020", Proceedings of Arab and Middle East International Conference and Fair on Solid Waste Management, Cairo, Egypt, 8–6 December.

Allenby, B.R. (1999), *Industrial Ecology, Policy Framework and Implementation*, Englewood Cliffs, NJ: Prentice Hall.

Al-Widyan, M.I., M.M. Al-Jalil, and N.H. Abu-Hamdeh (2002), "Physical durability and stability of olive cake briquettes", *Canadian Biosystem Engineering*, **44**, 3.41–3.35.

Anonymous (2001), "Recycling of Beverage Cartons: Little Packaging for a Lot of Product", *Internationale Papier Wirtschaft* (IPW), Vol. 5.

Aseptic Packaging Council (2005), "Question and Answer", http://www.aseptic.org/

Asian Institute of Technology (2006), School of Environment Resources and Development, "Cleaner Production (CP)", www.serd.ait.ac.th/cp/cp.html

ASTM Standards Source (1998a), "ASTM 695M-91, Standard Test Method for Compressive Properties of Rigid Plastics (Metric)", Philadelphia.

ASTM Standards Source (1998b), "ASTM 790M-93: Standard Test Methods for Flexural Properties of Unreinforced and Reinforced Plastics and Electrical Insulating Materials (Metric)", Philadelphia.

ASTM Standards Source (1998c), "ASTM D1525-96: Standard Test Method for Vicat Softening Temperature of Plastics", Philadelphia.

ASTM Standards Source (1998d), "ASTM D1525-96: Standard Test Method for Vicat Softening Temperature of Plastics", Philadelphia.

ASTM Standards Source (1998e), "ASTM D2240-97: Standard Test Method for Rubber Property – Durometer Hardness", Philadelphia.

ASTM Standards Source (1998f), "ASTM D543-95: Standard Practices for evaluating the Resistance of Plastics to Chemical Reagents", Philadelphia.

ASTM Standards Source (1998g), "ASTM D570-95: Standard Test Method for Water Absorption of Plastics", Philadelphia.

ASTM Standards Source (1998h), "ASTM D792-91: Standard Test Methods for Density and Specific Gravity (Relative Density) of Plastics by Displacement", Philadelphia.

ASTM Standards Source (1998i), "ASTM C410-60: Standard Specification for Industrial Floor Brick", Philadelphia.

ASTM Standards Source (1998j), "ASTM C902-95: Standard Specification for Pedestrian and Light Traffic Paving Brick", Philadelphia.

ASTM Standards Source (1998k), "ASTM C936-96: Standard Specification for Solid Concrete Interlocking Paving Units", Philadelphia.

Ayres, R.U. (1989), "Industrial Metabolism", in: J.H. Ausubel, and H.E. Sladovich (eds), *Technology and Environment*, pp. 23–49, Washington, DC: National Academy Press.

Bain, R.L. (1993), "Electricity from biomass in the United States: Status and future directions", *Bioresource Technology*, **46**, 86–93.

Bakker, P., *et al.* (2003), "Development of Concrete Breakwater Armour Units", 1st Coastal Estuary and Offshore Engineering Specialty Conference of the Canadian Society of Civil Engineering, Canada, June.

Bhattacharya, S.C., M.S. Ram, P. Wongvicha, and S. Ngamkajornvivat (1989), "A survey of uncarbonized briquettes and biocoal markets in Thailand", *Regional Energy Resources Information Center (RERIC) International Energy Journal*, **11**, June 1989.

Berger, K.R. (2005), "A Brief History of Packaging", University of Florida, Institute of Food and Agricultural Sciences, http://edis.ifas.ufl.edu/AE206

Biermann, C.J. (1993), *Essentials of Pulping and Paper Making*, Academic Press Inc.

Boateng, A.A. (1990), "Incineration of rice hull for use as a cementious material: The Guyana experience", *Cement and Concrete Research*, **20**, 795–802.

Bollinger, A. and M. Braungart, "Cradle to Cradle: Redesigning the Relationship between Industry and Nature", http://www.erscp2004.net/downloads/papers/andrewbollinger.pdf

BookRags Inc. (2005), "Aluminum", http://www.bookrags.com/sciences/earthscience/aluminum-woes-01.html

Bourdeau, L., P. Huovila, R. Lanting, and A. Gilham (1998), "Sustainable Development and the Future of Construction: A Comparison of Visions from Various Countries", CIB W82 Report Publication, No. 225.

Bovis Lend Lease (2001), "Bovis Lend Lease (UK) Wins Gold at the Green Apple Awards", http://bovislendlease.es/llweb/bll/main.nsf/all/20011029

Brecht, J.K. (2003), Hòrt 601 lecture notes. Vegetable Crops Department, University of Florida. http://www.aggiehotticulture.tamu.edu/syllabi/601/601nutri.pdf

Braungart, W. and M. McDonough (2002), *Cradle to Cradle: Remaking the Way we Make Things*, North Point Press.

Brazil Magazine (2006), "Brazil Starts World's First Recycling Plant for Carton Packaging", http://www.brazzilmag.com/content/view/2433/49/

Bregman, J.I. (1999), *Environmental Impact Statements*, Lewis Publishers.

Brewster, J.A. (2001), "Industrial Ecology and its Relationship to Cleaner Production", International Conference on Cleaner Production, Beijing, China, September, Paper 9 of 30.

Burcharth, H.F. and S.A. Hughes (2003), "Coastal Engineering Manual: Types and Functions of Coastal Structures", United States: US Army Corps of Engineers (USACE), (EM 1110-2-1100 Part V-Chapter 2).

CAPMAS, Control Agency for Public Mobilization And Statistics (1999), *Statistical Yearbook*, Cairo, Egypt.

CEPI, The Confederation of European Paper Industry (1999), *Annual Statistics*.

Chandrasekhar, S., P.N. Pramada, P. Raghavan, K.G. Satyanarayana, T.N. Gupta, (2002), "Microsilica from rice husk as a possible substitute for condensed silica fume for high performance concrete", *Journal of Materials Science letters*, **21**, 1245–1247.

Charlier, P. and G. Sjoberg (1995), "Recycling of Aluminum Foil from Post-Consumer Beverage Cartons", Proceedings of the TMS Fall Meeting, 3rd International Symposium on Recycling of Metals and Engineered Materials, pp. 676–683.

China Internet Information Center (2005), "China to Draft Law Promoting Circular Economy", http://china.org.cn/english/China/136793.html

Cichonski, T.J. and K. Hill (eds) (1993), *Recycling Sourcebook*, Washington, Thomson Gole.

CIRIA, Construction Industry Research and Information Association (1993), *Environmental Issue in Construction: A Preview of Issues of Initiatives Relevant to the Building, Construction and Relevant Industries*, London.

Cleaner Production in China (2006), "Cleaner Production Concepts", www.chinacp.com/eng/cp_concepts.html

Connecticut Metal Industries (2005), "Plastic Recycling", http://ctmetal.com/plastic%20more%20info.htm

Cooper, A.W., M. Solvi, and M. Calmes (1986), "Blast Furnace Slag Granulation", *Iron and Steel Engineer*, July, Vol. 63, 46–52.

Cordon, W.A. (1979), *Properties, Evaluation, and Control of Engineering Materials*, New York: McGraw-Hill.

Côté, R.P. and E. Cohen-Rosenthal (1998), "Designing Eco-Industrial Parks: A Synthesis of Some Experience", *Journal of Cleaner Production*, Vol. 6, 181–188.

CPA, Clean Production Action (2006), "Clean Production Action", www.cleanproduction.org/AAbase/default.htm

Dadd, D.L. (2004), "Textile Recycling by Worldwise", Dezignare Interior Design Collective, http://www.dezignare.com/newsletter/recycling textiles.html

Daly, H.E. (1977), *Steady State Economics*, San Francisco: Freeman.

Dasgupta, A. (1983), "Anaerobic Digestion of Solid Wastes of Cane Sugar Industry", Ph.D. Dissertation, University of Miami, May.

Dasgupta, A. and S.M. El-Haggar (2003), "Industrial Hazardous Waste Treatment and Disposal Guidelines", Report submitted to USAID, EEPP, June.

Deepchand, K. (2001), "Commercial Scale Cogeneration of Bagasse Energy in Mauritius", *Energy for Sustainable Development*, Vol. V, No. 1, March.

DETR, Department of the Environment Transport and the Regions (2000), "Building a Better Quality of Life: A Strategy for More Sustainable Construction", London.

DiChristina, M. and M. Henkenius (1999), "Construction-Waste Disposal: It's a Grind", *Popular Science*, Vol. 12, November.

Dietmar, R. (2003), "Approaches to a Circular Economy in Germany", Proceedings of Workshop on Circular Economy in Guiyang, August.

EcoAccess Environmental Licenses and Permits (2006a), "Clinical or Related Waste Treatment and Disposal", http://www.epa.qld.gov.au/publications/p00783aa.pdf/Clinical_or_related_waste_treatment_and_disposal.pdf

EcoAccess Environmental Licenses and Permits (2006b), "Defining Clinical Waste", http://www.epa.qld.gov.au/publications/p00786aa.pdf/Defining_clinical_waste.pdf

EEAA, Egyptian Environmental Affairs Agency (1994), "Comparing Environmental Health Risks in Cairo", *Project in Development and Environment*, Vol. 2, C.

EEAA, Egyptian Environmental Affairs Agency (1996), "EIA Guidelines", October.

Egyptian Standard for Cement Tiles (1974), *Book of Egyptian Standards*, Cairo, Egypt.

Ehrenfeld, S. (1997), "Industrial Ecology: A New Framework for Product and Process Design", *Journal of Cleaner Production*, 5(1–2), 87–95.

El-Haggar, S.M. (1996), "Recycling of Solid Waste", 3rd Annual AUC Research Conference, The American University in Cairo, Environmental Protection for Sustainable Development, 21–22 April.

El-Haggar, S.M. (1998), "Recycling of Solid Waste in Greater Cairo", Ain Shams 1st Conference on Engineering and Environment, Cairo, Egypt, 9–10 May.

El-Haggar, S.M., M.F. Hamouda, and M.A. Elbieh (1998), "Composting of Vegetable Waste in Subtropical Climates", *Int. J. Environment and Pollution*, Vol. 9, No. 4.

El-Haggar, S.M. and R. Baher (1999), "Design, Manufacturing and Testing of a Waste Paper Recycling System", *Int. J. Environment and Pollution*, Vol. 11, No. 2, 211–227.

El-Haggar, S.M. (2000a), "The Use of Cement By-Pass Dust for Sewage Sludge Treatment", The International Conference for Environmental Hazardous Mitigation, ICEHM2000, Cairo University, Egypt, 9–12 September.

El-Haggar, S.M. (2000b), *A Guide to Environmental Impacts of Industrial and Development Projects*, 1st edition, Cairo: Nah'det Misr Co.

El-Haggar, S.M. (2001a), "Paper Recycling in Egypt", International Symposium for Recovery and Recycling of Paper, University of Dundee, UK, 19–20 March.

El-Haggar, S.M. (2001b), "Environmentally Balanced Rural Waste Complex for Zero Pollution", Enviro 2001, The 3rd International Conference for Environmental Management and Technologies, Cairo, Egypt, 29–31 October.

El-Haggar, S.M. (2001c), "New Cleaner Production Hierarchy for Zero Pollution", Enviro 2001, The 3rd International Conference for Environmental Management and Technologies, Cairo, Egypt, 29–31 October.

El-Haggar, S.M. and M. Toivola (2001), "Mixing Plastic Waste with Sand as Raw Material for Eco-Building Products", WASTE 2001, The Middle East Congress and Exhibition of Recycling and Waste Management, 28 February–2 March.

El-Haggar, S.M., S. El-Attawi, and A. Mazen (2001), "Recycling and Deinking of Newspaper Mixed with Magazine", International Symposium for Recovery and Recycling of Paper, University of Dundee, UK, 19–20 March.

El-Haggar, S.M. (2002), "Zero Pollution for Sustainable Development in Egyptian Industries", International Congress on Sugar and Sugar Cane By-Products, Diversification, Havana, 17–22 June.

El-Haggar, S.M. (2003a), "Solid Waste Management Using Cleaner Production Technologies", Chapter 7 from the book entitled *Environmental Balance and Industrial Modernization*, Cairo: Dar El-Fikr El-Araby.

El-Haggar, S.M. (2003b), "Reaching 100% Recycling of Municipal Solid Waste Generated in Egypt", The International Symposium for Advances in Waste Management and Recycling, Dundee, Scotland, UK, 9–11 September.

El-Haggar, S.M. (2003c), "Industrial Ecology Using an Integrated CP–EMS Model for Sustainable Development", International Symposium on Advances in Waste Management and Recycling, Dundee, Scotland, UK, 9–11 September.

El-Haggar, S.M. (2003d), "6-R Golden Rule for Industrial Ecology", Enviro 2003, 4th International Conference on Environmental Technologies", 30 September–2 October.

El-Haggar, S.M. and G.H. Salama (2003), "Industrial Ecology to Approach Zero Pollution", Cairo 8th International Conference on Energy and Environment, Egypt, 4–7 January.

El-Haggar, S.M. and E.M. El-Azizy (2003), "Environmental Impact Assessment", Basis and Mechanisms of Sustainable Development Series (1), Dar El-Fikr El-Araby Book Co.

El-Haggar, S.M. (2004a), "Industrial Ecology for Renewable Resources", International Conference for Renewable Resources and Renewable Energy: A Global Challenge, Terista, Italy, 10–12 June.

El-Haggar, S.M. (2004b), "Cradle-to-Cradle for Industrial Ecology", Advances in Science and Technology of Treatment and Utilization of Industrial Waste, CMRDI and University of Florida, US-Egypt Joint Fund, Cairo, Egypt, 6–10 June.

El-Haggar, S.M. (2004c), "Solid Waste Management: Alternatives, Innovations and Solutions", Fundamentals and Mechanisms for Sustainable Development Series (3), Dar El-Fikr El-Araby Book Co.

El-Haggar, S.M., B.E. Ali, S.M. Ahmed, and M.M. Hamdy (2004a), "Increasing Nutrients Solubility from Natural Rocks During Composting of Organic Wastes", *Minia Journal of Agricultural Research and Development*, Vol. 24, No. 1, 71–88.

El-Haggar, S.M., B.E. Ali, S.M. Ahmed, and M.M. Hamdy (2004b), "Solubility of Natural Rocks During Composting", Second International Conference of Organic Agriculture, Cairo, Egypt, 25–27 March.

El-Haggar, S.M., S.M. Ahmed, and M.M. Hamdy (2004c), "Production of Compost from Organic Agriculture Enriched with Natural Rocks", *Egyptian Journal of Applied Sciences*, Vol. 19, No. 7B, 784–799, July.

El-Haggar, S.M. (2005), "Rural and Developing Country Solutions", Chapter 13 in *Environmental Solutions*, Nelson Nemerow and Franklin Agardy (eds), Elsevier Academic Press, pp. 313–400.

El-Haggar, S.M., I.O. Adeleke, and M. Gadallah (2005), "Briquetting of Solid Wastes from Cane Sugar Industry", Cairo 9th International Conference on Energy and Environment, Sharm El-Sheikh, 14–17 March.

El-Haggar, S.M., M.M. El Gowini, N.L. Nemerow, and N.T. Veziroglue (2005), "Sugarcane–Briquetting–Fertilizer: Environmentally Balanced

Industrial Complex", The International Hydrogen Energy Congress and Exhibition, Istanbul, Turkey, 13–15 July.

El-Haggar, S.M. and D.A. Sakr (2006), "Environmental and Technological Management System: ISO14001Plus", Cleaner Production Technologies Series, Dar El-Fikr El-Araby Book Co.

El-Madany, I.M. (1999), "Integrated and Sustainable Management for Municipal Solid Wastes", Cairo: Proceedings of Arab and Middle East International Conference and Fair on Solid Waste Management, 8–16 December.

El-Raey, M. (1997), "Utilization of Slag Produced by Electric Arc Furnace at Alexandria National Iron and Steel Company", Institute of Graduate Studies and Research, Alexandria University, Egypt.

Elliott, S. (2000), "Don't Waste Time", *International Construction*, 15–17 March.

Emmanuel, J., C. Hrdika, P. Głuszyński, R. Ryder, M. McKeon, R. Berkemaier, and A. Gauthie (2004), "Non Incineration Medical Waste Treatment Technologies in Europe", Health Care Without Harm, http://www.noharm.org/details.cfm?type=document&id=919

Environmental Agency for England and Wales (2001), "Waste Minimisation: An Environmental Good Practice Guide for Industry", United Kingdom.

Epstein, E. (1997), *The Science Of Composting*, Technomic Publishing Co., Inc.

Erkman, S. (1997), "Industrial Ecology, a Historical View", *Journal of Cleaner Production*, 5(1–2), 1–10.

European Aluminum Foil Association (2005), "Sustainable Energy Generation Coupled with Recovery and Recycling", http://www.alufoil.org/media/lnfoil16English.pdf

Evans, J.W. and R. Stevenson (2000), "A Practical Handbook for Policy Development and Action Planning for the Promotion of Cleaner Production", UNEP 6th International High-level Seminar on Cleaner Production, Montreal, Canada.

Faborode, M. and O. O'Callaghan (1987), "Optimizing the Compression/briquetting of Fibrous agricultural materials", *Journal of Agricultural Engineering Research*, **38**, 245–262.

Featherstone, W.B. and K.A. Holliday (1998), "Slag Treatment Improvement by Dry Granulation", *Iron and Steel Engineer*, July, 42–46.

Ferguson, J., N. Obe, N. Kermode, C.L. Nash, W.A. Sketch, and R.P. Huxford (1995), *Managing and Minimizing Construction Waste: A Practical Guide*, London: Thomas Telford.

Foster, A. (1996), "How to Make a Heap from Slag", *The Engineer*, Vol. 282, April 22.

Frosch, R.A. (1994), *Physics Today*, Vol. 47, Issue 11, November.

Frosch, R.A. (1995), *Environment*, Vol. 37, Issue 10, December.

Garner, A. (1999), "Industrial Ecology: An Introduction", National Pollution Prevention Center for Higher Education, University of Michigan, November.

Gavilan, R. (1994), "Source Evaluation of Solid Waste in Building Construction", *Journal of Construction Engineering and Management*, Vol. 120, No. 3, 536–552.

Georgescu-Roegen, N. (1971), *The Entropy Law and the Economic Process*, Cambridge, MA: Harvard University Press.

Gertler, Nicholas (1995), Industrial Ecosystems: Developing Sustainable Industrial Structures. http://www.sustainable.doe.gov/business/gertler2.html

Gibbs, D. and P. Deutz (2004), "Implementing Industrial Ecology: Planning for Eco-Industrial Parks in USA", *Geoforum*, 36(4), 452–464.

Graedel, T. and B.R. Allenby (1995), *Industrial Ecology*, Englewood Cliffs, NJ: Prentice Hall.

GreenBlue (2003), "Introduction to Cradle-to-Cradle Principles", http://www.greenblue.org/edesign/Intro_C2C.pdf

Hecker, T. (1993), "10 Common Problems with Recycle Operations", *Pit & Quarry*.

Heckett MultiServ (1997), "Reprocessing Steel Plant Fines", *Steel Times*, Vol. 225, No. 1, January, 19–24.

Heeres, R., W. Vermulen, and F. De Walle (2004), "Eco-Industrial Park Initiatives in the USA and The Netherlands", *Journal of Cleaner Production*, 12(8–10), 985–996.

Hill, B. and C. Pulkinen (1988), "A study of Factors affecting pellet durability and pelleting efficiency", Saskatchewan dehydrator Association.

Holliday, L. (ed.) (1966), *Composite Materials*, Elsevier Publishing Company: Amsterdam.

Hyvärinen, E. (2001), "Paper Recovery in Europe – Industrial Views", Proceedings of the International Symposium on Recovery and Recycling of Paper, pp. 1–8.

Ibrahim, Y. (2006), "Utilization of Municipal Solid Waste Rejects in Producing Wave Dispersion and Shoreline Erosion Protection Structures (Breakwaters)", M.Sc. Thesis, The American University in Cairo.

IFIA, International Fertilizer Industry Association (2002), *Fertilizers Indicators*, 2nd edition, October.

IGES, Institute for Global Environmental Strategies (2005), "Special Feature on the Environmentally Sustainable City, Waste Management and Recycling in Asia," *International Review for Environmental Strategies*, Vol. 5, No. 2, 477–498.

Indigo Development (2006), "China Seeks to Develop a 'Circular Economy' (CE)", http://www.indigodev.com/Circular1.html

Indigo Development (2005), "Eco-Industrial Park Handbook for Asian Developing Nations", http://www.indigodev.com/Handbook.html

Indigo Development (2006), "Eco-Industrial Parks", http://www.indigodev.com/Ecoparks.html

Indigo Development (2005), "Industrial Ecology Methods and Tools for Analysis and Design", http://www.indigodev.com/Tools.html

Indigo Development (2003), "The Industrial Symbiosis at Kalundborg, Denmark", http://www.indigodev.com/Kal.html

Jackson, T. (1993), *Clean Production Strategies: Developing Preventive Environmental Management in the Industrial Economy*, Florida: Lewis Publishers.

Japan Environmental Agency (1998). *Life Cycle and Problems of Waste*. Tokyo.

Jain, A., T.R. Rao, S.S. Sambi, and P.D. Grover (1994), "Energy and chemicals from rice husk", *Biomass and Bioenergy*, **7**, 285–289.

Jelinski, L.W., T.E. Graedel, R.A. Laudise, D.W. McCall, and C.K.N. Patel (1992), "Industrial ecology: concepts and approaches", *Proceedings of the National Academy of Science*, **89**, 793–797.

Khedari, Joseph, Charoenvai, Sarocha, Hirunlabh, Jongjit (2003), "New insulating particleboards from durian peel and coconut coir", *Building and Environment*, **38**, 435–441.

Johansson, A. (1992), *Clean Technology*, Florida: Lewis Publishers.

Kilby, E. (2001), "Current Statistics on Recovered Paper", International Symposium for Recovery and Recycling of Paper, University of Dundee, UK, 19–20 March.

Kourany, Y. and S.M. El-Haggar (2000), "Utilizing Slag Generated from Iron and Steel Industry in Producing Masonry Units and Paving Interlocks", 28th CSCE Annual Conference, 7–10 June, London, Ontario, Canada.

Kourany, Y. and El-Haggar, S.M. (2001), "Using Slag in Manufacturing Masonary Bricks and Paving Units", *TMS Journal*, September, 97–106.

Lankford, W.T., N.L. Samways, R.F. Craven, and H.E. McGaannon (1985), *The Making, Shaping and Treating of Steel*, Pittsburgh: Herbrick and Held.

Lardinos, I. and A. van de Klundert (1995), *Plastic Waste: Options for Small-Scale Resource Recovery*, Urban Solid Waste Series 2. Amsterdam: The TOOL Publications.

Lichts, F.O. (2004), "International Sugar and Sweetener Report", Vol. 136, No. 29, 5 October.

Mahmoud, M.Y. (2003), "Specifications of Natural Mineral Fertilizers Used in Producing Organic Fertilizers", Al-Ahram Company for Minerals and Natural Fertilizers, 1st Organic Farming Conference and Exhibition, Cairo, Egypt.

Mannan, M.A. and C. Ganapathy (2002), "Engineering properties of concrete with oil palm shell as coarse aggregate", *Construction and Building Materials*, **16**, 29–34.

Martin, S., A. Keith, A. Weitz, A. Cushman, A. Sharma, and R. Lindrooth (1996), "Eco-Industrial Parks: A Case Study and Analysis of Economic, Environmental, Technical, and Regulatory Issues Final Report", Economics Research Triangle Institute.

McKerlie, K., N. Knight, and B. Thorpe (2005), "Advancing Extended Producer Responsibility in Canada", *Journal of Cleaner Production*, 1–13.

McKinney, R. (1991), "Recycling Fiber – The State of the Art in Europe", World Pulp and Paper Technology.

Merck Index (1952), "Phosphorus", Merck & Co. Inc.

Metro Solid Waste Department, Portland, Oregon, USA (1993), "Characterization of Construction Site Waste".

Mohan, C. (2000), "IREDA Power Packed Opportunities from Biomass", Indian Renewable Energy Development Agency (IREDA) Limited, New Delhi.

Morris, S., "Project Waste Management Master Specification", Susan Morris Specifications Limited under contract to Greater Vancouver Regional District, 2001, http://www.gvrd.bc.ca/buildsmart/pdfs/WasteManagement Spec.pdf.

Morsy, M.M. and M.S. Saleh (1996), "Industrial and Non-Industrial Solid Waste Utilization in Road Building and as Ballast for Railroads", Transportation and Communications Research Council Report, Scientific Research and Technology Academy, Egypt.

Mostafa, A., S.M. El-Haggar, and A. Gad El-Mawla (1999), "Matching of an Anaerobic Animal Waste Digester with a Dual-Fuel Generator Unit", *International Journal of Environment and Pollution*, Vol. 12, No. 1, 97–103.

Nagao, Y., *et al.* (1989), "Development of New Pavement Base Course Material Using High Proportion of Steel Making Slag Properly Combined with Air-Cooled and Granulated Blast Furnace Slags", Nippon Steel Technical Report No. 43.

Nagy, C.Z. (2002), "Oil Exploration and Production Wastes Initiatives", Department of Toxic Substances Control, Hazardous Waste Management Program, and Statewide Compliance Division, May.

Natural Resources – Canada (2002), "Energy Efficient Recycling of aluminum", http://www.nrcan.gc.ca/es/etb/cetc/cetc01/htmldocs/factsheet_energy-efficient_recycling_of_aluminum_e.html

Nayak, B.B., B.C. Mohanty and S.K. Singh (1996), "Synthesis of Silicon Carbide from Rice Husk in a DC Arc Plasma Reactor", *Journal of American Ceramics Society*, Vol. 79, No. 5, 1197–2000.

NCDENR (1997), North Carolina Department of Environment and Natural Resources – Division of Pollution Prevention and Environmental Assistance, "Process Optimization at a Sugar Manufacturing Facility", http://www.p2pays.org/ref/10/09353.htm

Nemerow, N.L. (1995), *Zero Pollution for Industry*, NY: John Wiley and Sons Inc.

Ndiema, C.K.W., P.N. Manga, and C.R. Rottoh (2002), "Influence of die Pressure on Relaxation Characteristics of briquette biomass", *Energy Conversion and Management*, **43**, 2157–2161.

NewCity (2006), "Industrial Ecology: From Theory to Practice", http://www.newcity.ca/Pages/industrial_ecology.html

Newel, J. (1990), "Recycling Britain", *New Scientist*, 46, September.

Nielsen, L.E. and R.F. Landel (1994), *Mechanical Properties of Polymers and Composites*, 2nd edition, Revised and Expanded, New York: Marcel Dekker.

Nimityongskul, Pichai, and Daladar, Telesforo (1995), "Use of coconut husk ash, corn cob ash and peanut shell ash as cement replacement", *Journal of Ferrocement*, **25**, 35–44.

NMFRC (2005), National Metal Finishing Resource Center, "Pollution Prevention and Control Technologies for Plating Operations: Section 7 – Off-Site Metals Recycling", http://www.nmfrc.org/bluebook/sec73.htm

Nour, M.H. (2002), "Impact of Lignin from Non-Wood Pulping Processes on Kraft Wastewater Treatment and Receiving Water Quality", M.Sc. Thesis, The American University in Cairo, Egypt.

ODEQ (Oregon Department of Environmental Quality) (1997), "Report on Mixed Solid Waste Processing Facilities", http://www.deq.state.or.us/wmc//solwaste/mrfreport.html#three

OECD, Organisation for Economic Co-operation and Development (2001), *Environmental Outlook 2001*, France: OECD.

Ojewole, G.S. and O.G. Longe (2000), "Evaluation of the productive and economical efficiencies of cowpea hull and maize offal inclusion in layers' ration", *Nigerian Journal of Animal Production*, **27**, 35–39.

Olah, O. (2004), "Waste Management of Tetra Pak Products in Hungary and Sweden", M.Sc., Royal Institute of Technology, Stockholm.

Otubusin, S.O. (2001), "The effect of different combinations of industrial and agricultural waste as supplementary feeds on tilapia fingerlings production in floating net-hapas", *Nigerian Journal of Animal Production*, **28**, 108–111.

Okpala, D.C. (1990), "Palm kernel shell as lightweight aggregate in concrete", *Building and Environment*, **25**, 291–296.

Ogazi, J.N., J.A.I. Omueti, and J.A. Moore (2000), "Waste utilization through organo-mineral fertilizer production in south western Nigeria, Animal, agricultural and food processing wastes", *Proceedings of the Eight-International Symposium*, Des Moined, Iowa, USA 9–11 October 2000, pp. 640–647.

Parkin, S. (2000), "Contexts and Drivers for Operationalizing Sustainable Development", *Civil Engineering*, **138**, 9–15.

Parkin, S. (2000), "Sustainable Development: The Concept and the Practical Challenge", *Civil Engineering*, **138**, 3–8.

Peck, S. (1998), "Industrial Ecology: From Theory to Practice" Environmental Studies Association of Canada Meeting at Learned Societies Conference in St. Catherines, Ontario.

Peck, S., C. Callaghan, and R. Côté (1997), "EIP Development and Canada: Final Report", Peck & Associates.

Peng, C.L., D.E. Scorpio, and C.J. Kibert (1997), "Strategies for Successful Construction and Demolition Waste Recycling Operations", *Construction Management and Economics*, 49–58.

Planet Ark (2005), "The History of Cartons", http://www.planetark.org/campaignspage.cfm/newsid/72/newsDate/12/story.htm

Prigogine, I. (1955), *Thermodynamics of Irreversible Processes*, Springfield, IL: Charles Thomas.

Raghupathy, R., R. Viswanathan, and C.T. Devadas (2002), "Quality of paper boards from arecanut sheath", *Bioresource Technology*, **82**, 99–100.

Redefining Progress and Earth Day Network (2002), "Sustainability Starts in Your Community", http://www.rprogress.org/newpubs/2002/ciguide.pdf

Research and Markets, "Global Waste Management Market Report 2004", http://www.researchandmarkets.com/reportinfo.asp?report_id=72031&t=e&cat_id=

Riddick, K. (2003), "Producer Responsibility in the EC", Proceedings of the International Symposium on Sustainable Waste Management, pp. 79–84, September.

Rodushkin, I. and A. Magnusson (2005), "Aluminum Migration to Orange Juice in Laminated Paperboard Packages", *Journal of Food Composition and Analysis*, Vol. 18, 365–374.

Russell, Bill (1990), "Straw particleboard", *The Proceedings of the 24th Washington State University International Particleboard/Composite Materials Symposium*, April 3–5 1990, Pullman, W.A. USA.

Saikkuu, L. (2006), *Eco-Industrial Parks: A Background Report for the Eco-Industrial Park Project at Rantasalmi*, Finland: Publications of Regional Council of Etela-Savo.

Sampanthrajan, A., N.C. Vijayaraghavan, and K.R. Swaminathan (1992), "Mechanical and thermal properties of particle boards made from farm residues", *Bioresource Technology*, **72**, pp. 249–251.

Sarma, A. and A. van der Hoek (2004), "A Need Hierarchy for Teams: ISR Technical Report", UCI-ISR-04-9, Department of Informatics, Donald Bren School of Information and Computer Sciences, University of California Irvine, October.

Sarraf, M. and B. Larsen (2002), "Report No. 25175-EGT: Arab Republic of Egypt, Cost Assessment of Environmental Degradation", World Bank, METAP, 29 June.

Sarraf, M., K. Bolt, and B. Larsen (2004a), "Syrian Arab Republic Cost Assessment of Environmental Degradation", World Bank, METAP, Final Report, 9 February.

Sarraf, M., B. Larsen, and M. Owaygen (2004b), "Cost of Environmental Degradation – The Case of Lebanon and Tunisia", World Bank, METAP, Environmental Economic Series, Paper No. 97, June.

Sawiries, Y.L., S.M. El-Haggar, and A.M. Ghanem (2001), "Environmentally Balanced Municipal Solid Waste Complex for Zero Pollution", Enviro 2001, The 3rd International Conference for Environmental Management and Technologies, Cairo, 29–31 October.

SEAM (1998), Support for Environmental Assessment and Management Programme, "Reduction of Milk Losses at Misr Company for Dairy and Food, Mansoura – Egypt", http://www.seamegypt.org/downloads%5C14.pdf

SEAM (1999a), Support for Environmental Assessment and Management Programme, "Combining Preparatory Processes: A Low Cost High Productivity Solution", http://www.seamegypt.org/downloads%5C7.pdf

SEAM (1999b), Support for Environmental Assessment and Management Programme, "Oil and Fats Recovery at Tanta Oil and Soap Company, Tanta – Egypt", http://www.seamegypt.org/downloads%5C20.pdf

SEAM (1999c), Support for Environmental Assessment and Management Programme, "Recovery of Cheese Whey for Use as an Animal Feed", http://www.seamegypt.org/downloads%5C15.pdf

SEAM (1999d), Support for Environmental Assessment and Management Programme, "Sulphur Black Dyeing: A Cleaner Production Approach", http://www.seamegypt.org/downloads%5C8.pdf

SEAM (1999e), Support for Environmental Assessment and Management Programme, "Water and Energy Conservation at Edifna Company for Preserved Food, Alexandria – Egypt and Kaha Company for Preserved Food, Kaha – Egypt", http://www.seamegypt.org/downloads%5C17.pdf

SEAM (1999f), Support for Environmental Assessment and Management Programme, "Water and Energy Conservation at El Nasr Company for Spinning and Weaving, Mahalla El Kobra – Egypt", http://www.seamegypt.org/downloads%5C10.pdf

SEAM (1999g), Support for Environmental Assessment and Management Programme, "Waste Minimisation at Sila Edible Oil Company, Fayoum – Egypt", http://www.seamegypt.org/downloads%5C19.pdf

Shackel, B. (1990), *Design and Construction of Interlocking Concrete Block Pavements*, NY: Elsevier Applied Science.

Shen, Kuo C. (1991), "Direct conversion of biomass and wastes to panel boards", *Symposium paper: Energy from Biomass and Wastes*, 1991, pp. 571–582 and 16th IGT Conference on Energy from Biomass and Wastes, March 25–29, 1991, Washington, DC, USA.

Singh, S.K. (2002), "Synthesis of SiC from Rice Husk in a Plasma Reactor", *Bulletin of Material Science*, Vol. 25, No. 6, November, 561–563.

Singh, S. and V. Vijayalakshmi (2006), "Marble Slurry – A New Building Material", Technology Information, Forecasting, and Assessment Council, http://www.tifac.org.in/news/marble.htm

Smart Growth (2000), "Burnside Eco-Industrial Park – Halifax, Nova Scotia, Canada", http://www.smartgrowth.org/casestudies/ecoin_burnside.html

Smart Growth (2000), "Fairfield Ecological Industrial Park – Baltimore, Maryland", http://www.smartgrowth.org/casestudies/ecoin_fairfield.html

Smith, M., A. Manning, and M. Lang (1999), "Research on the Reuse of Drill Cuttings Onshore", CORDAH Research Limited, November.

Sujirote, K. and P. Leangsuwan (2003), "Silicon carbide formation from pre-treated rice husks", *Journal of Material Science*, **38**, 4739–4744.

Sun, L. and K. Gong (2001), "Silicon Based Materials from Rice Husks and their Applications", *Industrial Engineering Chemistry Review*, Vol. 40, 5861–5877.

Suriyage, G. (2005), "Milk Cartons", Recycled Paper Craft, http://www.recycledpapercraft.com/milkcar.htm

Susan Morris Specifications Limited (2001), "Project Waste Management Master Specification", Greater Vancouver Regional District, http://www.gvrd.bc.ca/buildsmart/pdfs/wastemanagementspec.pdf

Sustainable Agri-Food Production and Consumption Forum (2001), "Good Practices – Possible Solutions: Cleaner Production", http://www.agrifood-forum.net/practices/cp.asp

Swamy, R.N. (1993), "Concrete with Slag: High Performance and Durability without Tears", 4th International Conference on Structural Failure Durability and Retrofitting, Singapore, pp. 206–236.

Szekely, J. (1995), "A Research Program for the Minimization and Effective Utilization of Steel Plant Wastes", *Iron and Steelmaker*, 25–29, January.

TAPPI, Technical Association for the Worldwide Pulp, Paper and Converting Industry (2005), "How is Paper Recycled", http://www.tappi.org/paperu/all_about_paper/earth_answers/EarthAnswers_Recycle.pdf

Tchobanoglous, G., H. Theisen, and S. Vigil (1993), *Integrated Solid Waste Management: Engineering Principles and Management Issues*, New York: McGraw-Hill, Inc.

TecEco Pty Ltd. Sustainable Technologies (2006), "Life Cycle Analysis", www.tececo.com/sustainability.life_cycle_analysis.php

Tetra Pak (2001), "What Happens to Used Beverage Cartons?", http://www.tetrapak.com/docs/What_happens-eng_0002.pdf

The Aluminum Association Inc. (2004), "In-Depth Information/Recycling Process: Aluminum – A Great Economic Story", http://www.aluminum.org/Template.cfm?Section=In-depth_information&NavMenuID=758

The ASEC Group (2006), "The By-Pass Dust: The Cement Producers' Nightmare", http://www.asec-egypt.com/group/by_pass_dust.asp

The Cardinal Group Inc. (2005), "Canadian Eco-Industrial Network", http://www.greenroofs.ca

The European Aluminium Association (2005), "Recycled Aluminium", http://www.aluminium.org/home.htm

Thakur, S.K. and K.D.N Singh (2000), "Efficiency of agricultural waste and weeds for biogas production, journal of research", *Birsa Agricultural University*, **12**, 11–15.

The Government of the Hong Kong Special Administrative Region – Environment, Transports and Work Bureau (2003), "Proposed Clinical Waste Control Scheme", http://www.etwb.gov.hk/boards_and_committees/ace/2002ace/paper122002/index.aspx?langno=1&nodeID=282

Tremper, G. (1996), "The History and Promise of Jojoba", Armchair World, http://www.armchair.com/warp/jojoba1.html

Tripathi, Arun K., P.V.R. Iyer, and T.C. Kandal (1998), "A techno-economic evaluation of biomass briquetting in India", *Biomass and Bioenergy*, **14**, 479–488.

UNEP, United Nations Environmental Program (1997), "Papers from an Executive Seminar on: The Role of Information Technology in Environmental Awareness-Raising, Policy-Making, Decision-Making, and Development Aid", 3 September.

UNEP, United Nations Environment Program (2000), "Clean Production Assessment in Meat Technology, Denmark: COWI Consulting Engineers and Planners".

UNEP, United Nations Environmental Program (2000/2001), "Egypt Human Development Report – Chapter 4: The Environment and Sustainable Development".

UNEP, United Nations Environmental Program – Production and Consumption Branch (2001), "Cleaner Production – Key Elements", http://www.uneptie.org/pc/cp/understanding_cp/home.htm

UNEP, United Nations Environmental Program – Production and Consumption Branch (2001), "Cleaner Production Assessment in Industries", http://www.uneptie.org/pc/cp/understanding_cp/cp_industries.htm

UNEP, United Nations Environmental Program – Production and Consumption Branch (2001), "Extended Producer Responsibility", http://www.uneptie.org/pc/cp/tools/epr.htm

UNEP, United Nations Environmental Program – Production and Consumption Branch (2001), "The Evolution of an Industrial Park: The Case of Burnside, Halifax, Canada", http://www.uneptie.org/PC/ind-estates/casestudies/Burnside.htm

UNEP, United Nations Environmental Program – Production and Consumption Branch (2001), "The Industrial Symbiosis in Kalundborg, Denmark", http://www.uneptie.org/pc/ind-estates/casestudies/kalundborg.htm

UNEP, United Nations Environmental Program – Production and Consumption Branch: Environmental Management of Industrial Estates – (2001), "An Eco-Industrial Networking Exercise in Naroda Industrial Estate, Ahmedabad, India", http://www.uneptie.org/pc/ind-estates/casestudies/Naroda.htm

UNEP, United Nations Environmental Program (1997), "Cleaner Production at Pulp and Paper Mills: A Guidance Manual Publication in Cooperation with the National Productivity Council, India".

United Nations 96th Plenary Meeting (1987), "Report of the World Commission on Environment and Development", 11 December.

USEPA, US Environmental Protection Agency (May 1999), "Environmental Fact Sheet: Source Reduction of Municipal Solid Waste", http://www.epa.gov/garbage/pubs/envfact.pdf

USEPA (2005), United States Environmental Protection Agency, "Municipal Solid Waste Generation, Recycling, and Disposal in the United States: Facts and Figures for 2003", http://www.epa.gov/garbage/pubs/msw05rpt.pdf

USEPA, US Environmental Protection Agency (2006a), "Climate Change and Waste: Reducing Waste can make a Difference", http://www.epa.gov/mswclimate/folder.htm

USEPA, United States Environmental Protection Agency (2006b), "Project XL: Baltimore Development Corporation Proposal", http://www.epa.gov/projectxl/balt/030496.htm

USNA, United States Naval Academy (2006), "Breakwaters", http://www.usna.edu/naoe/courses/en420/bonnette/breakwater_design.html

Uwe G., Professor at Fraunhofer-Institut Elektronenstrahl und Plasma-technik (2005), Private Communication.

Vogler, J. (1984), *Small Scale Recycling of Plastics*, UK: Intermediate Technology Publications.

Wernick, I.K. and J.H. Ausubel (1997), "Industrial Ecology: Some Directions for Research", Pre-Publication Draft, Program for the Human Environment.

Whisperwave Dispersion Technologies Inc. (2005), "Floating Breakwaters Provide Erosion Control Protection and Wave Attenuation", http://www.whisprwave.com/floating.htm

Wikipedia (2005), "Sodium Aluminate", http://en.wikipedia.org/wiki/Sodium_aluminate

Wikipedia (2006a), "Industrial Ecology", http://en.wikipedia.org/wiki/Industrial_ecology

Wikipedia (2006b), "Sulfuric Acid", http://en.wikipedia.org/wiki/Sulfuric_acid

Williams, P.T. (1998), *Waste Treatment and Disposal*, John Wiley & Sons.

Wogrolly, E., M. Hofstatter, and E. Langshwert (1995), "Recycling of Plastic Waste", *Advances in Material Technology Monitor*, Vol. 2, No. 4, 1–6, Vienna: United Nations Industrial Development Organization (UNIDO).

World Bank (2004), *Analysis of Environmental Performance in Egypt*, USA: WB.

World Bank (2005), "*Arab Republic of Egypt – Country Environment Analysis (1992–2002)*", USA: WB.

Xiaofei, P.E.I. (2006), "Overview of the Circular Economy in China", http://www.jk.sh.cn/websb04/downloads/Overview%20of%20the%20Circular%20Economy%20in%20China.doc

Yussefi, M. and H. Willer (2003), *The World of Organic Agriculture*, Nuernberg Messe: Statistics and Future Prospects.

Index

4Rs Golden Rule, 264
7Rs Golden Rule (Regulation, Reduce, Reuse,
    Recycle, Recovery, Rethinking and
    Renovation), 12–13, 125–6
  capital investment, 13
  construction waste, 280–9
  industrial ecology, 281–3
  mitigation of construction materials, 283–9
100% recycling system, 157

ABBC see Animal fodder, briquetting, biogas
    and composting technologies
Abrasion resistance: construction materials,
    209–11
Achievements:
  oil and soap industry, 71
  sulfur black dyeing, 56
  textile industry, 62
Administration: eco-industrial parks, 107, 115
Adoption regulations, 23
Advantages of incineration, 10
Aeration see Oxygen supply
Agenda 21, 86
Agglomerator machines, 169, 199–200
Agricultural waste management, 223–60
  animal fodder technology, 226–7
  biogas, 232–3
  briquetting, 227–32
  case studies, 243–60
  composting, 233–4
  construction industry, 234–6
  Egypt, 254–5, 258
  environmentally balanced rural waste
      complex, 240–3
  household municipal solid waste, 242–3
  integrated complexes, 239–43
  introduction, 223–4
  output technology matching diagram, 226
  power generation, 236–7
  silicon carbide, 237–9
  solid waste, 151–3

technologies, 224–6
  utilization techniques, 256–9
Ahmedabad, India: eco-industrial park, 114–17
Air injection flotation cell, 162
Air injection screening devices, 186
Alcohol production, 120–1
Alternative fuel technologies, 358–9
Aluminum foundries, 326–39
  aluminum recycling, 327–32
  annealing furnaces, 334–9
  cleaner production, 328–32
  opportunity assessment, 328–9
  polishing processes, 330–1
  process description, 327–8
  process modification, 329–32
  rolling process, 328, 330
  smelting furnaces, 328, 329, 331–4
  spinning process, 328, 330
Aluminum recycling, 179, 183, 188, 327–32
Analysis:
  alternative fuel technologies, 358–9
  bentonite, 247, 251–2
  cement bypass dust, 310
  construction waste, 290–1
  dolomite, 247, 250–1
  feldspar, 247, 250
  fertilizers, 247–52
  rock phosphate, 248–50
  sugarcane industry ash, 353
  sulfur, 247, 251
Animal farming, 366–7
Animal feed see Cheese whey recovery
Animal fodder:
  organic fertilizer case study, 258–9
  sugarcane industry, 358–60
  technology, 226–7
Animal fodder, briquetting, biogas and
    composting (ABBC) technologies,
    225–34, 240
  organic fertilizer case study, 259–60
  see also individual technologies

Animal manure, 247–8
Annealing furnaces, 334–9
  combustion/control systems, 337–8
  design, 335–9
  diagrams, 335–8
  doors, 336–7
  process, 328
  upgrading, 330
Apartment construction waste example, 280
Armor units *see* Breakwaters
Artificial ecology, 91
Aseptic packaging, 182
Ash:
  fly ash, 102, 105
  peanut shell mortars, 235
  sugarcane, 353
Asia: recycling, 3
Asnaes Power Station, Denmark, 100, 101–5
Asphalt:
  companies, 108–9
  concrete, 323
  mixtures, 316
Assembly processes: wood furniture
    industry, 75, 77
Assessment *see* Opportunity assessment
Audit *see* Opportunity assessment
Australia: Queensland, 295–6
Autoclave, 298–9
Awareness: sustainable development
    proposed framework, 126

Bagasse, 353, 358–60
Bagasse-cachaza gasification, 356–60
Balanced complex: sugarcane industry, 355,
    357
Barriers:
  cleaner production, 24
  industrial ecology, 88–91, 132–3
Baseline activities: Brownsville
    eco-industrial park, 109–10
Basic oxygen furnaces (BOF), 317–252, 317,
    323, 324–5
Beach revetments, 219
Beet manufacturing facility:
  cleaner production, 39–41
  cost/benefit analysis, 41
  opportunity assessment, 40–1
  process description, 39–40
  techniques for CP, 40–1
  see also Sugarcane industry
Benefits:
  cleaner production, 23–4, 38
  eco-industrial parks, 92–3, 106, 114
  government, 93
  intangible, 6
  social, 93
  sulfur black dyeing, 56
  textile industry, 62
  see also Achievements; Cost/benefit
    analysis; Environmental benefits

Bentonite analysis, 247, 251–2
Benzene: chemical resistance to, 213
Beverage cartons, 184
BF *see* Blast furnaces
Biogas (biomass gasification), 232–3
  comparative analysis, 359
  organic fertilizers, 257–8
  sugarcane industry, 356–60
Biological systems *see* Natural ecosystems
Biological treatment: definition, 4
Blast furnaces (BF), 317–19, 323
Bleaching, 65, 161
Blow molding machines, 171–2
BOF *see* Basic oxygen furnaces
Boilers, 33, 36
Bone recycling, 172–3
Breakwaters, 217–19
Bricks:
  cement bypass dust, 314
  drill cuttings, 344
  off-site recycling, 314
  properties, 202–4
Brine disposal/utilization, 364–5
Briquetting, 227–32, 256–7
  alternative fuel technology, 358–9
  comparative analysis, 358–9
  process, 228, 230–1
  quality parameters, 229–30
  sugarcane industry, 354–6
  technology, 231–2, 354–6
Brownsville eco-industrial park, 107–14
Bruce Energy Centre, Canada, 120–1
Burnside eco-industrial park, Canada, 117–20
Business barriers: industrial ecology, 90
Bypass dust, 313, 314–16

C&D *see* Construction and demolition
    wastes
CAA *see* Competent Administrative
    Authority
Cachaza gasification, 356–60
Calcium, 252
Canada: eco-industrial parks, 96–7, 99, 117–21
Canal revetments, 219
Cane *see* Sugarcane industry
Capital investment: 7Rs Golden Rule, 13
Carbon/nitrogen ratio: composting, 190
Cartons *see* Beverage cartons; Packaging
    materials
Case studies:
  agricultural management, 243–60
  aluminum foundries, 326–39
  apartment construction, 280
  beet manufacturing, 39–41
  cement industry, 308–17
  cheese whey recovery, 41–9
  cleaner production, 30–84
  construction and demolition waste, 279–92
  drill cuttings, 339–46
  eco-industrial parks, 97–124

eco-rural parks, 254–60
edible oil company, 71–3
Egypt, 31–9
Finsbury Pavement Building, 289–91
food sector industry, 31–51
granite industry, 346–50
iron and steel industry, 82–4, 317–26
marble industry, 346–50
milk loss reduction, 49–51
oil and soap industry, 67–73
organic fertilizer, 243–54
petroleum sector, 339–46
preserved food companies, 31–9
restaurant construction waste, 280
rural management, 243–60
spinning industry, 62–7
sugarcane industry, 350–62
sulfur black dyeing, 51–6
Sydney Olympic Village, 291–2
textile industry, 51–67
tourist industry, 362–9
weaving industry, 62–7
wood furniture industry, 73–82
CBOT *see* Chicago Board of Trade
CCC *see* Clark County Code
Cement industry, 308–17
  bricks, 314
  bypass dust, 310, 314
  cement manufacture, 9
  cleaner production opportunities, 313–16
  eco-industrial parks, 312
  methodology, 309–11
  off-site recycling, 314
  on-site recycling, bypass dust, 313
  sewage sludge, 314–16
  sustainability approach, 311–13
  waste disposal, 312
Ceramic glass, 314
CETP *see* Common effluent treatment plant
Cheese whey recovery:
  animal diets, 45–6
  cleaner production, 41–9
  cost/benefit analysis, 46–8
  factory outcomes, 48
  farms, 48
  process description, 41–3
  prototype phase, 47–8
  sustainability, 48–9
  waste analysis, 42–3
Chemical analysis *see* Analysis
Chemical disinfection, 299–300
Chemical properties: marble slurry, 347
Chemical pulping processes, 159
Chemical reagent resistance, 212–13
Chemical spraying, 79, 80–1
Chemical treatment: definition, 4
Chicago Board of Trade (CBOT): waste
  exchanges, 89
China, 94–5, 302
Circular economy initiative, 93–7

Clark County Code (CCC), 261
Cleaner production (CP), 21–84
  aluminum foundries, 328–32
  assessment *see* Cleaner production,
    opportunity assessment
  barriers, 24
  beet manufacturing, 39–41
  benefits, 38
  case studies, 30–84
  cement industry, 313–16
  cheese whey recovery, 41–9
  dairy industry, 41–51
  edible oil company, 71–3
  environmental management systems,
    129–31
  environmental reform, 141
  equipment modification, 26–7
  experimental phases, 43–4
  food sector industry, 31–51
  good housekeeping, 26
  granite/marble industry, 348–50
  guidelines, 23
  industrial ecology, 91, 95
  input material change, 27–8
  iron and steel industry, 82–4
  management systems model, 131
  methodology, 29–30
  milk loss reduction, 49–51
  Naroda eco-industrial park, 115
  obstacles/solutions, 24
  off-site recycling, 28
  oil and soap industry, 67–73
  on-site recycling, 28
  opportunity assessment, 29–30
    aluminum foundries, 328–9
    beet manufacturing facility, 40–1
    cheese whey recovery, 43
    drill cuttings, 341–5
    edible oil company, 72
    granite/marble industry, 348–50
    milk loss reduction, 50
    oil and soap industry, 69
    preserved food companies, 34–7
    sulfur black dyeing, 52
    textile industry, 57–8, 63–4
    wire rod production, 83
    wood furniture industry, 79–80
  paper mill example, 27
  pilot phases, 44–6
  preserved food companies, 31–9
  process control, 26
  production modification, 28–9
  promotion, 22–3
  prototype phase, 47–8
  soap industry, 67–73
  sulfur black dyeing, 51–6
  sustainable development proposed
    framework, 126
  techniques, 25–9
    beet manufacturing facility, 40–1

Cleaner production (*contd*)
  techniques (*contd*)
    cheese whey recovery, 43–6, 49
    edible oil company, 72–3
    milk loss reduction, 50–1
    oil and soap industry, 69–71
    preserved food companies, 37
    sulfur black dyeing, 53–4
    textile industry, 58–60, 64–6
    wire rod production, 83–4
  technology change, 27
  textile industry, 51–67
  wire rod production, 82–4
  wood furniture industry, 73–82
Cleaner Production Programme (UNEP
    1989), 22–4
Cleaning stage:
  paper making and recycling, 161
  plastic waste recycling, 164–7
  wood furniture industry, 78
Clinical solid waste management, 293–306
  amounts, 151
  autoclave, 298–9
  chemical disinfection, 299–300
  China, 302
  clinical waste types, 294
  costs, 305–6
  current experience, 302–3
  disinfection, 297, 298–302, 305–6
  disposal, 297–8
  electron beam techniques, 303–6
  Europe, 302
  handling waste, 296–7
  Health Care Without Harm Organization,
    297
  incineration, 298
  interim storage, 297
  methodology, 294–5
  microwave disinfection, 300–1
  minimization techniques, 296
  screw feeding, 301
  sorting wastes, 296
  systems, 295–8
  waste
    categories, 294
    definition, 293
    treatment, 297
    usage, 305–6
Closed loop economy, 93, 104, 122–3
Coding systems: plastic waste, 168
Collection procedures:
  contractor's responsibilities, 268
  plastic waste recycling, 164–7
  reuse technique, 276
Co-location: Brownsville eco-industrial park,
    112, 113
Color stripping/bleaching: paper making, 161
Combined desize and scour and bleach, 58–60
Combustion systems: furnaces, 337–8
Common effluent treatment plant (CETP), 115

Communities:
  eco-industrial parks, 92
  see also Complexes
Comparative analysis: alternative fuel
    technologies, 358–9
Competent Administrative Authority (CAA),
    271
Complexes:
  sugarcane industry, 355, 357
  see also Communities
Composite packaging materials, 180–7
Composting, 189–94
  agricultural waste management, 233–4
  carbon/nitrogen ratio, 190
  composting plant, 195
  fertilizer mixtures, 246
  moisture content, 190
  organic fertilizer case study, 257
  oxygen supply, 191
  processes, 190–4
  rural waste management, 233–4
  temperature, 191
Compressive strength:
  construction materials, 205–7
  slag, 322, 324–5
Concrete:
  palm kernel shells, 235
  products, 343
  slag compressive strength, 322
Conservation: natural resources, 1–2
Construction and demolition (C&D) wastes,
    261–92
  4Rs Golden Rule, 264
  7Rs Golden Rule, 280–9
  amounts, 151
  case studies, 279–92
  Competent Administrative Authority, 271
  disposal, 270–2
  Egypt, 262–3
  flow diagrams, 266, 271
  guidelines, 263–72
  management, 261–92
  manifesto monitoring procedure, 272
  recovery techniques, 270
  recycle techniques, 270, 271
  reduction at source, 264–70
  reuse techniques, 270
  waste definitions, 262, 280
Construction industry, 234–6
Construction materials:
  abrasion resistance, 209–11
  bricks, 202–4
  compressive strength, 205–7
  density/water absorption, 211–12
  Durometer hardness, 208–9
  flexural properties, 207–8
  leaching tests, 214
  mechanical properties, 205–9
  mix-design matrix, 204
  morphology analysis, 204–5, 214–15

properties, 202–15
service properties, 209–14
tests, 213–15
water leaching tests, 214
Construction phase:
contractor's responsibilities, 267–70
designer's responsibilities, 267
owner's team's responsibilities, 266–7
reduction at source, 266–7
Construction wastes *see* Construction and
demolition wastes
Consumption: sugarcane industry, 351
Continuous granulation systems, 321
Contract formulation phase, 265–6, 274–5
Contractor's responsibilities:
collection procedure, 268
construction phase, 267–70
execution phase, 269–70
material handling, 268
procurement, 267
site layout, 267
storage, 268
waste management personnel, 268–9
Control systems: annealing, 337–8
Conveyor belts, 155
Cooling tower installations, 36
Corporate organization ecology barriers, 90
Cost/benefit analysis:
beet manufacturing facility, 41
cheese whey recovery, 46–8
edible oil company, 73
milk loss reduction, 51
oil and soap industry, 69–71
preserved food companies, 37, 38
spinning industry, 66–7
sulfur black dyeing, 54
textile industry, 60, 66–7
weaving industry, 66–7
wire rod production, 84
wood furniture industry, 80–2
Costs: electron beam accelerators, 305–6
Cow dung *see* Animal manure
CP *see* Cleaner production
Cradle-to-cradle approach, 16–18, 125, 130–2,
344–6
see also 7Rs Golden Rule
Cradle-to-grave concept, 14–15, 17, 125
Crushers: glass, 156
Cultural reform, 137
Current practice, 1–19
7Rs Golden Rule, 12–13
cradle-to-grave concept, 16–18
incineration, 6–10
landfills, 11–12
life cycle assessment/analysis, 13–16
treatment, 3–6
waste management, 1–3
Cutting:
cartons/paper, 184
granite/marble, 346–7

DAF *see* Dissolved air flotation
Dairy industry, 41–51
Deep well injection, 364
Definitions:
eco-industrial parks, 92
industrial ecology, 86–8, 100, 311
sustainable development, 85–6
sustainable treatment, 4–6
Deinking paper, 161
Demolition waste management:
disposal, 278–9
guidelines, 273–9
management plan summary sheet, 278
recovery technique, 278
recycle technique, 277–8
reduce technique, 273–5
reuse technique, 276–7
see also Construction and demolition
wastes
Denmark, 18, 92, 97–8, 100–7
Novo Nordisk, 101–4, 105
Density: construction materials, 211–12
Desalination plants, 363
Designer's responsibilities, 267
Design phase, 265
Desize and scour combination, 58–60
Development:
eco-rural parks, 254–60
industrial ecology, 280–9
rejects recycling, 197–8, 201–2
Disadvantages:
incineration, 10
landfills, 11–12
Disinfection:
chemical, 299–300
clinical waste, 297, 298–302, 305–6
irradiation, 302
microwave, 300–1
plasma systems, 302
usage, 305–6
see also Sterilization
Disposal:
cement industry, 312
clinical solid waste management, 297–8
construction and demolition waste, 270–2
demolition waste management, 278–9
Dissolved air flotation (DAF), 5
Documentation: reuse technique, 277
Dolomite analysis, 247, 250–1
Doors: furnaces, 336–7
Drain cleaning, 151
Drill cuttings, 339–46
approach, 341, 344–6
cleaner production, 341–5
concrete products, 343
cradle-to-cradle approach, 344–6
economic assessment, 345–6
fluids, 340–1
methodology, 341
mud, 340, 342

Drill cuttings (*contd*)
  opportunity assessment, 341–5
  recycle opportunity, 343–4
  reuse opportunity, 343–4
  treatment, 342–3
  waste generated, 340–1
  waste utilization, 345
Dry granulation systems, 321
Drying kilns, 74, 77
Durometer hardness, 208–9
Dyeing *see* Sulfur black dyeing; Textile industry

EAFs *see* Electric arc furnaces
EBRWC *see* Environmentally balanced rural waste complex
Eco-Efficiency Centre, Burnside, 118–19
Eco-industrial parks (EIPs), 91–3, 255–6
  benefits, 92–3, 106, 114
  case studies, 97–124
  cement industry, 312
  definitions, 92
  Denmark, 18
  recycling initiatives, 93–7
  scenarios, 109–14
Economic evaluation:
  drill cuttings, 345–6
  eco-industrial parks, 92, 106, 114
  heavy oil, 361
  natural gas, 361
  sugarcane industry, 360
  textile industry, 62
Economic growth commitments, 85
Economic indicators, 113
Economic instruments, 23
Economic reform, 137
Eco-rural parks (ERPs), 240, 254–60
Ecosystems *see* Industrial ecology; Natural ecosystems
Eco-tourist village, 368
Edible oil company, 71–3
Education role: eco-industrial parks, 119
Effluent treatment: dyeing industry, 55, 56
Egypt:
  agricultural waste quantities, 254–5, 258
  case studies, 41–67
  construction waste, 262–3
  dairy industry, 41–9
  edible oil company, 71–3
  EIA approach, 144
  fats recovery, 67–71
  granite industry, 346
  iron and steel industry, 319–20, 323
  municipal solid waste amounts/types, 151–3
  oil and soap industry, 67–73
  preserved foods, 31–9
  recycling, 161–3
  spinning/weaving, 62–7
  sugarcane industry, 350–5
  textile industry, 51–67

EIA *see* Environmental impact assessment
EIPs *see* Eco-industrial parks
EIS *see* Environmental impact statement
Electric arc furnaces (EAFs), 317, 323–5
Electricity consumption: preserved food companies, 33
Electron beams:
  accelerator costs, 305–6
  sterilization, 304–6
  technology, 303–4
Emissions:
  chemical spraying, 79, 81
  wood furniture industry, 76
EMS *see* Environmental management systems
End-of-pipe treatment, 3
Energy conservation:
  preserved food companies, 34–6, 37, 38
  spinning/weaving, 62–7
Energy consumption:
  oil and soap industry, 68
  preserved food companies, 32–3
Energy flow analysis: eco-industrial parks, 104–6
Energy linkages, 120–1
Energy recovery, 3, 182–3
Energy saving measures *see* Energy conservation
Energy sources: alternative fuel technology, 358
Environmental awareness, 134
Environmental benefits:
  eco-industrial parks, 92, 114
  sulfur black dyeing, 55
  textile industry, 61–2
Environmental degradation, 135, 136
Environmental impact assessment (EIA), 139–47
  contents, 142–3
  EIS process approach, 144–7
  mechanisms, 141–7
  process diagram, 145
  screening list approach, 143–4
Environmental impact statement (EIS) process, 144–7
Environmentally balanced rural waste complex (EBRWC), 240–3
Environmental management systems (EMS):
  cleaner production, 129–30
  concepts, 127–9
  environmental reform, 139, 140
  model, 128
  reform structure diagram, 140
  sustainable development proposed framework, 126–30
Environmental reform, 135–41
  cleaner production, 141
  economic reform, 137
  EMS, 139, 140
  political reform, 136–7

proposed structure, 137–41
regulations, 137–9
social/cultural reform, 137
structure diagram, 138
sustainable development, 125–48
Environment Outlook report (OECD), 153
EPR *see* Extended producer responsibility
Equipment modification, 26–7
ERPs *see* Eco-rural parks
Europe, 302
Evaporation ponds, 364
Execution phase:
    contractors, 269–70
    reduce technique, 275
Experimental controls: water, 45
Experimental phases: cleaner production, 43–4
Extended producer responsibility (EPR), 13–16
External assistance provision, 23
Extrusion: plastic, 170–2

Fabric quality, 54, 55, 59, 60–1
Facilitators: sustainable development, 134–5
Factory outcomes: cheese whey recovery, 48
Fairfield Ecological Industrial Park, USA, 121–4
Farming:
    animal farming, 366–7
    cheese whey recovery, 48
    fish farms, 365
    organic farming, 243–54, 367
    tank farms, 108, 111
    worldwide organic, 246
Fats recovery, 67–71
Feldspar analysis, 247, 250
Fertilizers:
    composting mixtures, 246
    eco-industrial parks, 102, 105
    quality, 252–4
Financial barriers: industrial ecology, 90
Finishing processes: furniture industry, 75–6,
        77–8
Finsbury Pavement Building case study, 289–91
Fish farms, 365
Flame furnaces, 200–1
Flexural properties: construction materials,
        207–8
Floating breakwaters, 218
Flotation cells: air injection, 162
Flow analysis: materials and energy, 104–6
Flow diagrams:
    construction and demolition waste, 266,
        271
    environmentally balanced rural waste
        complex, 241, 242
Fluidized-bed incineration, 10
Fly ash, 102, 105
Foam glass, 175–9
    mechanism, 177–8
    process, 178–9
Food sector industry:
    beet manufacturing facility, 39–41

Bruce Energy Centre, 120–1
case studies, 31–51
cheese whey recovery, 41–9
cleaner production, 31–51
dairies, 49–51
milk loss reduction, 49–51
preserved food companies, 31–9
waste recycling, 189–94
see also Preserved food companies
Forced aeration, 193
Formulae: sustainable development, 133–4
Fossil fuels: alternatives, 359
Foundries: aluminum, 326–39
Fumes, 324–5
Furnaces, 317, 318–19, 324–5
    annealing, 328, 330
    basic oxygen furnaces, 317–252, 317, 323,
        324–5
    blast furnaces, 317–19, 323
    doors, 336–7
    electric arc furnaces, 317, 323–5
    flame furnaces, 200–1
    heating/mixing types, 201
    iron and steel industry, 318–19
    smelting furnaces, 328, 329, 331–4
    types, 200–1
Furniture industry *see* Wood furniture
        industry
Future sustainability, 1–19
    7Rs Golden Rule, 12–13
    cradle-to-grave concept, 16–18
    incineration, 6–10
    landfills, 11–12
    life cycle assessment/analysis, 13–16
    treatment, 3–6
    waste management, 1–3

Garbage collection, 367
Gasification, 302, 356–60
    see also Biogas
Gas supply: eco-industrial parks, 102, 104–5
GDP *see* Gross domestic product
Germany, 93
Glass:
    crushers, 156
    foam glass, 175–9
    history, 180
    manufacture, 174–5
    off-site recycling, 314
    recycling, 173–5
    waste amounts, 152
Golden Rule *see* 4Rs Golden Rule;
        7Rs Golden Rule
Good housekeeping, 26
Government benefits, 93
Granite/marble industry, 346–50
    cleaner production, 348–50
    grani layers, 220
    grinding, 347
    industrial process, 346–8

Granite/marble industry (*contd*)
    opportunity assessment, 348–50
    polishing, 347
    processes, 346–8
    sawing, 346–7
    slurry, 347–9
    trimming, 347
    wastewater recycling, 349–50
Granulation systems, 321
Grate-type incineration, 9–10
Green farming *see* Organic farming
Grinding granite/marble, 347
Gross domestic product (GDP): MSW
        increase, 1–2
Guidelines:
    construction waste management, 263–72
    demolition waste management, 273–9
Gyproc, 100, 101–2, 104–5

Handling wastes: clinical, 296–7
Healthcare facilities *see* Clinical solid waste
        management
Health Care Without Harm Organization, 297
Health, safety and environmental (HSE)
        issues, 6
Heavy metal content, 214
Heavy oil, 359, 361
History:
    glass, 180
    Kalundborg eco-industrial park
        structuring, 101–4
Horizontal axis shredder, 168
Hose nozzle installation, 36
Hospital waste *see* Clinical solid waste
        management
Household municipal solid waste, 242–3
Housing, 117–18, 123
HSE *see* Health, safety and environmental
        issues
Hydrapulper schematic, 186
Hydraulic presses, 156
Hydrochloric acid use, 188

IA *see* Industrial audit
IE *see* Industrial ecology
Implementation: aluminum foundries,
        329–32
Improvement:
    boiler efficiency, 36
    water collection systems, 36
Incineration, 6–10
    advantages, 10
    clinical solid waste management, 298
    disadvantages, 10
    fluidized-beds, 10
    grate-type, 9–10
    liquid feed, 7–8
    municipal solid waste management, 194–6
    rotary kilns, 8–9
India, 114–17

Indicators: sustainable development, 133–4
Indirect flame furnaces, 200–1
Industrial audit (IA), 126, 127–9
Industrial ecology (IE), 85–124
    7Rs Golden Rule, 281–3
    barriers, 88–91, 132–3
    circular economy initiatives, 93–7
    definitions, 86–8, 100, 311
    development, 280–9
    eco-industrial parks, 91–3, 97–124
    leadership, 106
    life cycle analysis, 132
    recovery, 282
    recycling, 93–7, 282
    reducing, 281–2
    regulations, 281
    renovation, 282–3
    rethinking, 282
    reusing, 282
    sustainable development proposed
        framework, 126, 132–3
Industry:
    definition, 307
    Naroda eco-industrial park, 115
    see also individual industries
    waste management, 307–86
Infectious categories, 294
Informational barriers, 89
Information technology (IT), 134–5
Injection molding machines, 171
Input material change, 27–8
Inspection covers *see* Manholes
Installation:
    cooling towers, 36
    hose nozzles, 36
    temperature controllers, 35
    water meters, 36
Intangible benefits, 6
Integrated complexes, 239–43
Integrated waste management (IWM), 2
Interim storage, 297
Interlocks, 202–4
Iron in soil, 252
Iron and steel industry, 317–26
    cleaner production, 82–4
    Egypt, 319–20, 323
    furnaces, 317, 318–19
    slag, 320–6
    solid waste materials, 320
    waste materials, 320
    wire rod production, 82–4
    worldwide problems, 318–20
Irradiation: disinfection, 302
Irrigation: salt tolerable plants, 365
ISO standard, 15
IT *see* Information technology
IWM *see* Integrated waste management

Japan, 93–4
Johannesburg Summit, 2002 , 86

Kalundborg eco-industrial park, Denmark, 92, 97–8, 100–7
  administration, 107
  history of structuring, 101–4
  materials and energy flow analysis, 104–6
Knitted fabrics, 59
Kraft (chemical) pulping processes, 159
Kyoto bleaching units, 65

Laminated plastic recycling, 187–9
Landfills, 11–12, 196
LCA *see* Life cycle assessment/analysis
Leaching tests, 214
Leadership, 106
Legislation, 93–7
von Liebig, Justus, 233
Life cycle assessment/analysis (LCA), 13–16
  cradle-to-cradle concept, 125, 130–2
  cradle-to-grave concept, 14–15, 17, 125
  industrial ecology, 132
  ISO standard, 15
  new basis, 17
  OECD, 15
  sustainable development proposed framework, 130–2
  traditional, 14
Lignocellulostic material, 235
Liquid feed incineration, 7–8

Machining: wood furniture industry, 77
Management plan summary sheet, 278
Manholes, 215–16
Manifesto monitoring procedure, 272
Manufacture *see* Production
Marble industry, 346–50
  processes, 346–8
  slurry properties, 347
  see also Granite/marble industry
Market barriers, 89
Marshall Test, 323
Materials:
  construction waste, 263
  flow analysis, 104–6
  generation, 154
  handling, 268
  municipal solid waste, 154
  open industrial systems, 86–7
  recovery, 3
    municipal solid waste, 154
    packaging materials, 182–3
  substitution, 64
  TECTAN, 183–4
  textile industry, 64
  volume reduction, 2
  see also Packaging materials
Mechanical properties: construction materials, 205–9
Mechanical pulping processes, 158–9
Metal waste amounts, 152
Methodology:

cement industry, 309–11
clinical solid waste management, 294–5
drill cuttings, 341
organic fertilizer case study, 246–52
tourist industry, 363
Microwave disinfection, 300–1
Milk:
  loss reduction, 49–51
  packaging materials, 180–1
Minimization techniques, 296
Mining granite/marble, 346
Mitigation of construction materials:
  7Rs Golden Rule, 283–9
  recovery techniques, 286
  recycling techniques, 287–8
  reducing techniques, 283, 284–5
  regulations, 283
  renovation techniques, 288–9
  rethinking techniques, 288–9
  reusing techniques, 283, 285–6
Mix-design matrix: construction materials, 204
Mixing: plastic waste recycling, 170–2
Moisture content: composting, 190
Morphology analysis, 204–5, 215
MSW *see* Municipal solid waste
Mud, 340, 342
Mudcake, 353
Municipal solid waste (MSW):
  100% recycling system, 157
  amounts, 151–3
  composition, US, 151
  conveyor belts, 155
  Egypt, 151–3
  GDP, 1–2
  household waste, 242–3
  incineration, 194–6
  landfill, 196
  management, 149–96
    1995–2020 , 153
    incineration, 194–6
    landfill, 196
    OECD, 149–50
    rejects, 194–6
    transfer stations, 155–8
  material generation, 154
  organic fertilizer case study, 259
  recovery, 154
  rejects recycling, 194–222
  transfer stations, 155–8
Municipal wastewater sludge (MWWS), 315–16
Municipal wastewater treatment plants, 365–6
MWWS *see* Municipal wastewater sludge

Naroda eco-industrial park, India, 114–17
National Environmental Policy Act (NEPA) 1969 , 139–40, 149
Natural composting, 191

Natural ecosystems, 87–8, 91
Natural gas, 359, 360–1
Natural resource conservation, 1–2
Natural rocks, 247, 248–52
NEPA *see* National Environmental Policy
    Act (1969)
Netherlands, 123
Networking, 115–16
Novo Nordisk, Denmark, 101–4, 105

Obstacles to cleaner production, 24
OECD *see* Organization for Economic
    Cooperation and Development
Off-site recycling:
    cement bypass dust, 314
    cleaner production, 28
    road pavement layers, 316
Oil and fats recovery, 67–71
Oil refinery, Denmark, 101–5
Oil and soap industry:
    achievements, 71
    cleaner production, 67–73
    cost/benefit analysis, 69–71
    edible oil company, 71–3
    energy consumption, 68
    oil and fats recovery, 67–71
    process description, 68–9
    service units, 68
    wastewater generation, 69
    water consumption, 68–9
Oil supply:
    Brownsville eco-industrial park, 108–9
    Kalundborg eco-industrial park, –1, 103
On-site recycling:
    cement bypass dust, 313
    cleaner production, 28
Open industrial systems, 86–7
Opportunity assessment, 29–30
    aluminum foundries, 328–9
    beet manufacturing facility, 40–1
    cheese whey recovery, 43
    cleaner production, 29–30
    drill cuttings, 341–5
    edible oil company, 72
    milk loss reduction, 50
    oil and soap industry, 69
    preserved food companies, 34–7
    sulfur black dyeing, 52
    textile industry, 57–8, 63–4
    wire rod production, 83
    wood furniture industry, 79–80
Organic farming, 243–54
    tourist industry, 367
    worldwide, 246
Organic fertilizer case study, 243–54
    ABBC approach, 259–60
    animal fodder, 258–9
    biogas, 257–8
    composting, 257
    fertilizer quality, 252–4

    methodology, 246–52
    mixtures, 246
    MSW, 259
    objectives, 245–6
    waste utilization techniques, 256–9
Organic waste amounts, 152
Organization for Economic Cooperation and
    Development (OECD):
    Environment Outlook report, 153
    life cycle assessment/analysis, 15
    municipal solid waste management, 149–50
    total waste generated, 149–50
Output technology: agricultural/rural waste,
    226
Ovens *see* Annealing furnaces; Incineration
Owner's team's responsibilities:
    construction phase, 266–7
Oxygen supply: composting, 191

Packaging materials:
    air injection screening devices, 186
    aluminum, 183
    aseptic packaging, 182
    energy recovery, 182–3
    hydrapulpers, 186
    material recovery, 182–3
    milk, 180–1
    paper pulp separation process, 184–7
    recycling, 180–7
    thermal compression, 183–4
Palm kernel shells, 235
Paper:
    color stripping/bleaching, 161
    making and recycling, 159–63
    mill example, 27
    processes, 161
    product recycling, 161–3
    pulping machines, 162
    pulp separation process, 184–7
    recycled products, 161–3
    recycling, 158–63
    stripping/bleaching, 161
    utilization, 160
    waste amounts, 152
Paper cartons *see* Packaging materials
Parks *see* Eco-industrial parks
Partnerships:
    Brownsville eco-industrial park, 108–9, 112
    Kalundborg eco-industrial park, –1
    Naroda eco-industrial park, 116
Passive composting, 192–3
Pathological categories, 294
Pavement layers, 316
Peanut shell mortars, 235
Pelletizing machines, 169–70
Personnel: contractor's, 268–9, 276
Petroleum industry *see* Drill cuttings
Pharmaceutical waste, 294
    see also Clinical solid waste management
Phosphates *see* Rock phosphate

Physical treatment: definition, 4
Pilot phases: cleaner production, 44–6
Pith (sugar cane), 248, 249
Planning phase:
reduce technique, 274
reduction at source, 264–5
wood furniture industry, 75, 77
Plants: irrigation, 365
Plasma reactor technology, 238–9
Plasma systems (disinfection), 302
Plasterboard recycling, 291
Plastic waste amounts, 152
Plastic waste recycling, 163–72, 187–9
cleaning stage, 164–7
coding systems, 168
collection, 164–7
extrusion, 170–2
mixing, 170–2
process flow diagram, 164
product manufacture, 170–2
reprocessing, 167–70
separation, 164–7
sorting, 164–7
tests, 165–6
Polishing processes, 330–1, 347
Political reform, 136–7
Political support, 279
Pollution control, 3
react and treat rule, 25
see also Zero pollution
Pollution minimization, 73–82
Polyester/polyurethane spraying emissions, 79
Polyethylene recycling, 188
Polymer composites, 349
Poultry manure, 248
Power generation, 236–7
Preserved food companies, 31–9
boiler fuel consumption, 33
cleaner production, 31–9
cost/benefit, 37, 38
electricity consumption, 33
energy consumption, 32–3
energy saving measures, 34–6, 38
process description, 32
steam usage, 33
water consumption, 32–3
water saving measures, 36–7, 38
water sources/discharges, 33
see also Food sector industry
Presses: hydraulic, 156
Process control, 26
Processes:
aluminum foundries, 327–8
beet manufacturing facility, 39–40
bone recycling, 173
briquetting, 228, 230–1
cheese whey recovery, 41–3
composting, 190–4
edible oil company, 71–2
foam glass production, 178–9

milk loss reduction, 49
oil and soap industry, 68–9
paper making and recycling, 161
paper pulp separation, 184–7
plastic waste recycling, 164
preserved food companies, 32
spinning/weaving, 63
wood furniture industry, 74–6, 77–8
Procurement: contractor's responsibilities,
267
Production:
glass, 174–5
modification, 28–9
sugarcane, 351
wood furniture industry, 77
see also Cleaner production
Productivity: sulfur black dyeing, 55
Products:
development, 198, 201–2
manufacture, 170–2
plastic waste recycling, 170–2
rejects recycling, 198, 201–2
reuse, 2
Promotion: cleaner production, 22–3
Properties:
bricks, 202–4
construction materials, 202–15
granite/marble slurry, 347, 349
Proposed structure: environmental reform,
137–41
Prototype phase, 47–8
Provision:
CP implementation guidelines, 23
external assistance, 23
support measures, 23
Pulping processes, 162
kraft type, 159
mechanical, 158–9
Pyrolysis technology, 238, 302

Quality:
briquetting, 229–30
fertilizers, 252–4
Queensland, Australia, 295–6

Radioactive categories, 294
React and treat rule, 25
Recovery techniques:
construction and demolition waste, 270,
278, 286
energy, 3
industrial ecology, 282
materials, 3
paper utilization, 160
steam condensate, 36
Recycling:
aluminum, 179, 183, 188, 327–32
Asia, 3
bones, 172–3
composite packaging materials, 180–7

Recycling: (*contd*)
  construction and demolition waste, 270,
    271, 277–8
  drill cuttings, 343–4
  Egypt, 161–3
  food waste, 189–94
  glass, 173–5
  granite/marble wastewater, 349–50
  industrial ecology, 282
  laminated plastics, 187–9
  mitigation of construction materials,
    283–6, 287–8
  municipal solid waste rejects, 197–222
  packaging materials, 180–7
  paper products, 161–3
  plasterboard, 291
  plastics, 163–72, 187–9
  polyethylene, 188
  products, 2
  textiles, 179–80
  tin cans, 179
  waste paper, 158–63
  wastewater, 349–50
  see also Off-site recycling; On-site
    recycling; Rejects recycling
Recycling economy initiative, 93–7
Reducing techniques:
  construction materials, 283, 284–5
  contract formulation phase, 274–5
  demolition waste management, 273–5
  execution phase, 275
  industrial ecology, 281–2
  material volume, 2
  planning phase, 274
  tender formulation phase, 274–5
Reduction at source:
  construction and demolition waste, 264–70
  construction phase, 266–7
  contract formulation phase, 265–6
  design phase, 265
  planning phase, 264–5
  regulations development, 125
  tender phase, 265–6
Refineries, –5, 108–9, 111
Refining process: paper, 161
Regional strategies barriers, 90–1
Regulations:
  adoption, 23
  barriers, 90
  environmental reform, 137–9
  industrial ecology, 90, 281
  mitigation of construction materials, 283
  reduction at source, 125
  sustainable development proposed
    framework, 127
  see also 7Rs Golden Rule
Rejects recycling:
  agglomeration machines, 199–200
  breakwaters, 217–19
  bricks, 202–4

  construction materials, 202–15
  development, 197–8, 201–2
  manholes, 215–16
  morphology analysis, 204–5
  municipal solid waste, 194–222
  products, 198, 201–2
  road ramps, 222
  silica-plast products, 198
  table topping, 219–21
  technologies, 198–201
  wheels, 221–2
Remote partners, 108–9, 112
Renovation techniques:
  industrial ecology, 282–3
  mitigation of construction materials, 288–9
Replacement of defective items, 35
Reprocessing plastic waste, 167–70
Repulping, 161, 184
Resistance to chemical reagents, 212–13
Resource recovery projects, 116
Responsibilities *see* Contractor's
  responsibilities; Designer's
  responsibilities; Extended producer
  responsibility; Owner's team's
  responsibilities
Restaurant construction waste example, 280
Rethinking techniques, 282, 288–9
Reuse techniques:
  collection procedures, 276
  construction and demolition waste, 270,
    276–7
  construction materials, 283, 285–6
  documentation, 277
  drill cuttings, 343–4
  industrial ecology, 282
  waste management personnel, 276
  work activities, 276–7
  see also Recycling
Revetments, 219
Rice husks, 235, 237
Road map to sustainable development, 147–8
Road pavements, 316
Road ramps, 222
Rock phosphate, 247, 248–50
Rolling process: aluminum, 328, 330
Rotary kiln incineration, 8–9
Rural community technologies, 224–6
Rural waste management *see* Agricultural
  waste management

Safety and environmental (HSE) issues, 6
Salt tolerable plants, 365
Sanding processes: furniture industry, 75, 77
Sand-plast manholes, 216
Savings:
  cooling tower installation, 36
  defective steam trap replacement, 35
  hose nozzle installation, 36
  leaking steam valve replacement, 35
  spinning, 54–5

steam pipeline insulation, 34
sulfur black dyeing, 54, 55
temperature controller installation, 35
textile industry, 61
water collection system improvement, 36
water meter installation, 36
weaving, 54–5
see also Energy conservation
Sawing processes:
 furniture industry, 75, 77
 granite/marble, 346–7
Scenarios: eco-industrial parks, 109–14
Scrape recycling, 348–9
Screening list approach, 143–4
Screening pulp, 161, 185
Screw feeding, 301
Scrubbing systems, 330–1
SEAM *see* Support for Environmental
  Assessment and Management
Separation: plastic waste recycling, 164–7
Service properties:
 abrasion resistance, 209–11
 construction materials, 209–14
 density/water absorption, 211–12
Service units: oil and soap industry, 68
Sewage sludge, 314–16
Shredders, 168
Shrimp harvesting, 364–5
Silica-plast rejects products, 198
Silicon carbide, 237–9
Site layout: contractors, 267
Slag:
 asphalt concrete, 323
 BF slag, 318–19, 323
 BOF, 318–19, 320–2, 323
 compressive strength, 322, 324–5
 iron and steel industry, 320–6
Sludge amounts, 151
Slurry:
 cement bound, 349
 granite/marble, 348–9
 polymer composite, 349
 recycling, 348–9
Smelting furnaces, 331–4
 design, 332–4
 diagrams, 333
 process, 328
 upgrading, 329
Soaking, 184
Social benefits, 93
Social/environmental reform, 137
Sodium hydroxide, 189, 213
Solar ponds, 364
Solid fuel *see* Briquetting
Solid waste:
 agricultural, 151–3
 Burnside eco-industrial park, 117
 iron and steel industry, 320
 management systems, 117
 sugarcane industry, 351

wood furniture industry, 79–80
 see also Municipal solid waste; Slag
Solutions to cleaner production, 24
Sorting:
 clinical waste, 296
 plastic waste, 164–7
Spinning aluminum, 328, 330
Spinning and weaving industry:
 cost/benefit analysis, 66–7
 energy conservation, 62–7
 process description, 63
 savings, 54–5
 sulfur black dyeing, 52
 water conservation, 62–7
 see also Textile industry
Spraying: chemicals, 79, 80–1
Statoil refinery, Denmark, 101–5
Steady state biological systems, 87–8
Steam:
 Bruce Energy Centre, 120–1
 condensate recovery savings, 36
 Kalundborg eco-industrial park, 103, 105
 pipeline insulation, 34
 preserved food company usage, 33
 sterilization, 298–9
 supply, 103, 105, 120–1
 trap replacement, 35
 valve replacement, 35
Sterilization, 35, 298–9, 304–6
Stone companies, 108–9, 111
Storage:
 clinical waste, 297
 contractor's responsibilities, 268
 textile industry facilities, 64
Stripping paper, 161
Structure: environmental reform, 137–41
Sugar beet *see* Beet...
Sugarcane industry, 350–62
 alternative fuel technologies, 358–9
 animal fodder, 358–60
 ash composition, 353
 bagasse composition, 353
 balanced complex, 355, 357
 biogas, 356–60
 briquettes, 354–6
 cane consumption, 351
 economic evaluation, 360
 heavy oil, 359, 361
 natural gas, 359, 360–1
 production, 351
 solid waste generation, 351
 solutions, 354–61
 traditional energy form, 351–4
 traditional fossil fuel alternatives, 360–1
 see also Beet...
Sulfur, 104, 105, 247, 251
Sulfur black dyeing, 51–6
 achievements, 56
 benefits, 56
 chemical substitutes, 53

Sulfur black dyeing (*contd*)
  cost/benefit analysis, 54
  effluent treatment, 55, 56
  environmental benefits, 55
  fabric quality, 54, 55
  productivity improvement, 55
  savings, 54, 55
  working conditions, 55–6
Sulfuric acid, 188–9, 213
Support for Environmental Assessment and
    Management (SEAM), 31, 52
Support measure provision, 23
Sustainability:
  agricultural waste management, 223–60
  cement industry, 311–13
  cheese whey recovery, 48–9
  clinical solid waste management, 293–306
  construction and demolition waste
      management, 261–92
  industrial ecology relationship, 87–8
  industrial waste management, 307–86
  municipal solid waste management,
      149–96
  rural waste management, 223–60
Sustainable development:
  definitions, 85–6
  environmental reform, 125–48
  facilitators, 134–5
  formulae, 133–4
  indicators, 133–4
  road map, 147–8
  tools, 133–4
Sustainable development proposed
      framework, 126–33
  environmental management systems,
      126–30
  industrial audit, 126, 127–9
  industrial ecology, 126, 132–3
  life cycle assessment, 130–2
  regulations, 127
Sustainable human economic systems *see*
      Life cycle assessment
Sustainable treatment: definition, 4–6
Sydney Olympic Village construction waste
      study, 291–2
Symbiosis:
  Brownsville eco-industrial park, 109, 111
  criteria for, 106–7
  Kalundborg eco-industrial park, –4

Table topping, 219–21
Tank farms, 108, 111
TAPPI report: paper recycling, 161
Technical barriers: industrial ecology, 88–9
Technique combination *see* Integrated waste
      management
Technologies:
  ABBC, 225–34, 240
  briquetting, 231–2, 354–6
  changing, 27

cleaner production, 27
construction industry, 234–6
electron beams, 303–4
rejects recycling, 198–201
rural communities, 224–6
silicon carbide, 238–9
TECTAN (packaging material), 183–4
Temperature:
  composting, 191
  sterilizer control, 35
  Vicat softening, 213–14
Tender phase: reduce techniques, 265–6, 274–5
Tests:
  construction materials, 213–15
  plastic waste recycling, 165–6
  Vicat softening temperature, 213–14
Textile industry:
  benefits/achievements, 62
  bleaching units, 65
  cleaner production, 51–67
  combined desize and scour and bleach,
      58–60
  cost/benefit analysis, 60, 66–7
  desize and scour combination, 58–60
  economic benefits, 62
  Egypt, 51–67
  energy conservation, 65–6
  environmental benefits, 61–2
  fabric quality, 59, 60–1
  knitted fabrics, 59
  material substitution, 64
  preparatory processes, 56–62
  recycling, 179–80
  savings, 61
  scour and bleach, 58–60
  spinning, 54–5, 62–7
  storage facilities, 64
  waste amounts, 152
  water conservation, 65–6
  weaving, 54–5, 62–7
  working conditions, 61–2
  see also Spinning and weaving industry
Thermal compression: packaging
      materials, 183–4
Thickness loss: abrasion resistance, 210
Tin can recycling, 179
Tiverton, Canada, 120–1
Tools:
  sustainable development, 133–4
  sustainable human economic systems, 125
Total waste generated, 149–50
Tourist industry:
  animal farming, 366–7
  approach, 363
  case study, 362–9
  desalination plants, 363
  garbage collection, 367
  methodology, 363
  organic farming, 367
  wastewater treatment plants, 365–6

Traditional energy form, 351–4
Traditional fossil fuel alternatives, 360–1
Traditional life cycle assessment/
    analysis, 14
Traditional sugar mills, 352
Transfer stations, 155–8
Transportation, 122
Treatment, 3–6
    biological treatment, 4
    chemical treatment, 4
    drill cuttings, 342–3
    municipal wastewater, 365–6
    physical treatment, 4
    sustainable treatment, 4–6
    see also Effluent treatment
Trimming granite/marble, 347
Tunisia, 39–41

United Nations Environment Program
    (UNEP), 21, 22–3, 86
United States (US):
    eco-industrial parks, 95–6, 98–9, 107–14,
        121–4
    MSW composition, 151
Usage:
    clinical solid waste, 305–6
    disinfection, 305–6
    steam, 33
    waste paper, 160

Vermi-composting, 193–4
Vicat softening temperature, 213–14

Waste management:
    construction and demolition wastes,
        261–92
    contractor's responsibilities, 268–9
    current practice, 1–3
    future sustainability, 1–3
    personnel, 268–9, 276
    reuse technique, 276
    see also Agricultural waste management;
        Clinical solid waste management
Wastes:
    Burnside eco-industrial park, 117
    cheese whey recovery, 42–3
    drill cuttings, 340–1, 345
    eco-industrial parks, 91–2, 104–6, 115–16
    exchanges, 89
    iron and steel industry, 320
    Naroda park, 115–16
    paper recycling, 158–63
    utilization techniques, 256–9, 345
    wood furniture industry, 73–82
    see also Municipal solid waste; Solid
        waste

Wastewater:
    oil and soap industry, 69
    recycling, granite/marble, 349–50
    tourist industry, 365–6
    treatment plants, tourist industry, 365–6
Water:
    absorption, 211–12
    collection system improvement, 36
    conservation, 62–7
    consumption, 32–3, 68–9
    discharges, 33
    as experimental control, 45
    leaching tests, 214
    meter installations, 36
    oil and soap industry, 68–9
    preserved food companies, 32–3, 36–7, 38
    saving measures, 36–7, 38
    settling scrubbing systems, 330–1
    supply, –5, 112
    textile industry, 65–6
WCED *see* World Commission on
    Environment and Development
Weaving industry *see* Spinning and weaving
    industry
Web addresses xix
Wheels, 221–2
WHO classification: clinical waste, 294
Wire rod production, 82–4
Wood furniture industry:
    assembly processes, 77
    cleaner production, 73–82
    cleaning processes, 78
    cost/benefit analysis, 80–2
    drying processes, 74
    emissions, 76
    finishing processes, 77–8
    machining, 77
    pollution, 76–9
    polyester/polyurethane spraying, 79
    process description, 74–6
    process materials inputs, 77–8
    sanding processes, 77
    solid waste, 79–80
    waste, 76, 79
Work activities: reuse technique, 276–7
Working conditions: textile industry, 55–6,
    61–2
World Commission on Environment and
    Development (WCED), 86
Worldwide environmental problems, 318–20
Worldwide organic farming, 246
Worldwide usage: waste paper, 160

Yeast production, 105

Zero pollution, 12–13, 88